iPhone
Unlocked

Everything You Need to Know to Get Cracking in iOS 14

DAVID POGUE

Simon & Schuster Paperbacks

New York London Toronto Sydney New Delhi

Simon & Schuster Paperbacks
An Imprint of Simon & Schuster, Inc.
1230 Avenue of the Americas
New York, NY 10020

First Simon & Schuster trade paperback edition January 2021

SIMON & SCHUSTER PAPERBACKS and colophon are registered trademarks of Simon & Schuster, Inc.

For information about special discounts for bulk purchases, please contact Simon & Schuster Special Sales at 1-866-506-1949 or business@simonandschuster.com.

The Simon & Schuster Speakers Bureau can bring authors to your live event. For more information or to book an event, contact the Simon & Schuster Speakers Bureau at 1-866-248-3049 or visit our website at www.simonspeakers.com.

Manufactured in the United States of America

1   3   5   7   9   10   8   6   4   2

Library of Congress Cataloging-in-Publication Data has been applied for

ISBN 978-1-9821-7664-8
ISBN 978-1-9821-7665-5 (ebook)

ALSO BY DAVID POGUE

How to Prepare for Climate Change

Mac Unlocked

Windows 10: The Missing Manual

Pogue's Basics: Tech

Pogue's Basics: Money

Pogue's Basics: Life

Abby Carnelia's One and Only Magical Power

The World According to Twitter

Opera for Dummies

Classical Music for Dummies

Magic for Dummies

Contents

Part Three: The iCommunicator

Part Six: Appendixes

Introduction

These days, the iPhone is barely recognizable as a telephone. It gets far more use as a camera, alarm clock, GPS navigator, web browser, and email terminal, not to mention a music player, radio, boarding pass, fitness tracker, TV, voice recorder, weather station, notepad, newspaper, ebook reader, calendar, stock ticker, calculator, address book, text communicator, and so on.

Heck, lots of people use the iPhone's flashlight more often than they make phone calls with it.

Incredibly, though, the iPhone has resisted becoming a seething, sloshing mass of chaotic features. That's mostly because of Apple's traditional ace in the hole: Unlike its competitors (*cough* Android *cough*), Apple is the exclusive maker of both the phone's hardware and its software. It can therefore design things to work harmoniously, minimizing complexity and maximizing the "things just work" quotient.

Apple has also blessed most of its apps (that is, its software programs) with roughly the same design, so you have less to learn. The buttons are in consistent places, and they work consistently from app to app.

Every year, Apple introduces a new version of iOS, the operating system and apps of the iPhone. The book in your hands covers the 14th version—iOS 14.

> **NOTE:** More specifically, it covers iOS 14.2.1, a version with a few changes since the original iOS 14 release. (For example, version 14.2 included 100 new emoji symbols. Stop the presses!) But everything should work as described here no matter which version of iOS 14 you have—and no matter which iPhone model you're using.

How This Book Was Born

Hello there, I'm David.

I wrote my first iPhone book, *iPhone: the Missing Manual*, in 2007. It managed to cover every feature, app, and setting of the iPhone in 278 pages.

But each year, Apple introduced a new iOS version with more features; each year, the book got thicker. Eventually, it was 720 pages long, weighed over 2 pounds, and *still* didn't cover everything—I'd started offloading entire chapters as downloadable PDF files.

Eventually, it hit me: You want to learn to use your phone, yes—but good instruction isn't just walking through a thousand features and telling you what each one does. It's also *curation*. It's telling you which features are even *worth* knowing about and explaining when you might use them. It's distinguishing between features that Apple is excited to promote—and truly inspired features that never got any marketing love.

That's the idea behind *iPhone Unlocked*: to teach you the features you'll actually find valuable, in the right order, with the right emphasis. (As for the title: It's a little pun. Of course you want to unlock the power within your iPhone—but you also have to unlock it with your password, face, or fingerprint before you can even begin.)

Seven Fundamentals of the iPhone

It's great that Apple has seen to it that iOS generally uses consistent design elements and operational methods in order to make life easier for you. Before

---

THIS BOOK'S ARROW SHORTHAND

As a side effect of iOS's growing complexity, it now takes more steps to find things.

Over and over, the full instructions to find a certain setting might be, "Tap the **Settings** app; in the list that opens, tap **Accessibility**. On the Accessibility screen, tap **Display & Text Size**. On this screen, tap **Larger Text**; drag the slider to make the phone's type bigger on the screen."

That's a lot of verbiage. So in this book, you'll see a shorthand that looks like this: "Adjust the text size in **Settings→ Accessibility→Display & Text Size→ Larger Text**."

Here's hoping you can interpret that notation. Without it, *this* book might be 720 pages long, too. ✦

---

you dive in, therefore, it's worth getting to know a few of those techniques, since you'll be encountering them over and over again. Learn once, use often.

Long-Presses Are a Thing

It doesn't take long to figure out how to *tap* things on the screen. Tap to open an app. Tap to take a photo. Tap to begin video playback.

What you may not guess is that many more functions are hiding behind every app's icon, every web link, every email in your inbox, every photo thumbnail. To see these additional options, you have to use another kind of press: a *long*-press, where you leave your finger down on an icon or button for about a second.

On the home screen, the result is a pop-up list of commands (or a command panel, described next). In Mail, you can long-press a message's name in your inbox to see a pop-up view of its contents. In Safari, you can long-press a web link to open a pop-up preview window of that web page. You get the idea.

Long-press to see a secret shortcut menu.

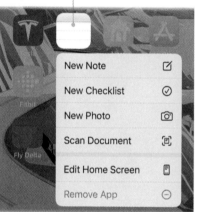

Long-pressing

Command Panels Are the New Menus

The iPhone doesn't have a menu bar like the one on a Mac or a PC. It does, however, have something very close: the command panel. These scrolling lists of options are hiding all over the phone. You'll see one when you long-press a text message in Messages, a place in Maps, a song in Music, a person or device in the Find My app, a file in Files, a link in Safari, and the ⬆ button in any app.

The main things to remember: Tap or drag the top of the panel to enlarge it to full-screen size—and tap the gray background to make it vanish again.

The command panel peeks up.

Swipe up to see more.

The command panel

You Can Configure Your Own Lists

Apple wants you to make your iPhone *your* iPhone. And that means giving you the tools to choose which icons appear in your Control Center, which options appear on the share sheet (page 203), which apps appear in your Messages "apps drawer" (page 262) which widgets appear on your Today screen (page 102), and so on.

If these terms mean nothing to you at the moment, don't worry; you'll learn about them. For now, it's worth meeting the standard list configurator, which you'll encounter in various corners of iOS.

The basic idea: This screen lists all the modules or buttons you can install. The ones you've already installed appear in the top half of the list; in the bottom half, you can see additional options you haven't yet installed.

The list configurator

To uninstall anything that's currently listed at the top, tap its little ⊖. (Tap Remove to confirm.) At that point, it flies down into the lower part of the list—the Uninstalled.

And to install anything in the lower list, tap its ⊕.

You can also drag the ☰ up or down to specify the order of these elements. For example, the top-to-bottom order on this configurator list determines the sequence of the tiles that appear in the Control Center (page 52).

You'll Be Zooming In and Out a Lot

Your little iPhone screen may be bright and clear and gorgeous, but it's still a *little* iPhone screen. Compared with the screen on a regular computer, it doesn't have a lot of real estate.

For that reason, life with the iPhone means zooming into things a lot (to read them more clearly) or zooming out (to get a big-picture view of the whole page).

Most often, the technique for doing that is *pinching and spreading.* That is, place two fingers on the glass—usually your thumb and index finger—and spread them apart (to magnify the image) or pinch them together (to shrink it down again). It makes more sense if you imagine that the screen is printed on a stretchy sheet of rubber.

Spread two fingers to zoom in.

Pinch and spread

In some apps, you can also *double-tap* (two fast taps in the same spot) to zoom in. That's a quick way to enlarge photos, maps, PDF files, and so on.

No matter how you zoom in, you're now looking at one small *part* of the map, photo, web page, or document. You're peeking at it through a magnifying glass. At this point, you need a way to pan around—to move the magnifying glass around on the image. You do that by *dragging* one finger across the screen.

Swipes and Taps Work the Same Way Everywhere

No mouse or keyboard came with your iPhone. Therefore, you're expected to do everything you'd do on a computer—point, click, type, scroll—using nothing but touches on the iPhone's glass screen.

Tapping is easy: Bop an onscreen object lightly and crisply with one finger. It's gotta be the pad of your finger; pens, pencils, toothpicks, and fingernails don't work.

> **TIP:** Some people tap with a stylus—an inkless pen with a special tip, available by the thousand online.

Swiping means dragging your finger across the screen. Most often, you'll swipe horizontally across something in a list, right to left, to delete it.

Flicking means—well, flicking your finger fast across the screen. That's how you scroll big lists, like the messages in your inbox. The quicker you flick, the faster the list spins. You'll love flicking; the lists on the phone are animated as though they have their own inertia, slowing to a stop after a moment. (You can interrupt the scroll at any time, either with another flick or by tapping.)

The Same Markup Tools Appear Everywhere

Your iPhone's touchscreen isn't so different from the fancy graphics tablets that professional illustrators use. You can use it to make drawings, sketches, and signatures right on the screen—including (and especially) onto photos and PDF documents. There's no easier way to sign a PDF contract.

To pull this off, open the standard markup tools by tapping Ⓐ. It appears, for example, in the Messages, Mail, Notes, and Photos apps, as you'll read in this book.

Your artist's toolkit includes a **Pen** (solid lines that get thicker as your finger slides faster); a **Highlighter** (translucent lines, like...a highlighter); a **Pencil** (a more textured, graphitey marking than the Pen, and no width variance); an **Eraser** (long-press to choose either **Pixel Eraser**, which erases pieces of lines, or **Object Eraser**, which deletes entire lines and shapes); and the **Selection Pen** (drag to enclose lines or shapes—and then drag to move the chunk you've selected).

> **TIP:** You can long-press the **Pen, Highlighter**, or **Pencil** tools to choose a starting line thickness and to specify the opacity of the marks you're making.

There's also a surprisingly powerful **Ruler** tool. Change its position and angle by putting two fingers on the ruler and then twisting or sliding them. At that point, you can make perfect straight lines by "pressing your finger against" the ruler as you draw.

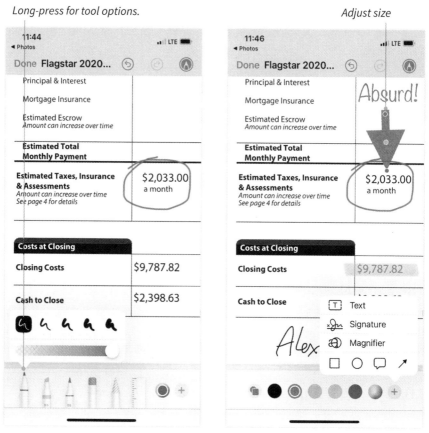

Long-press for tool options.

Adjust size

Markup tools

Apple ran out of room for the last seven tools, believe it or not; they're hiding in the ⊕ button.

When you tap **Text**, you get a text box. Double-tap it to type something; tap the text box and then tap AA to choose font and paragraph formatting. You can drag to move the box, adjust the blue handles to reshape it, or twist two fingers to rotate it.

You can tap **Signature** to insert a handwritten signature—great for signing documents electronically! To teach the phone what your signature looks like, tap **Add or Remove Signature** and then +; use your finger to draw your signature. (These signatures sync automatically across your various Apple machines—see page 398.)

The **Magnifier** creates a superimposed, round, magnified area on your photo or document, which is useful for calling your viewer's attention to some small detail. The blue handle adjusts the circle's size; the green one changes the degree of magnification. (Drag inside the circle to move it.)

The **Square, Circle, Speech Bubble**, and **Arrow** tools superimpose these shapes on your photo. To change line thickness and specify whether it's solid or hollow, tap . Drag blue dots to change size, or green ones to change the shape, angle, or direction of the arrows and speech bubbles.

Throughout your markup adventures, you can tap the Undo button (⊚) over and over again to rewind your steps.

> **TIP:** If you have a 3D Touch iPhone (the iPhone 6S, 7, 8, X, and XS families), you can press harder as you're drawing to get fatter lines.

Apple Is Obsessed with Data Privacy

It's a wonderful thing that the company takes such great care with your information. That's a far cry from Facebook and Google, whose business models are based on selling data about you to advertisers.

Eventually, though, you may get a little sick of iOS's nagging interruptions, seeking your approval for security-related actions. Are you sure the app you've just opened is allowed to use your camera or microphone? Is it OK for a weather app to know your location? Was it really you who just logged into your iCloud.com account on a different computer?

In the end, though, you may well come to appreciate Apple's zeal. Most people feel better not being tracked.

In iOS 14, you'll find more privacy and security features than ever.

What's New in iOS 14

These days, it's awfully hard for Apple (or Google) to invent massive, paradigm-changing new features for their phones. Once you've added the big-ticket game changers like voice recognition, autocorrect, water resistance, and the ability to locate the phone, how many huge ideas are left?

That's why iOS 14's catalog of new features is fairly scattershot. It's a long list of nips and tucks across the acres of software. That's not to say that some of these features aren't welcome, though. Here's a summary of what's new:

- **Widgets.** For the first time ever, your home screens can harbor more than just app icons. Now you can install *widgets:* miniwindows that sit right there among the apps to provide quick glances at current information like weather, stocks, news, photos, and your calendar.

  Android phones have had widgets for years—and now, for those who wish them, they're available on the iPhone, too. Apple starts you off with a choice of 19, but app makers can also offer their own widgets.

  In iOS, you can save space by *stacking* several widgets in the same home-screen space, and then flick through them at will. Page 101 tells all.

- **App Library.** The apps-on-a-home-screen organizational concept may have seemed like a good idea back when the App Store listed only 500 apps. But nowadays, most people's home screens are a disaster. Sure, you organized the apps on your first or second screens—but after that, you probably let the apps lie wherever they landed when you download them.

  Now there's a new home screen, off to the right of all others, that lists *all* your apps. You can view them either by category or in a tidy alphabetical list—but the point is that you never have to wonder which home screen holds a certain app.

  And now that you have the App Library, you may never again need home screens 3, 4, 5, 6, and so on—so in iOS 14, Apple gives you the option of *hiding* certain screens completely.

- **Sleep tools.** The iPhone is ready to help you get better sleep. It's prepared to help you start winding down before bedtime (with a certain playlist or a nice, boring book); keep your lock screen dark and quiet during sleeping hours; and wake you gently after a number of hours of sleep you specify.

- **FaceTime picture-in-picture.** When you're on a video call and need to root around on your phone to look something up, the call no longer blacks out your screen. Instead, the other person appears in a small inset window that you can move, resize, or temporarily park off the edge of the screen.

- **Better texting.** The Messages app now lets you "pin" up to nine of your most frequent correspondents above the list of conversations, so you don't have to go scrolling through all your chats to resume one of the most important ones.

You now have *inline replies,* too—you can respond to a *particular* text message, even an old one. It appears indented under the original message.

You can now choose both a name and a photo to represent a group chat. And Memojis—the cartoon characters that you design to look like you (or your dream alter ego)—are more inclusive. They now offer new ages, headwear styles, face coverings, and hairstyles, including, for the first time, a man bun. How did we *live* before this?

- **More compact comms screens.** Incoming phone calls, incoming FaceTime calls, and Siri responses now appear as compact banners rather than taking over your full screen.

Incoming call

- **Camera improvements.** The app now lets you shoot pictures with much less recovery time between shots (how does four shots per second sound?). A new exposure control lets you lock in a brightness setting for your whole shooting session. And you can now change video resolution or frame rate right before you shoot, without having to duck back into Settings.

- **More info in Maps.** The app now includes Guides, which are folders full of points of interest in each city, as recommended by the editors of big-name travel guides. You can now ask for navigation instructions for bike riding; you'll be shown how hilly your route is and warned about the presence of stairs. And if you have an electric car, Maps can automatically incorporate charging stops into your route.

- **Photo sorting.** Now the Photos app lets you view subsets of your photo collection with a couple of taps—only the shots you've edited or favorited, for example. You can sort albums by oldest or newest first.

- **Smarter reminders.** Apple's Reminders to-do app has been getting awfully robust in the past few versions, but now it's even robustier. For example, you can split up tasks by delegating them wirelessly to other people, who receive notifications of their new burdens. There are new automated reminder suggestions and smarter smart lists.

- **Translate.** This new app (page 367) works something like the Google Translate website: It offers instant translation between two languages. You can either type out what you want translated or you can *speak* your piece in English (or whatever your language is). The phone speaks the translation aloud. When your conversation partner replies (in their native tongue), you hear the English rendition.

- **AirPods switching.** If you own AirPods (Apple's wireless, detached white earbuds), you'll enjoy this one: As you move from one Apple machine to another (Mac, iPhone, iPad), the AirPods switch automatically, too.

- **Backtrack menu.** The more you use the iPhone, the more you realize how much of navigation consists of *drilling down*. Open **Settings**, then tap **Accessibility**, then tap **Zoom**, then tap **Touch**, then tap **Zoom Controller**, then tap **Double-Tap**, and suddenly you're six screens deep.

  But now, instead of tapping the back button (<) over and over again, you can long-press it. A menu lists all the screens you crossed to get here, so you can jump directly to any intermediate point.

Long-press...   ...and backtrack to any previous screen.

The backtrack menu

This trick is most useful in Settings, of course—but in fact it's available in *all* Apple's apps: Mail, Notes, Safari, Reminders, and so on.

- **Security cameras.** If you have a home security camera that uses Apple's HomeKit standard, the Home app can now learn to identify faces, so you won't be alarmed when a family member comes into the picture.

- **Sign language recognition.** When you're on a group FaceTime call and somebody starts using sign language, the app is smart enough to make that person's window big enough to see.

- **Safari upgrades.** The iPhone's web browser is faster than ever. When a foreign-language site pops up, you can translate it in place with one tap.

Above all, though, Apple has put work into security and privacy features. You can now view a privacy report for every website you visit, letting you know just how many scammy cross-site trackers each page has tried to clip onto you (and which Safari has blocked). Safari also alerts you if one of your passwords was involved in a corporate data breach, so you can change it before anything bad happens.

- **App Clips.** Has this happened to you? You're trying to use an electronic parking meter, or rent a bike or scooter, or order food—and it turns out you need an app for that. And you're on cellular. And it's going to take eight minutes, plus creating an account and all that jazz.

  Apple has a solution. Software companies can now create *App Clips,* which are stripped-down *parts* of apps that download fast and contain only enough code to get you going.

- **Sound recognition.** The phone can now recognize—and notify you about—important sounds in the background, like doorbells, crying babies, running water, or fire alarms. Apple figures this feature will help you if you have trouble hearing.

- **People Detection.** The Magnifier app has a wild new feature, available on the iPhone 12 Pro models: It knows if there are people near you, and exactly how far away they are. Great if you're sight-impaired and want to know when it's your time to move forward in line (especially in the time of social distancing).

- **Back taps.** Just the craziest, weirdest, most useful minor feature to come along in years: You can double-tap the *back* of your phone to trigger some action. You get a long list of choices for what action the tap performs—scrolling, going to the home screen, opening the Control Center, and so on. And you can set up a different action for *triple*-tapping.

- **Emoji search.** Finally!

- **Default browser and email apps.** You're no longer forced to use Safari and Mail when you tap links in Twitter, in email, and on web pages.

- **Voice Memos organization.** The app now lets you put your recordings into folders or mark them as favorites. The app can even take a stab at reducing room echo or background noise.

- **More weather info.** The Weather app can now show you the intensity of rain or snow that's coming in the next hour. It also lets you know if it'll be much hotter, colder, or rainier tomorrow.

- **Increased privacy.** Apple continues its quest to ensure that no app, ever, can spy on you without your awareness. Now, whenever any app is using your phone's camera or microphone—even the Camera app or Voice Memos—a tiny "LED" dot indicator appears at the top of the screen, just to keep you informed.

Similarly, you no longer have to choose between giving an app access to your precise location (for example, a GPS navigation app) and no location info at all. Now you can give it access to your *approximate* location, which should be enough for, for example, a weather or air-quality app, or a TV app that needs to know what region of the country you're in.

To turn this feature on for an app, open **Settings→Privacy→Location Services.** Tap the app's name and then turn off **Precise Location.** Now the app will know where you are only within a few miles, and the phone will update your location only a few times an hour.

With this quick summary, the following pages, and an optimistic attitude, you should have no problem diving into iOS 14—and unlocking your iPhone's full potential.

PART ONE

Meet the Machine

CHAPTER ONE

The Hardware in Your Hand

The very first iPhone, which Steve Jobs unveiled onstage in 2007, was a Model T by today's standards. There was so much it was missing! It didn't have a front camera, or a flash for the back camera. It couldn't shoot video. It couldn't cut, copy, or paste. It had no GPS. You couldn't send pictures as text messages. You couldn't dial by voice. There was no autocomplete or autocorrect. There was no app store, either—you got 16 apps, and you were happy.

Technology has surged ahead since that day, and yet, incredibly, the essential layout of the iPhone hasn't changed much at all. It still has a multitouch screen on the front, earpiece at the top, camera on the back, microphone and charging connector at the bottom. The volume buttons and silencer switch are still in the same places on the left edge.

And you still have to recharge the thing pretty much every day.

Whether you've got a shiny new (and shiny expensive) iPhone 12 Pro Max, or an iPhone 6s that's been soldiering on since 2015, the fundamental hardware design is the same.

The tour is now departing; please remain seated at all times.

On, Off, and Asleep

When your iPhone comes from the factory, it's turned off; Apple didn't want it to arrive with a dead battery.

You turn it on by pressing the side button for a few seconds. When the logo appears, you can let go. It takes about 15 seconds for the full startup process to bring you to the lock screen, where you're supposed to enter your password, supply your fingerprint, or use face recognition to unlock your iPhone.

> **NOTE:** On the iPhone SE, the power button is on the top edge instead. And it has a different name: the top button.

That ecstatic moment of unboxing a new iPhone is one of the few times you'll ever arouse the iPhone from its fully off state. For the rest of its life, when you're not actually looking at the iPhone, you aren't supposed to turn it off. You're supposed to put it to *sleep*.

To do that, just give the side button a quick click.

> **TIP:** The iPhone also goes to sleep on its own if you ignore it long enough. You can specify how long that takes in **Settings→Display & Brightness→Auto-Lock**.

When the iPhone is asleep, the screen goes black and the machine uses very little power. But everything you've been doing is still open in the phone's memory; all your apps are still running. Your music keeps playing, your audiobook keeps reading, your GPS keeps guiding you, your email keeps downloading.

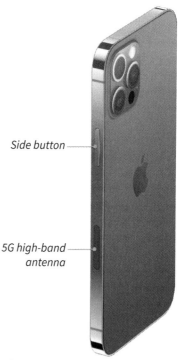

Side button

5G high-band antenna

The right side

Waking the phone once again doesn't take much. Just tapping the screen, pressing the home button (on the older models), or even picking up the phone brings the screen to life. (Technically, it brings up the lock screen; see page 34.)

The lesson here: When you're finished using the phone for now, don't shut it down; put it to sleep. It costs you almost no battery power, and the phone will be immediately ready to go next time you need it.

That side button, by the way, has a thousand uses. It's involved in commanding Siri, taking screenshots, silencing an incoming call, and so on. But for your very first iPhone lesson, it's enough to understand that you press it to put the phone to sleep—or to wake it up again.

The Left-Side Buttons

On the skinny left edge of the phone, three more buttons await your inspection. At top: the silencer switch.

It's a little flipper. When you use your fingernail or the pad of your thumb to click it toward the back of the phone, so a little patch of orange appears in the flipper socket, the phone will make no unbidden sounds. You'll never be embarrassed by the phone ringing, a text-message chime, or some social-media notification. That's probably what you want in a meeting, in a theater, or in a church. (Flip it forward, toward the screen, when you want to hear the rings, chirps, and chimes again.)

Note, by the way, that the silencer switch doesn't silence everything. It does not silence the following types of sounds:

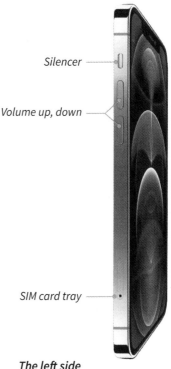

Silencer

Volume up, down

SIM card tray

The left side

- **Alarms.** Even when you've got the silencer on, your phone still rings for any alarms you've set. Apple figures that if you've gone to the trouble of setting an alarm, you really want to hear it. You wouldn't want to oversleep, now, would you?

- **Sounds you play.** If *you* request a sound—if you play a video, listen to a podcast, play some music, play back a voicemail, start an audiobook— you'll hear it, even if you have "silenced" the phone. Apple calculates that if you've tapped a ▶ button, you probably want to hear something.

Below the silencer switch are two other buttons: volume up and volume down. Mostly, of course, you'll use them to adjust the phone's speaker volume. But there's a little more nuance to them than that:

- **When you're not on a call,** they adjust the volume of sound effects: the phone-ringing sound, alarms, the voice of Siri, and so on.

> **NOTE:** That statement is true as long as you haven't turned off **Settings→Sounds & Haptics→Change with Buttons**.

- **When a call comes in,** press one of these buttons to make the ringing or vibrating stop.

- **When you're on a call,** these buttons control the speaker or earbud volume.

- **When you're playing music or podcasts,** they adjust the playback volume. They work even when the screen is off.

- **When you're taking a photo or video,** you can press one of these volume keys to snap the photo or start/stop the video.

The Screen

If Apple is going to design a computer that's all screen and no keys, it had better provide one amazing screen. And sure enough: Every year, the iPhone screen gets better, brighter, and higher resolution.

The iPhone 12 Pro Max, for example, is composed of 3,564,384 tiny pixels. If you tried to count them, one pixel per second, it would take you almost six weeks—with no eating, sleeping, or bathroom breaks. (But don't bother. They're smaller than the human eye can detect.)

You're probably aware that it's a touchscreen; you tap buttons and icons directly on the glass with your finger. In fact, though, it's a *multitouch* screen, meaning that it can detect the touch of more than one finger simultaneously. You can zoom into a photo or a map, for example, by spreading two fingers apart on the glass.

Over the years, Apple has worked with Corning Glass, the biggest inventor of industrial glass formulations, to develop glass that's more resistant to shattering when the phone slips from your hand. Apple says the iPhone 12 family's screens are four times tougher than previous iPhones' glass, in part because Corning has introduced tiny particles of ceramic to the liquid goop that it cooks into glass.

Well, any little bit is welcome—but the screen can still break. Be careful out there.

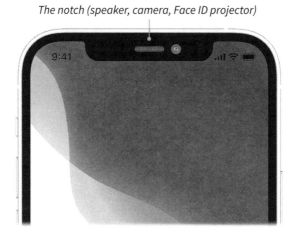

The notch (speaker, camera, Face ID projector)

The screen and the notch

The Cameras

These days more than ever, the name "iPhone" seems like an anachronism. According to the latest research, conducted by parents glancing over at their kids on the couch, making phone calls is about the *last* thing young people do with their smartphones. A more appropriate name for this thing would be iTexter, iSnapchat—or iCamera.

In fact, the iPhone is the most-used camera model on the planet. On Flickr.com, a photo-hosting site, nine out of the ten most-used cameras are iPhone models.

The modern iPhone may have up to four cameras: one on the front, for taking selfies, and as many as three on the back, depending on your model. A smartphone is too skinny to include a traditional zooming lens, so Apple has equipped some models with two or three *non*-zooming lenses, one apiece at different levels of zoom.

Also crammed into the back camera panel:

- **A very bright flash,** which you'll probably use mostly as a flashlight.

- **A tiny microphone,** which lets the iPhone record sound when you're shooting video, and also provides noise cancellation when you're making a phone call.

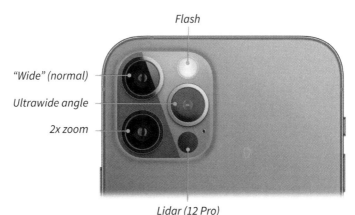

Flash

"Wide" (normal)

Ultrawide angle

2x zoom

Lidar (12 Pro)

The cameras on the back

- **On the iPhone 12 Pro models, a lidar sensor.** Lidar (for *light detection and ranging*) blasts out a spray of infrared-light dots and measures the reflections, detecting the precise distance of each one. Add together enough of these distance measurements, and you've got a camera that can build a mesh—a 3D contour map—of things in front of it. In that way, the lidar sensor works like the Face ID camera on the front of the phone (page 37)—but the lidar has a much longer range, about 16 feet.

In the non-phone world, lidar helps keep self-driving cars, robots, and drones from crashing. On the iPhone 12 Pro, lidar helps the camera focus faster in low light, allows the Measure app to gauge people's height (page 351), and makes Portrait mode photos(page 171) possible in light that would ordinarily be too dim.

The sensor is also a blessing to augmented-reality apps, which turn the phone into a viewer for the world around you—and superimpose graphics on it. As you move the phone, the simulated objects' perspective and sizes change exactly as though they were really in front of you.

AR lets you preview a piece of furniture in the actual room where you're considering putting it, or visualize a different wallpaper in your house, or just measure things without a tape measure. And, of course, AR opens up a new world of games, in which characters, vehicles, and weapons seem to live in the real world with you.

As these apps are updated to take advantage of the iPhone's lidar, they'll need less time to scan the room before showing the results, and they'll be better at handling occlusions (where one object is in front of another).

The Home Button (Maybe)

Apple has been quite clear on one design point: The home button is going away. The oldest models capable of running iOS 14 still have a home button: the iPhone 6s, 7, and 8 families, as well as the iPhone SE (2016 and 2020).

All newer models feature a screen that extends from the top to the bottom of the phone, paving right over the strip where the home button used to be. Instead of recognizing your fingerprint to unlock the phone, these phones use Face ID (face recognition).

If your iPhone has a home button, you can press it to wake the phone, hold it down when you want to speak to Siri, and touch it to unlock the phone (instead of entering your password). That's right: The fingerprint reader is embedded in the home button itself.

Home button + Touch ID fingerprint sensor

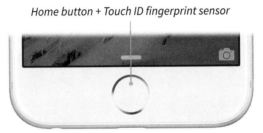

The home button

Speakers and Microphones

At the top center of your screen, there's a small slit in the glass. (You may not be able to see it, but you can feel it with your thumb.) Behind it is the speaker that produces the sound when you're on a phone call.

There's another speaker on the bottom edge, which is what you hear in speakerphone mode. On the latest models, the top and bottom speakers can play in stereo, which is great when you're holding the iPhone sideways and watching a movie.

The bottom edge also contains two microphones. Why two? To help cancel out ambient noise, making your voice clearer to the people you call. (The microphone on the back panel, on the other hand, provides noise cancellation for *your* benefit, so what *you* hear on calls is cleaner.)

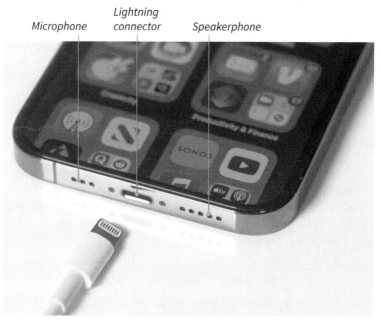

Microphone   *Lightning connector*   *Speakerphone*

The bottom edge

Connections

There's only one place to plug anything into the modern iPhone, and Apple is doing its darndest to eliminate even that.

It's the Lightning connector on the bottom edge. That's where you plug in the white USB charging cable that came with your iPhone—although Apple really, really wants you to use a wireless or magnetic charging panel instead. It's also where you plug in the white earbuds that came with all iPhones until the iPhone 12 family—although Apple really, really wants you to use wireless earbuds like the AirPods. (Apple eliminated the headphone jack starting with the 2016 iPhones.)

> **NOTE:** You *can* plug headphones into the modern iPhone's Lightning connector; you just need to buy a little adapter to do it (about $8).
>
> The only real inconvenience with this setup is that you can't charge your phone with the Lightning jack while your headphones are plugged in there. If that bothers you—well, you have yet another adapter cord to buy.

Charging (and Battery Tips)

If you ever have the delightful opportunity to see an iPhone taken apart, you'll be astonished to see what's inside. It's mostly battery. That single component occupies about three-quarters of the iPhone's interior space.

And even though it's so big, you *still* have to recharge the phone every night. That should give you some indication of how much power the *other* elements draw—the screen, processor, and wireless features.

In any case, these days, the term "battery life" has two different meanings, and you're expected to be the manager of both of them.

Hours per Charge

First, "battery life" can mean how many hours of life you get out of each charge. That depends primarily on what you're doing with the phone. If you've got GPS navigation or Personal Hotspot (page 78) going, the battery will be dead in a couple of hours. If all you're doing is listening to music or reading an ebook, it can last a couple of days.

If you want to get the most out of a charge, adopt techniques like these:

- **Turn on Low Power Mode.** In this special economy mode, the iPhone stops performing a lot of background tasks—downloading email, checking for app updates, and listening for you to say "Hey Siri" (page 139), for example. You get fewer animations, the processor slows down, and the battery icon at the top of the screen turns yellow to remind you what's going on. If you're willing to use the phone in this hobbled state, you can get an extra three hours per charge.

    You can turn on Low Power Mode manually in Settings→Battery, or in the Control Center (page 52). But the iPhone also invites you to turn it on—once when your battery charge hits 20% remaining, and again at 10%.

    You don't really have to think about turning *off* Low Power Mode. Once the phone is charged enough, Low Power Mode turns off automatically.

- **Dim the screen.** The screen is the number-one consumer of battery power. The dimmer it is, the longer the battery lasts. Use the Control Center to make this adjustment (page 52).

If you're desperate, you can also turn off Wi-Fi, Bluetooth, and cellular data (all in the Control Center) to save a few more drops of battery juice—or just turn on airplane mode (page 54).

You can also keep tabs on which of your apps seem to be the hungriest battery gluttons; open **Settings→Battery** and scroll down.

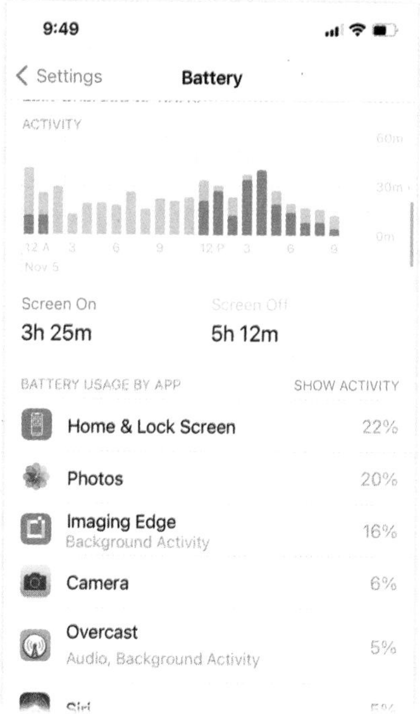

9:49
Settings
Battery
ACTIVITY
60m
30m
0m
12 A
3
6
9
12 P
3
6
9
Nov 5
Screen On
3h 25m
Screen Off
5h 12m
BATTERY USAGE BY APP
SHOW ACTIVITY
Home & Lock Screen
22%
Photos
20%
Imaging Edge
Background Activity
16%
Camera
6%
Overcast
Audio, Background Activity
5%

Battery usage by app

Some of the apps in this list bear a **Background Activity** label. That's a warning that these apps are connecting to the internet without your awareness, in the background, and eating into your battery charge along the way. You're welcome to turn off this background activity for any apps that don't deserve the privilege: Open **Settings→General→Background App Refresh** and switch them off.

> **NOTE:** You may also see the words **Low Signal**. Nothing eats up the phone's power faster than hunting for a cellular signal—when, for example, you're on a plane or out at sea and forgot to turn on airplane mode. In these situations, the phone directs more power to its antennas in hopes of finding a signal.

Charges per Battery

Here's the other way of defining "battery life": how many times you can recharge your battery before it starts *holding* less charge.

Every lithium-ion battery has a limited life span. After about 500 charges, it holds only about 80% of its original capacity. That's chemical reactions for ya.

Once your phone can't even make it through a day without recharging, you can take it in to an Apple Store and get the battery replaced for $50 or $70, depending on the model. But wouldn't it be better to postpone that dreaded day? There is a way.

It turns out that lithium-ion batteries don't like to sit at 100% full. The more time they spend at that level, the faster they degrade. Unfortunately, if you charge your phone when you go to bed, it charges fully in about three hours and then spends *the rest of the night* at 100%. After many nights like that, its longevity will shrink.

The iOS feature called Optimized Battery Charging studies your charging patterns. It might discover that you usually plug the phone in at midnight, and then unplug it at 7:30 a.m. when you get up.

Once it has a pretty good grasp of your routine, Optimized Battery Charging limits the *immediate* charging to 80%, where it sits for most of the night—and then completes the charge just in time for you to get up at 7:30. This ingenious charging pattern minimizes the time the battery spends at 100% full.

"But wait!" you cry. "What about when I'm traveling? What if I have to get up early? This feature might send me out into my day without a full charge!"

First, Optimized Battery Charging doesn't kick in when you're traveling, period. (In fact, this feature requires Location Services to be turned on in **Settings→Privacy→Location Services**; the feature needs to know when you are, in fact, away from home.)

As for getting up super early: You can simply turn this feature off for a night in **Settings→Battery→Battery Health**. You can also override it when you see it happening. When the charge has reached 80%, the phone displays a notification bubble that says **Scheduled to finish charging by 7:30 AM** (or whatever).

Long-press that notification and tap the **Charge Now** button that appears; you've just told Optimized Battery Charging to step aside. Your phone will charge to 100% right away.

Optimized charging notification

Maybe you hate the whole idea of your phone trying to guess when it should complete your nightly charge. Maybe you trade in your phone every year anyway, and so you don't really care how many years the battery will last. In that case, you can turn the whole feature off in Settings→Battery→Battery Health.

How to Charge Your Phone

There's still no such thing as *truly* wireless charging, where your phone is in your pocket, receiving some cool kind of electrical waves from a transmitter in the corner of the room. The electronics industry knows perfectly well that that would be awesome, and they're working on it.

In the meantime, you have to recharge your phone by connecting it to something:

- **Plug in the white cable that came with it.** The smaller end is what Apple calls a Lightning connector; you shove it into the bottom of the phone (see "The bottom edge" on page 24). You don't have to pay attention to which way is up.

  The other end goes into a USB jack—of a computer or a car, for example.

  NOTE: No matter how you charge, check to make sure there's a tiny lightning bolt on the battery icon in the upper-right corner. That's your indication that the phone is charging properly, and that there's no problem with the cable, the wall outlet, or the laptop you're plugged into.

  Or, when speed is of the essence, you can plug the far end of this cable into a power outlet, courtesy of the little wall-outlet adapter that *used* to come with every iPhone. The phone charges faster when it's plugged into the wall than into a computer.

  TIP: On iPhones that offer "fast charging"—the iPhone 8 and later models—you can get faster charging with higher-wattage power bricks. The standard Apple white cube adapter is a 5-watt charger, which charges fairly slowly. But 18- and even 30-watt chargers are available, which go from 0% to 100% in about an hour and 40 minutes— one hour faster than the 5-watt charger.

- **Set it down on a Qi charger.** If you have an iPhone 8 or later, a special luxury awaits: You can charge your phone each night just by setting it down onto what looks like a plastic drink coaster or easel. You're spared the agony and frustration of plugging in a physical cable.

These things are called Qi charging pads (pronounced "chee," the Chinese word for life force). Qi is not an Apple technology—many phone brands can use these pads and stands, which is why they cost only about $12. Some of these Qi chargers are like tiny beds for your phone; some prop the phone at an angle; some are integrated into mounting brackets that connect to the windshield or vents of your car.

Qi charger

In any case, the Qi charger plugs into a power outlet. You set the phone down, screen up, onto the pad, and presto: The phone begins to charge.

It's not as fast as plugging in the Lightning cable; after two hours of Qi charging, your battery might be at 40% or 50%, depending on the wattage of the pad. It would be at 80% using the cable.

- **Slap a MagSafe charger onto it.** Despite their convenience, Qi chargers have one small drawback: If your phone isn't exactly centered on the pad, it doesn't charge. Nobody ever said life wasn't a struggle.

WHAT'S NO LONGER IN THE IPHONE BOX

If you buy one of the 2020 iPhones—the iPhone 12 family or iPhone mini—you might be startled to discover that the *box* is much smaller than iPhone boxes used to be. That's because there's not as much inside.

Apple has stopped including earbuds in the box. You no longer get the tiny white power brick, either—the wall wart that plugs into a wall outlet for charging the phone. (All you get is the Lightning cable, which must be plugged into a computer or other USB jack.)

Apple says it's thinking of the planet. Manufacturing and shipping all those parts produces as much greenhouse gas

as 450,000 cars a year. And after 13 years of iPhones, we the people already have 700 million of the earbuds and 2 billion Apple power bricks clattering around in our junk drawers.

Of course, if you're slightly more cynical, you might realize that if you *don't* have those parts lying around, you now have to buy them separately. Apple charges $20 apiece for the wired earbuds and the wall wart, $160 and up for the AirPods, and $40 for the MagSafe magnetic charger for the iPhone 12 family.

The bottom line: Apple's reasoning might be easier to believe in if the price for each iPhone were $40 lower. ✶

The charger magnetically snaps onto the iPhone…and lets you know it's working.

MagSafe charger

That's why, beginning with the iPhone 12 family, Apple introduced the MagSafe charger for iPhones ($40 for the charger, plus $20 for the required 20-watt USB-C adapter). It's not to be confused with the MagSafe connector that once made it easy to plug in MacBook laptop charging cables.

It's a disk, 2 inches across, that snaps solidly and magnetically onto the back of the iPhone 12 models. The magnets are strong enough to hold the weight of the phone when it's dangling from the MagSafe cable (although that's not recommended except when you're standing over a mattress). When you snap the magnet onto the back, a supercool charging indicator appears on the iPhone's screen.

NOTE: If you plug in the Lightning charging cable *and* put the phone on a Qi pad or MagSafe disk, the cable wins. And that's exactly what you would want, because the cable charges faster.

Unfortunately, the magnetism isn't strong enough to reach through a phone case—unless it's one of Apple's special MagSafe phone cases ($50). On the other hand, the strong magnets present a handy new option: a MagSafe *wallet* ($60), which is a leather pouch just big enough to hold a few cards and is supposed to cling to the back of your phone without glue, Velcro, or welding.

Clearly, getting into MagSafe requires a decent amount of disposable income, as well as an incredibly low tolerance for first-world problems.

But it is kind of cool.

> **TIP:** No matter how you charge your iPhone, you can use it while it's charging. But if speed is important, note that the phone charges faster if it's asleep, faster still in airplane mode, and fastest of all if it's powered fully off.

Antenna Gaps

The band around the edges of the iPhone is metal. It's attractive and rugged, but radio waves can't pass through it easily. You might consider that a downside on a device whose primary purpose is *transmitting radio waves*.

That's why Apple interrupts the metal band with thin strips of plastic—little slots that are enough to let the radio waves pass through. On the iPhone 12 models, there's a relatively huge window cut into the right edge, too (shown on page 18). That's for access to the 5G millimeter wave signal (page 72).

Steve Jobs would've hated these interruptions in the smooth, metallic finish of the iPhone—but then again, he also would have hated a phone that can't make calls or get on the internet.

SIM Card Slot

There's a final stop on your tour of the edges of the phone, which will probably interest you only once in your iPhone career. It's the SIM card slot (shown on page 19).

Apple may have made your phone, but a cellular carrier like Verizon, AT&T, or T-Mobile/Sprint provides your *service*. And they generally provide that in the form of a SIM card: a tiny memory card, the size of your pinkie fingernail, that stores your phone number and calling-plan details.

It sits in an equally tiny tray that pulls out from the right side of your iPhone (the left side, on iPhone 12 models). To eject the tray, push an unfolded paper clip—or the tiny, pointy tool that came with your phone—straight into the hole on the SIM card tray cover.

When you upgrade to a newer phone, you move this card from your old phone to your new one. When somebody dials your phone number, whichever phone contains your SIM card is the one that rings.

> **TIP:** When you travel overseas, and you want to remain reachable, you can rent a cheap local SIM card while you're there. That's usually much less expensive than using your own account and paying exorbitant international roaming fees.
>
> If you do decide to keep your own plan, don't forget to contact your carrier to let them know when and where you're going. If you don't, you may not be able to use your phone at all. As a bonus, your carrier may offer a temporary international plan that keeps the roaming fees below brain-exploding level. (Try not to be jealous of T-Mobile subscribers, for whom international texting and internet are free.)

ADVANCED SIM CARD TRICKS

Most people insert the SIM card into the phone and never think about it again.

You may wind up thinking about it, though, if you decide to switch from one carrier to another—for example, from AT&T to Verizon.

If you bought your iPhone directly from the cell carrier, it's probably a *locked* phone—imprisoned by software to work only with that carrier. They do that to make sure you can never leave their service while you still owe them money.

But when the time comes to switch carriers, contact their customer service and let them know you'd like them to unlock the phone. They can do it by remote control—assuming that you do not, in fact, owe them money.

(Now you understand why iPhone aficionados prefer to buy an *unlocked* phone in the first place—a phone that's not associated with any carrier in particular.)

Impressively enough, recent iPhone models (iPhone XS and later) have, in essence, two SIM cards: a traditional physical one, and also an electronic one (an eSIM).

These phones can have two phone numbers, each with a different ring sound. You can use one as your work number and one as your personal number. For each person in Contacts, you can specify which "line" to make calls from.

Not all cellular carriers offer this feature, so investigate before you buy. ✦

CHAPTER TWO

Eleven Settings to Change First Thing

It's not easy to design a smartphone that makes every customer equally happy: Danish software engineers, 7-year-olds in China, choir conductors in Brazil, college kids in Cincinnati…and you. Everybody is different. The design and settings that seem perfect to one person might seem unusable to you.

All Apple can do is make sure you can *change* every one of the thousand controls, buttons, sliders, and switches in the iPhone's Settings app—and then guess which factory settings will please the most people most of the time.

That's why, upon installing iOS 14 or getting a new iPhone, you should take a few minutes to change some of the settings to make this iPhone *your* iPhone. Along the way, you'll discover the locations of some of the iPhone's most important control centers.

To make most of these tweaks, you'll begin by opening the Settings app, whose icon on your home screen shouldn't be hard to find.

Set Your Password, Face ID, or Fingerprint

Apple has gone to extraordinary lengths to ensure that if you leave your phone sitting out somewhere, nobody else—in your family, in your workplace, or in society at large—can get in. When you're not using your phone, it locks.

There are good reasons for that setup. You may have personal stuff on the phone—emails, texts, and web bookmarks that probably shouldn't become public. You may also have financial stuff on it—passwords for your bank, insurance, and online store accounts, for example.

The Lock Screen

When you wake the iPhone, the first thing you see is the lock screen. Its fundamental purpose, of course, is to say: "OK, your phone is awake—but you shall not pass until you provide your fingerprint, faceprint, or passcode."

But that's not the lock screen's only purpose. It's also a convenient, one-glance status screen. It shows, for example, the time and date, which has led millions of people to decide they don't need to own a watch anymore.

The lock screen also displays notifications about things that took place while the phone was asleep: calls you missed, text messages and emails that came in, and so on (page 42).

This screen even gives you instant access to the camera: Swipe to the left. Now you're ready to take a shot.

Swipe down to see older notifications.

Swipe left for the camera.

Swipe right for your Today screen.

Notifications

Swipe up to reach the home screen.

The lock screen

If you swipe to the right, you open the Today screen, full of widgets that give you updates on the weather, news, stocks, your calendar, and so on (page 102).

The Control Center (page 52) and the Notification Center (page 43) are available here. And, maybe most importantly of all, the Flashlight (🔦) is here, too.

> **NOTE:** If you can believe it, when you turn on the Flashlight, its infinitesimal power switch—on the barrel of that tiny flashlight icon—actually *moves to the on position.* Somebody at Apple *really* cares about icon design.

The bottom line: Often, you don't need to unlock the phone at all. Just consulting the information on the lock screen might be enough.

The Passcode

All right, but what if you do need full access to your phone? Society's evildoers and snooping roommates are locked out, but how hard is it for *you* to get in?

The first and most basic line of defense is the passcode: a password that unlocks the phone.

Yes, the iPhone offers face or fingerprint recognition to make unlocking easier. But you must *also* set up a passcode; it will always be the fallback method of unlocking.

In fact, even if you've set up face or fingerprint recognition, the iPhone still demands your passcode every now and then—for example, after you've shut down the phone, if you've gone two days without using the phone, if you make several failed attempts to log in with your face or fingerprint, and so on.

The very first time you turned on your phone, the setup process (Appendix A) invited you to make up a passcode. But if you skipped that step, or if you want to change your passcode, here's how it goes:

1. **Open** Settings→ Face ID & Passcode.

   On phones that have home buttons, this panel is called Touch ID & Passcode instead.

   Either way, you now have to enter your existing passcode, if you have one.

2. **Tap** Turn Passcode On (or Change Passcode, if you already have one).

   The iPhone suggests that your passcode should be six digits long.

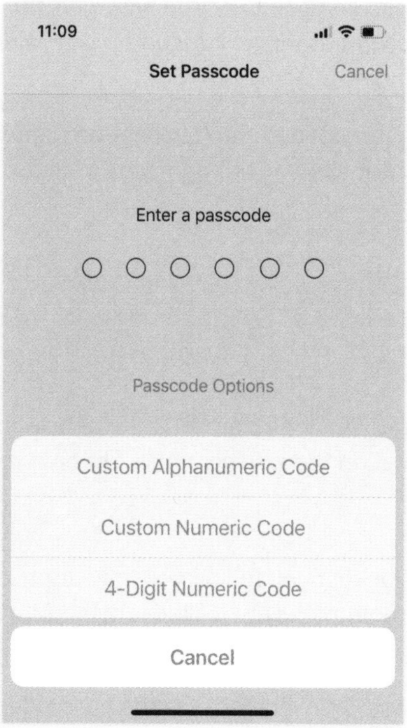

Set Passcode   Cancel

Enter a passcode

○ ○ ○ ○ ○ ○

Passcode Options

Custom Alphanumeric Code

Custom Numeric Code

4-Digit Numeric Code

Cancel

Creating a passcode

But there's some fine print on the screen: If you tap **Passcode Options**, you can choose instead a **Custom Alphanumeric Code** (any password, any length, using any letters, numbers, or symbols), **Custom Numeric Code** (all numbers, any length), or **4-Digit Numeric Code** (ATM style).

You have to retype the passcode, to rule out typos. Finally, you return to the Passcode screen.

3. **Set up the conditions for your passcode.**

For example, **Require Passcode** controls how quickly the iPhone requires your passcode again after the last time you entered it. The idea is to save your having to type it over and over when you're sitting at your desk and checking email every few minutes. (If you're using Touch ID or Face ID, **Immediately** is the only choice.)

Here, too, is where you can indicate which features and information bits are visible on the lock screen *without* entering your passcode: your **Notifications**, the **Control Center**, **Siri**, **Reply with Message** (answering text messages right from their notification bubbles on the lock screen), **Return**

Missed Calls, and so on. These features save you a lot of tedious unlocking for simple information checks—but if you worry about somebody sneaking a look when you're not around, you can turn them off selectively.

> **NOTE:** This screen also houses the nuclear option, **Erase Data**. If you turn this on, the iPhone *erases itself completely* if somebody makes 10 incorrect attempts to type in your passcode. Clearly, that person is not you and should not have access to your information.
>
> On the other hand, this could be a disastrous option if you're impaired after a night on the town.

You've now created (or changed) your passcode.

Next step: setting up the feature that spares you from having to type it 273 times a day.

Face ID

Passwords in general are a fussy inconvenience—and they're not even especially secure. All somebody has to do is spy as you unlock your phone, and presto: They now know your passcode.

Facial recognition is a different story. Once you've set up Face ID, the face-recognition feature of the iPhone X and later models, your face is the only one that unlocks the phone—period. Even if somebody goes to the trouble of making a life-size photo of you, or a mask, or even a flawless model of your head, it won't work—only your actual, living, breathing face can unlock

---

MAKING FACE ID WORK FOR YOU

A set of extremely useful options rests quietly in **Settings→Face ID & Passcode**. Once you've set up Face ID, they're worth exploring.

Require Attention for Face ID means you also have to be *looking* at the phone, eyes open, for Face ID to work. Now nobody can unlock the phone while you're asleep by holding it up to your face. On the other hand, you'll have trouble unlocking the phone if you're wearing sunglasses or can't open your eyes.

The **Attention Aware Features** have nothing to do with unlocking the phone. Instead, they use the iPhone's TrueDepth camera to detect *when you're looking at the screen*.

When you are, the screen doesn't dim to save battery power, it doesn't lock after five minutes (or whatever your autosleep setting is), and your morning alarm rings a little bit quieter. After all—the phone sees that you're already up. ✦

You can't actually see the infrared beams, but this Apple video shows what you'd see if you could.

The phone learns your face.

Move your head slowly to complete the circle.

How Face ID works

the phone. (Well, either yours or your identical twin's. Let's hope you have a good relationship.)

Face ID works by blasting 30,000 invisible infrared light dots and examining how they distort when they hit your face. Its TrueDepth camera system accurately measures the contours of your face, even if you change your hairstyle, glasses, makeup, skin color, or facial hair. Face ID *doesn't* recognize you when you're wearing a face mask, but hats and scarves don't faze it.

Face ID also still works if you gain or lose weight, or as you age; it continually updates its conception of your face over time. (You will, however, have to retrain it if you have radical plastic surgery and enter the Witness Protection Program.)

You can set things up so nobody can unlock your phone by holding it up to your face while you're asleep; see the box on the previous page. A law-enforcement official can't force you to unlock the phone, either; see page 69.

Face ID isn't only for unlocking the phone. At the top of the Settings→Face ID & Passcode screen, you'll see other features that facial recognition can make faster and easier for you: making purchases on Apple's online stores (iTunes & App Store), using Apple Pay, filling in your names and passwords on websites

(Password AutoFill), and confirming transactions in financial, insurance, airline, shopping, and password-locker apps.

> **NOTE:** Apple wishes you to know that the iPhone never transmits the scan of your face. Apple doesn't have it, and it's never on the internet. It's locked deep within a protected security chip on the phone itself.

To set up Face ID, open **Settings→Face ID & Passcode**. Enter your passcode to proceed (as you now know, you can't use Face ID without one).

Tap **Set Up Face ID**. The iPhone shows your current image and asks you to trace a circle in the air with your nose. Gradually, the tick marks around the circle fill in, as Face ID learns your face from every angle. Then you have to do it a second time.

> **TIP:** If you tap **Set Up an Alternative Appearance**, you can repeat the process to store a second facial model. Maybe you want your partner or child to be able to unlock the phone. Or maybe you just look really different sometimes.

From now on, whenever you wake the phone, just look straight at it to unlock it. The little white 🔒 opens to show that Face ID has worked. Now you can read all your notification messages right there on the lock screen.

But if your intention is to proceed to the home screen and actually *use* the phone, make a quick, short swipe up from beneath the screen to blow past the lock screen.

> **TIP:** If Face ID doesn't work, and the phone asks for your passcode, make *another* quick swipe up from the bottom of the screen. That's how you force the phone to attempt another Face ID scan.

Fingerprint Security (Touch ID)

Older iOS 14 iPhones, from the iPhone 6s through the iPhone 8, don't have that facial-recognition business. They have a home button, which doubles as a fingerprint reader called Touch ID.

It's pretty darn good: It can read your fingerprint from any angle, and nobody can fool it with a plastic finger or even a detached finger. Your phone can learn to recognize as many as five fingerprints, whether they're all yours or attached to other people you trust.

If you didn't set up your fingerprint during the iPhone setup process (Appendix A), you can teach it or retrain it at any time in **Settings→Touch ID &**

Passcode. Here again, you can't set up Touch ID unless you've already created a passcode; and here again, you'll need that passcode every time you try to add or change a stored fingerprint.

Tap **Add a Fingerprint.** As directed, put your finger right onto the home button.

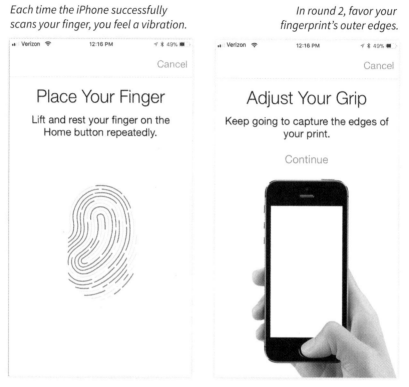

Touch ID setup

(Most people use their thumb or index finger.) You're supposed to touch the home button repeatedly, until the little fingerprint image on the screen darkens fully.

Now the phone tells you to Adjust Your Grip; tap **Continue** and repeat the process, this time tilting your finger slightly to give the sensor a better look at the edges of your finger.

> **TIP:** At this point, you can rename the fingerprint by tapping its current name (**Finger 1** or whatever it says). You can also delete a fingerprint by tapping its name and then hitting **Delete Fingerprint.**

From now on, you can simultaneously wake and unlock the phone by pressing the home button and *leaving your finger on it* for about a second.

Auto-Type Your Email Address

It's one of the greatest features on the iPhone, and maybe 2% of the world uses it: auto-typing. They're short bits of text—little abbreviations or codes that you make up—that magically expand into much longer phrases, saving you time and eliminating typos.

The classic example: Set up @@ so it types your complete email address into web forms, which is especially useful if yours looks like augustine.anastasia@ millenniumcorp.com.

But you might also set up the code *addr* to expand into your complete name, address, and phone number. Or maybe you do a lot of email work, and you find yourself typing the same phrases over and over. Set up one of these self-expanding abbreviations, and your fingers won't evolve into hideous claws from repetitive keyboard-tapping.

To create one of these expanding abbreviations, open Settings→General→ Keyboard→Text Replacement. Tap +. Now it's easy: Type or paste the expanded phrase you want into the Phrase box, and your abbreviation into the Shortcut box. Then tap Save.

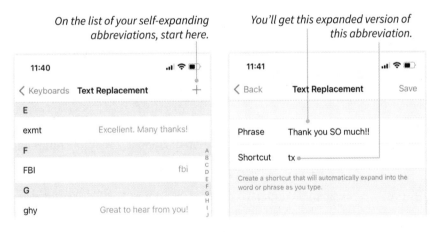

On the list of your self-expanding abbreviations, start here.

You'll get this expanded version of this abbreviation.

Set up auto-typing

The next time you type your abbreviation, in any app, marvel as iOS instantly swaps it out for the fully expanded version.

Adjust Auto-Sleep

As you may remember from Chapter 1, you're not supposed to shut your iPhone fully off every time you finish a task. Instead, you put it to sleep by clicking its side button.

But you can also set up the phone to go to sleep automatically if it notices that you haven't been doing anything for a while. To specify when that happens, open Settings→Display & Brightness→Auto-Lock. You can choose any interval from 30 Seconds to Never (which means the phone will stay on until you click the side button or its battery dies, whichever comes first).

Chill Your Notifications

Notifications are the message boxes that slide onto the top of your screen: alerts, messages, and warnings. They appear to let you know—well, everything.

You get a notification when an email comes in, when a text message arrives, when there's been a charge on your Apple Card, when there's a new Facebook or Twitter post, when a calendar appointment arrives, when you're entering or leaving Low Power Mode, when Maps needs your attention, when someone edits one of your shared Notes or Reminders…and on and on.

If you don't take control of the situation, you may discover that entire days go by and you've gotten no work done. You've been too busy dismissing notifications.

Processing Notifications

Here's what you can do when you see a notification appear:

- **Ignore it.** The iPhone can display two kinds of notification banners: *temporary*, which appear briefly and then disappear (great for things like incoming email), and *persistent*, which stay on the screen until you swipe them away (good for alarms and flight updates). Neither interrupts your work. You can keep on doing whatever you're doing.

- **Swipe up to dismiss it.** Flick a notification upward to get rid of it right now.

- **Tap to open it in its app.** For example, tap an email notification to open it in Mail, a News notification to read the full article, or an Uber notification to see where your driver is.

- **Long-press to process it on the spot.** You can act on some notifications right on the banner, without having to fire up any app. For example, you can

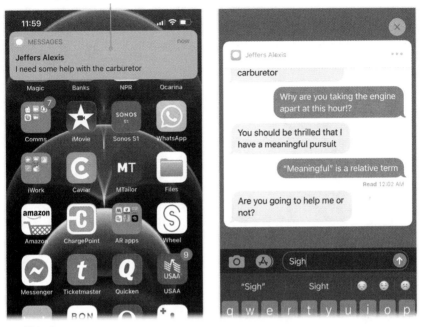

Notifications

reply to a text message, delete or mark an email as read, mark a Reminder as done, accept a calendar invitation, and so on.

The trick is to long-press the banner. It expands, revealing some options. A text-message notification becomes a miniature texting window so you can reply. A reminder offers buttons like **Mark as Completed**, **Remind me in 1 Hour**, or **Remind me Tomorrow**. An email expands to show you the whole thing and offers **Trash** and **Mark as Read** buttons.

All of this is very handy, because you can process each new development without having to switch into a different app.

The Notification Center

Just because you've taken care of a notification bubble doesn't mean it's gone forever. At any time, you can open the Notification Center: a collected list of all recent notifications, even ones you've already seen. To open it, swipe down from the top center of your screen.

It looks a lot like your lock screen, complete with the current date and time, and whatever wallpaper you've put there. In fact, you can even swipe to the left for the camera or to the right for the Today widgets (page 102), just as you can on the lock screen.

But your main business here is looking over all your recent notifications. To save screen space, iOS *groups* most notifications, stacking them up by app. Tap a stack to reveal all the notifications within it. At this point, you can process them one at a time, or you can tap the ⊗ above them (and then confirm with Clear) to clear them all at once.

Tap a stack of one app's notifications...   *...to make them spring apart.*

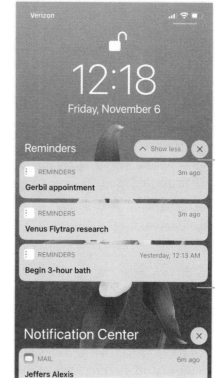

Expand-O-stacks

As usual, you can open a notification into its corresponding app by tapping it (an email opens into Mail, a text message opens into Messages, and so on).

But in the Notification Center, if you drag your finger leftward across the notification bubble, you reveal three new buttons:

- **Manage offers two big blue buttons** that can tone down the notifications from this particular app. If you tap Deliver Quietly, then this app's notifications won't chime, vibrate, display a bubble, sprout a number badge on its icon (⑰), or even show up on the lock screen. It's *almost* as though this notification didn't happen—but it does show up here, on the Notification Center.

The **Turn Off** button is a more definitive shutting up. Tap it, and then confirm your decision, to turn off this app's notifications completely. It's your way of saying, "When I want to know about this app's activities, I'll open it."

Swipe left to reveal these buttons. *If you tap **Manage**, you can quiet this app.*

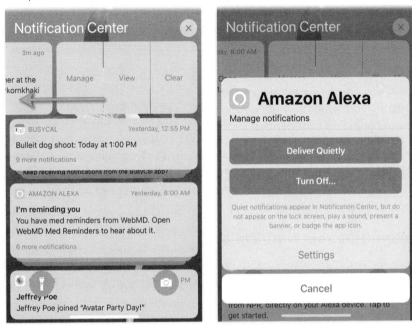

Managing notifications

If you silence enough apps using these buttons, your storms of notifications will melt down to just a drizzle. And the apps you *do* care about will have a better chance at getting your attention.

> **TIP:** If iOS notices that you consistently ignore the notifications from a certain app, it proactively offers you **Keep** and **Manage** buttons. Tap **Manage** to view your bubble-quieting options. The phone is trying to say: "I couldn't help noticing that you never look at these. You want me to silence this app?"

- **View** lets you act on a notification right in place, exactly as though you had long-pressed the incoming notification bubble itself.

- **Clear** deletes the notification.

Notifications on the Lock Screen

Believe it or not, notifications can appear in yet a third place: on your lock screen (page 34). Which is handy, really, because it means that when you wake your phone after a break, you can see everything you missed all in one place.

You can work through the notifications here exactly as you would in the Notification Center: Swipe left for Manage, View, and Clear buttons; tap to expand a stacked group.

But there's more to notifications on the lock screen, because the lock screen is visible to *anyone*, including people who pass by your desk while you're up getting coffee. Some of the notifications they find there might be deeply personal or highly confidential.

You might, in other words, want the lock screen to let you know *that* an app has tried to notify you—but you might not want the *contents* of the notification to be revealed until you, the rightful owner, have unlocked the phone. Read on to learn how to set this up.

Customizing Your Notifications

This is a chapter about setting up the iPhone to make it the most useful and the least annoying. A big part of that job is controlling which notifications pop up. If you have 65 apps, and each notifies you about every little thing, you'll lose your mind.

So open Settings→Notifications. Here you can individually tailor the notification behavior of each app.

When you tap an app's name, you get to make changes like this:

- **Allow Notifications.** This is the master on/off switch for this app. If you turn it off, you'll never be interrupted by notifications from this app.

- **Lock Screen, Notification Center, Banners.** These are the three places notifications can appear. You get to choose, for this app, where its notifications show up.

  For example, you might not want your job-search app to display notifications on your lock screen, where your boss might spot them—but you're OK if they appear as banners when you're *using* the phone.

  Conversely, you might turn Banners off for the News app, so you're not interrupted every single time a new article comes in. But you *do* want them to collect in your Notification Center, so when you finally have time

12:27	..ll 🛜 🔋	12:27	..ll 🛜 🔋
‹ Settings **Notifications**		‹ Notifications **Facebook**	

Master list of apps	Settings for this app
Siri Suggestions ›	Allow Notifications ⬤
Choose which apps can suggest Shortcuts on the lock screen.	
NOTIFICATION STYLE	ALERTS
📷 **Air Canada** › Banners, Sounds, Badges	Lock Screen · Notification Center · Banners
ⓐ **Airbnb** › Off	○ ✓ ✓
📷 **AirVisual** › Off	Banner Style Temporary ›
📷 **Akinator VIP** › Off	Sounds ⬤
📷 **Alaska** › Banners, Sounds, Badges	Badges ⬤

Notification settings

for a good read, you can just swipe down from the top of the screen to see what's going on in the world.

- **Banner Style.** Here's where you choose whether this app's notifications should be the Temporary kind (which vanish after a moment) or the Persistent kind (which stay on the screen until you tap to dismiss them).

- **Sounds.** When this app needs your attention, do you want it to play a little sound as the bubble appears?

- **Badges.** Do you want to see a little numeric counter (❷) on this app's home-screen icon, showing how many notifications have piled up?

- **Show Previews.** Do you want the *contents* of the notification to appear in the bubble—the actual text/email/Facebook message, the details of the Uber pickup, the name of your alarm or to-do item? If there's a risk that somebody might stroll by and read something embarrassing or incriminating, then maybe not.

For each app, you can choose to show the preview Always, Never (you'll have to tap the notification to read it in its app), or only When Unlocked (previews will appear only when you're actually using the phone—not on the lock screen).

- **Notification Grouping.** Remember how notifications can save space by stacking up? Here's where you can turn that clumping behavior Off, just for this app. By App makes this app's notification bubbles clump together. Or you can choose Automatic, which groups banners logically: separate stacks for each individual conversation in Messages, correspondent in Mail, news source in News, and so on.

- **Other settings.** Certain apps have special, unique controls. For Mail, for example, you can change the notification settings for each of your accounts (and for your VIPs). Photos offers individual notification settings for Memories, Shared Albums, and so on. It all makes sense.

You're not expected to get all your notification settings right on the day you first meet iOS 14. You *are* expected to remember that you can silence the notifications from individual apps that become annoying.

Set Up Do Not Disturb

A "Do Not Disturb" sign, of course, is the classic hotel-room doorknob hanger that you put out when you don't want anyone to barge in. On the iPhone, it's exactly the same idea: There are times when you don't want any notifications to appear, chime, or interrupt your work.

Do Not Disturb quiets all interruptions, regardless of any app-by-app notification settings you've established. (A ☾ appears on the status bar to help you figure out why your phone seems so morose.)

This is not the same as airplane mode, which turns off the phone's antennas. In Do Not Disturb, calls, texts, emails, and other communications still arrive; they just don't announce themselves.

Do Not Disturb is the condition people turn on when they're entering a meeting, a movie, or a bedroom.

Turning On Do Not Disturb

The quickest way to turn on Do Not Disturb is to use Siri. Say, "Do not disturb." Boom.

The next-quickest way is to open the Control Center (page 52) and tap the ☾. It lights up to show that it's activated.

> **TIP:** If you long-press ☾, you can make Do Not Disturb turn itself *off* again after an interval. For example, you can make it suppress notifications **For 1 Hour,** **Until tomorrow morning, Until I leave this location,** or at the time your current appointment ends (assuming the phone knows from your calendar that you are, in fact, at some event).
>
> This is a fantastic feature. Millions of people have missed important calls and texts because they turned on Do Not Disturb for a movie or a meeting, and then forgot to turn it off afterward. Never again!

Do Not Disturb on a Schedule

If you're the kind of person who sleeps at night—hey, it could happen—you can set Do Not Disturb to invoke itself automatically, according to a schedule. You might set it up so your phone doesn't ring or light up between the hours of 11 p.m. and 8 a.m., for example.

You set this up in Settings→Do Not Disturb→Scheduled.

New trick for setting times in iOS 14: Drag up or down directly on the digits.

Too fussy? Then long-press the digits to open the regular number keyboard.

Do Not Disturb settings

Do Not Disturb While Sleeping

There are two problems with scheduled Do Not Disturb.

First, the Do Not Disturb settings don't give you the opportunity to set different sleeping schedules for weekdays and weekends. Second, ideally, you would stop getting bombarded by notifications an hour or so *before* your bedtime, so you're not wound up and unable to fall asleep.

That's why Apple invented Sleep Mode, a new feature in iOS 14. It addresses both of those problems—but it's built into the Health app, not the Do Not Disturb settings. See page 423.

Emergency Bypass Options

In principle, Do Not Disturb is a spectacular feature. It has saved people millions of hours of sleep, and allowed millions of meetings, movies, and dates to proceed uninterrupted.

In a few situations, though, you may not want this blockade to be quite so impenetrable.

First of all, most people think of Do Not Disturb as a feature that keeps the phone quiet *when they're not using it*—when it's on the bedside table or in your pocket. If you're actually using the phone, you might not care that you're getting incoming notifications. That's why Settings→Do Not Disturb offers the While iPhone is Locked option. When the phone is *unlocked*, then it *does* ring and vibrate—because if the phone is awake, so are you.

But there's a second consideration: What if somebody is trying to reach you desperately in the middle of the night? What if somebody depends on you—an older parent, a younger child, a needy boss? If Do Not Disturb were truly impassable, you would be unreachable in an emergency.

Fortunately, in Settings→Do Not Disturb, you can allow certain callers and texters to break through:

- **Allow Calls From.** Here you can set up exceptions for Everyone (all calls and texts come through), No One (nobody gets through), or Favorites, meaning people you have designated as Favorites in the Phone app (page 223). That's probably the setting you want, since these are obviously the most important people in your life.

  If, on your Mac, you've created *groups* of people in Contacts, they show up here, too. You might create a group called Essential Underlings or Respected Overlords, and set up Settings→Do Not Disturb→Allow Calls From to let them through.

- **Repeated Calls.** This option lets *anybody* break through Do Not Disturb if they call more than once within three minutes. The assumption is that such a person is probably trying to reach you urgently. You wouldn't want Do Not Disturb to prevent you from finding out that somebody's gone missing, somebody's been hurt, or you've just been nominated for the Nobel Peace Prize.

Do Not Disturb While Driving

According to carefully controlled scientific studies, people are *terrible* drivers. Distracted driving kills 2,800 Americans a year, and no wonder: At highway speeds, glancing at your phone for five seconds means you're driving blind for the length of a football field.

It's a terrifying problem, and Apple is trying to do something about it.

When you turn on **Do Not Disturb While Driving**, your phone detects that you're driving and turns on Do Not Disturb automatically. Your phone goes dark and silent, with no notifications, sounds, or vibrations to distract you.

If anybody tries to text you, your phone silently sends an automated text response, like "I'm driving. I'll see your message when I get where I'm going." And then, a second later: "(I'm not receiving notifications. If this is urgent, reply 'urgent' to send a notification through with your original message.)"

Now the sender knows that you're a safe driver, that you'll attend to this message later, and that they can get through if this is really an emergency. If they text you the word "urgent," you see the original message. You can pull over to read it, or say to Siri, "Read my new messages."

> **TIP:** Not everybody who texts you gets these automatic responses; in fact, you can control who does. In **Settings→Do Not Disturb→Auto-Reply To**, you can choose **No One** (nobody gets an auto-reply), **Recents**, **Favorites**, or **All Contacts**. In other words, strangers never get the auto-replies—only people in your Contacts app.
>
> You can also tap **Auto-Reply** on this screen to edit the first auto-response message. (You're not allowed to edit the second reply, the one about using the word "urgent.")

During Do Not Disturb While Driving time, the only sounds your phone makes are alarms and timers you've set, music you're playing, and GPS navigation instructions.

Setting Up DND While Driving

When you tap Settings→Do Not Disturb→Activate, you're supposed to tell the phone how you want it to *know* when you're driving:

- **Automatically.** The phone uses its own sensors to figure out when it's in motion. It turns on Do Not Disturb automatically.

> **TIP:** Unfortunately, even the iPhone's superior sensors can't figure out *where you're sitting.* Do Not Disturb kicks in even if you're in the passenger seat, where it's perfectly safe to text.
>
> In that situation, you'll see a Do Not Disturb notification on your lock screen. Long-press it; tap **I'm Not Driving.** Now you can return to the important matter of texting your friends.

- **When Connected to Car Bluetooth.** Most modern cars have Bluetooth: a wireless connection between your phone and the car so that, for example, you can use the car's sound system for playing music from your phone or making phone calls. When your phone connects to the car this way, that's a pretty good sign that you're going to be driving, and therefore it should turn on Do Not Disturb.

- **Manually** means you'll turn DND While Driving on and off manually, using the Control Center (page 52).

- **Activate With CarPlay.** CarPlay, another high-tech Apple feature built into many recent car models, puts a version of the iPhone screen image onto the car's own dashboard touchscreen (details on page 433).

Your phone connecting to CarPlay is another great clue that you're about to be driving.

> **NOTE:** Here's another terrifying automotive statistic: 15% of teenagers admit they text while driving. If you are the parent of a teenager, therefore, you might want to consider setting up DND While Driving so it can't be turned off.
>
> It's part of the Screen Time feature described on page 439. Turn it on for your kid's phone, and set up a passcode. Finally, turn on **Settings→Screen Time→Content & Privacy Restrictions→Do Not Disturb While Driving.** Now your teenager will be protected by Do Not Disturb and can't turn it off without your Screen Time passcode.

Personalize the Control Center

The Settings app contains over 1,000 individual controls and switches. (Yes, your present author actually did count them all. It's been a long pandemic.)

Frankly, they're not all equally important. And trundling all the way to the Settings app every time you want to adjust the screen brightness or turn off Wi-Fi...it gets old fast.

That's why iOS offers the Control Center, a compact panel that delivers quick access to the most essential settings on your iPhone, like volume, brightness, Wi-Fi, and Do Not Disturb.

Some Control Center tiles are expandable.
Long-press the "wireless" cluster...

...to see these deeper controls.
Then you can long-press Wi-Fi...

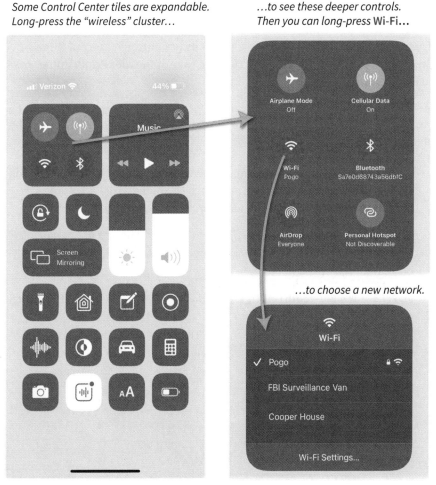

...to choose a new network.

Control Center

To open the Control Center, swipe downward from the right "ear" of your screen. (If your phone has a home button, swipe *up* from the bottom edge of the screen.) The Control Center slides into view, temporarily hiding whatever app you're using.

To close it again, tap any blank spot, or swipe up from the very bottom of the screen. (On a home-button phone, swipe down from the *top* of the screen.) You're right back in whatever you were doing.

> **TIP:** Although the Control Center is a quick way of making changes, it's not the fastest way. That would be Siri. You can say, "Make the screen brighter," "Turn on Do Not Disturb," "Turn on the flashlight," and so on. See Chapter 6.

The Core Tiles of the Control Center

The beauty of the Control Center is that it's customizable—sort of. A few of the items here are nonnegotiable. Here's a tour of the tiles Apple thinks are too important for you to remove.

> **TIP:** Most of the core Control Center tiles harbor secrets: You can *long-press* them to reveal further settings, as you'll learn on the following pages.

- **Airplane Mode (✈).** When you tap, the icon turns orange. You've just turned off all the phone's wireless features, in accordance with FAA regulations for phone use in flight. (You're also saving a lot of battery power.) Tap again to turn off airplane mode.

- **Cellular Data (ᵗᵖ).** This button controls your iPhone's connection to the cellular airwaves (page 71).

- **Wi-Fi (📶) and Bluetooth (✱).** Tap to turn your phone's Wi-Fi and Bluetooth off (white) or on (blue).

> **TIP:** If you long-press anywhere in this cluster of four buttons, a new panel opens. It's got the same four buttons *plus* icons for **AirDrop** (page 429) and **Personal Hotspot** (page 78).
>
> You can long-press on three of *these* buttons for even *more* options, as shown in the figure "Control Center" (previous page). Long-press the **Wi-Fi (📶)** button to specify *which* hotspot you want to join; long-press the **Bluetooth (✱)** icon to specify a Bluetooth gadget for connecting; and long-press **AirDrop (◎)** to specify who's allowed to send you files wirelessly. All of these used to require a tedious slog into the bowels of the Settings app.

- **Music.** These playback buttons control whatever app is playing sound at the moment: Music, Podcasts, Spotify, or whatever. They let you pause playback to chat with someone, skip a song you hate, or skip ads in a podcast, without ever interrupting what you're doing in an app. In fact, you can even use these controls on the lock screen.

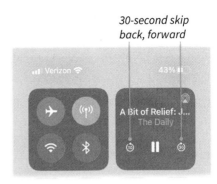

30-second skip back, forward

Podcast controls

Tap 🔊 to direct playback to a Bluetooth speaker, wireless earbuds, or an Apple TV or other AirPlay receiver (page 427). And if you long-press the tile, you reveal a few more goodies: album art, a scrubber bar for jumping around in the song, and a volume slider, for making volume adjustments faster than you could by pressing the volume buttons on the side of the phone.

- **Rotation lock** (🔄). In many apps, when you rotate the phone 90 degrees, the picture rotates with it. That's fantastic when you want to watch a video, for example.

  But sometimes, you don't *want* the screen to rotate; what if you're trying to read in bed, with your head sideways on the pillow? In that situation, tap this icon to lock the screen image in portrait (upright) mode.

- **Do Not Disturb** (🌙) is a quick way to turn on Do Not Disturb (page 48), so you won't be disturbed by incoming notifications. Long-press to schedule a time (for example, an hour from now, or tomorrow) when you want Do Not Disturb to turn itself back off again.

- **Brightness** (☀) controls the screen brightness. If you long-press, you reveal quick on/off switches for **Night Shift**, which makes the screen turn yellowish before bedtime (on the premise that the bluish light of electronic screens disrupts our melatonin production and makes it harder to sleep); **True Tone**, a feature of the iPhone 8 and later phones that adjusts colors on the screen so they look right in the current lighting conditions; and **Dark Mode**, described on page 87.

- **Volume** (🔊) controls the volume of your iPhone speaker, just like the buttons on the side of the phone. Long-press for a jumbo-size version of the same slider, which is better for making slight adjustments.

TIP: If you're the proud owner of AirPods Pro, the giant-slider screen also offers the on/off switch for **Noise Control** (background-noise cancellation) and **Spatial Audio** (a feature that, in some videos, creates a surround-sound effect).

The Home Controls

Below Apple's core icons, you're permitted to install a set of buttons that control whatever HomeKit-compatible smart-home gadgets you own (app-controllable light bulbs, security cameras, doorbells, and so on).

If, in Settings→Control Center, you turn on Show Home Controls, the Control Center automatically offers you tiles that control the home devices it thinks you might appreciate. Each one shows the status of the corresponding device, serves as an on/off switch, and offers more options if you long-press.

You also get a tile for Home Favorites, which is whatever set of gadgets you've established as your favorites in the Home app (page 337).

Design Your Own Control Center

Finally, at the bottom of the Control Center, Apple offers a set of smaller tiles. These, though, are optional; you're free to move or remove them in Settings→Control Center:

- **Flashlight (🔦)** turns on the iPhone's "flashlight"—the LED on the back panel that's otherwise the camera flash. This is one of the iPhone's most beloved features. You'll be flooded with gratitude in every dim restaurant, every time you can't find something in your bag, and every time you're trying to make your way along a dark pathway at night.

> **TIP:** Long-press to reveal a four-segment "slider" that controls the flashlight's brightness.

- **Timer (🕓)** opens the Timer tab in your Clock app, so you can set a count-down. It's handy for timing something in the oven, for timing mandatory music practice, or for seeing how many family members can hold a plank for two minutes.

  To specify how much time to count down, long-press the tile.

- **Calculator (🖩)** opens your Calculator app. That's a quick avenue to calcu-lating a tip, for example.

> **TIP:** If you need to copy the answer of your most recent calculation—when, for example, you need to remember how much you tipped—long-press to view a button called **Copy Last Result**.

- **Camera (📷)** opens the Camera app in a hurry, so you don't miss the magic moment.

But these are only teaser tiles! In fact, you can install as many as 25 of these buttons, if you don't mind turning the Control Center into a tower of tiles.

To look over your options, open **Settings→Control Center**. The Included Controls list indicates which ones you've already installed; the lower list, More Controls, awaits your inspection.

To install a tile onto your Control Center, tap the ⊕ next to its name. To remove one, tap the ⊖ and then confirm by tapping **Remove**. You can also drag the ≡ to specify the *order* of these tiles as they'll appear on the Control Center. See page 4 for more on using this type of list editor.

Tap enough ⊕ buttons in Settings, and you can create a very tall Control Center indeed.

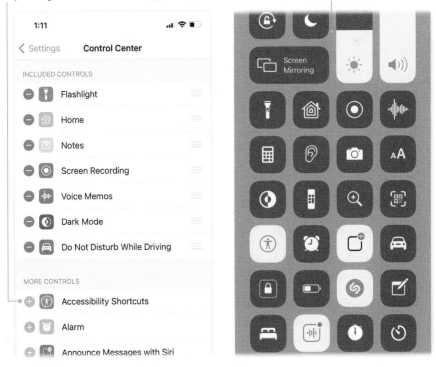

The Control Center catalog

Here's the complete list of what you can add to your Control Center:

- **Flashlight, Timer, Calculator, Camera, Home.** As described above.

- **Accessibility Shortcuts** opens a list of disability-accommodation features that you can use right now: Zoom, VoiceOver, AssistiveTouch, and so on. (They're all described starting on page 444.) This list matches the triple-click menu described on page 451.

- **Alarm** opens the Clock app to its Alarm tab. Here it's easy to set a new alarm—or preemptively turn off any upcoming ones.

- **Announce Messages with Siri** is an option exclusively for owners of Apple's wireless AirPods earbuds. When you're wearing your AirPods and a text message comes in, you hear a chime to get your attention—and then Siri reads the messages aloud; you don't have to lift a finger. Siri says, "A message from Alex Gorton says, 'I'm running late. So sorry.'"

  Siri listens for a few seconds in case you want to say, "Reply" or "Repeat that."

  > **TIP:** Long-press this button to turn on **Mute for 1 hour** (so Siri won't read you messages during an important meeting) or **Off for the day**.
  >
  > Note, too, that in **Settings→Notifications→Announce Messages with Siri→ Messages**, you can limit the announcements to people in your **Contacts**, or people you've corresponded with recently (**Recents**), or only people on your **Favorites** list (page 223).

- **Apple TV Remote** is an onscreen replica of your Apple TV remote control. It's perfect if you've lost your physical remote somewhere in the furniture.

- **Dark Mode**—what else?—turns dark mode on or off (page 87).

- **Do Not Disturb While Driving.** Here's a quick on/off switch for the feature that silences notifications on the road (page 51).

- **Guided Access** is the iPhone's kiosk mode, otherwise known as padded-walls mode. If you ever hand your phone to a toddler as an entertainment device, you'll need this. See page 449.

- **Hearing.** Here's the on/off switch for a supercool iPhone feature, called Live Listen, that's useful in noisy restaurants: You can set your iPhone down in front of whomever's talking. Its microphones transmit the audio directly to AirPods, or compatible hearing aids, that you're wearing. See page 425.

- **Home** is yet another way to access your Favorites among your smart-home devices, as you've set them up in the Home app (page 337).

- **Low Power Mode** is an on/off switch for the iPhone's battery-saving feature (page 25).

- **Magnifier** turns your iPhone into an illuminated magnifying glass. See page 63.

- **Music Recognition,** which Apple added in iOS 14.2, makes your phone listen to whatever music is playing around you—and, after only a few seconds, identify the song, album, and performer. It's just like the Shazam app! (Actually, it *is* the Shazam app, which Apple bought in 2018.)

- **Notes.** You're out and about, and suddenly you have to write something down: an idea, a recommendation somebody just mentioned, the phone number of a cute stranger. With one tap on this button, you're in your Notes app, ready to jot down the information.

> **TIP:** If you long-press the button, you get four additional commands: **New Note**, **New Checklist**, **New Photo**, and **Scan Document**. These shortcuts to specialized Notes features save you time when you're trying to quickly photograph, scan, or write something down.

If you've opted to make the Control Center available at the lock screen (in Settings→Face ID & Passcode→Allow Access When Locked), you might have yourself a little security problem. What if somebody picks up your phone, taps this tile on the Control Center, opens Notes, and starts reading your innermost thoughts and recipes?

Not to worry. Opening Notes from the lock screen doesn't show you any other notes but the one you've opened.

Meanwhile, in Settings→Notes→Access Notes from Lock Screen, you can further control what happens when you open Notes from the lock screen. Off means you can't open Notes at all. Always Create New Note means you get a new, fresh, empty Notes page each time. And Resume Last Note reopens whatever note you last opened—either the last one you Viewed in Notes App or the last one you Created on the Lock Screen.

> **TIP:** Even if you choose **Created on the Lock Screen**, you'll still get an empty, new note if a certain amount of time has elapsed since you last looked at your lock screen note. That's the purpose of the time intervals listed here. Yes, it's another security protection. Yes, this feature is getting complicated.

- **QR Code Reader** opens the Camera app to scan one of those square bar codes (page 438).

- **Screen Recording.** Your iPhone, believe it or not, can capture videos of everything happening on its own screen (page 437)—a useful feature for training or demo videos. This is a rare bird: an iOS feature that's available *only* in the Control Center. (Most of the other tiles are just shortcuts for Settings settings and apps.)

> **TIP:** Long-press the Screen Recording tile to open an options screen. Here you can turn the **Microphone** on—useful if you intend to narrate what you're doing in an onscreen demonstration. You may also see a list of apps here; that's so you can specify where to *send* the video once you're finished capturing it. (Most people just send it into the Photos app.)

- **Sleep Mode** is new in iOS 14 (see page 423). It incorporates three features, all designed to assist you with getting better sleep.

  First, Do Not Disturb can turn on well *before* bedtime, so you're not all riled up when you hit the pillow. Second, Sleep Mode can automatically run an app or play some music to get you in the mood. Finally, Sleep Mode can wake you in the morning with an alarm.

  Ordinarily, Sleep Mode kicks in automatically at the appointed time; this Control Center tile gives you manual control.

- **Sound Recognition** is the on/off switch for a new iOS 14 feature that's designed to help if you have trouble hearing. It notifies you if it detects certain sounds in the room: alarms, barking or meowing, car horns, someone ringing the doorbell or knocking, water running, a baby crying, and so on. See page 448.

- **Stopwatch** opens the Clock app to its Stopwatch tab (page 325). You're all ready to time somebody running a 100-yard dash or reciting all 50 state names.

- **Text Size.** Here's an infinitely more direct way to manipulate the type-size control that's ordinarily hiding in Settings→Accessibility→Display & Text Size→Larger Text. You get a vertical slider, which instantly changes the type size in most apps.

- **Voice Memos.** Tapping this tile opens the Voice Memos app (page 370), so you can tap the round, red Record button to begin recording audio.

But if you long-press this tile, a menu lists your three most recent recordings (for instant playback), as well as a **New Recording** button. Choosing **New Recording** is the quickest way to start recording audio on your phone, because it saves you the extra step of tapping the ⏺ button in the app.

- **Wallet** opens the Wallet app. If you have an Apple Card or some other credit card set up, here's where you can check your latest balance and recent transactions. If you long-press, you can choose to view *either* that credit card or your **Last Transaction**.

It may take you a few attempts to design the Control Center exactly the way you like it for maximum convenience. But it's worth taking the time; over the life of your iPhone, you'll save literally *minutes* of fiddling.

Get Your Languages Ready

In theory, your iPhone's menus, icons, and error messages already speak whatever language you chose the first time you set it up.

But in Settings, you can change the iPhone's language—or get it ready to switch among languages as you type. Polyglots, this one's for you.

To add a second language, open **Settings→General→Language & Region**. The **iPhone Language** indicates the iPhone's primary language (for menus, button labels, and so on).

But your job here is to tap **Add Language** and then choose from the long list of lovely languages. For each new language, iOS asks if you want to make it your new primary language or **Keep English** (or whatever your primary language is).

If you've added a few languages, you can now drag them up or down the list to prioritize them. Later, the ones at the top of the list here will also be at the top of the languages menu when you're tapping into your apps.

Now, not all languages use the 26-letter Roman alphabet. Some, like Japanese and Chinese, use different symbols; some, like Hebrew and Arabic, you read right to left. And some languages use the Roman alphabet but on a keyboard with a different layout from the U.S. one. (For example, pressing the ; key when you're typing in Swedish produces the ö symbol.)

Therefore, the second part of your language-installation job is to install a corresponding *input method*. Often that means a keyboard layout, but sometimes it's a completely different method of entering characters. For Japanese,

Tap to bounce between two keyboard layouts; hold to view the menu.

When he woke up, he knew that the machine had teleported him into Korea. The first thing he saw was a sign that said:

Keyboard Settings...

| English (US) |
| Emoji |
| Føroyskt |
| বাংলা |
| 한국어 |

Ok

U I O P

ㅓ J K L

ㅌ N M ⌫

return

When he woke up, he knew that the machine had teleported him into Korea. The first thing he saw was a sign that said:

하지만 이런 상황에서 복귀 절차를
이런 가운데 최초다 😑|

⊞ Aa ✓ 📷 Ⓐ ✕

" 😑 "

#123 ㅣ · ㅡ ⌫

ABC ㄱㅋ ㄴㄹ ㄷㅌ ↵

한글 ㅂㅍ ㅅㅎ ㅈㅊ

😊 → ㅇㅁ .,?! 간격

Multilingual typing

you might want to select a Kana keypad. For Simplified or Traditional Chinese, you can draw your symbols onto the screen with your fingertip.

In **Settings→General→Keyboards→Add New Keyboard**, the iPhone proposes, under **Suggested Keyboards**, input methods that match the languages you've chosen so far. Tap the one you want—or, if you're an oddball, one of the **Other iPhone Keyboards**—and you're on your way.

Now then. Suppose you're typing along, and you suddenly feel that *je ne sais quoi*, that feeling of *ennui*, that sense that English-only documents are your *bête noire*.

Just choose the new language's name from the ⊕ key below the standard iPhone keyboard. (If you long-press it, you get a menu of input methods; if you just tap it, you hop back and forth between the most recent two you've used.)

In other words, that ⊕ key lets you freely mix languages and alphabets within the same document; you never have to switch out to some Settings panel to make the change.

Turn On Insta-Zoom

There are all kinds of reasons you might want to magnify what's on your screen. Zooming in on tiny type is one obvious example, but sometimes you'll see a tiny headshot icon on social media and want a closer look. Or maybe you're looking over a scan of a document, and you're having trouble making out a couple of the words.

The iPhone can magnify the screen to exactly the level you find useful. But until you turn it on, that feature lies dormant. Here's how you get it ready.

Open **Settings→Accessibility→Zoom**. Turn on **Zoom**. Immediately, a rectangular magnifying lens appears; anything inside it appears at 200% of actual size. You can move the "lens" around the screen by dragging its white handle.

Drag to move the lens; tap to change settings.

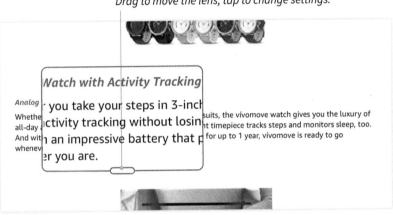

Magnifying lens

If using the magnifying lens feels like peering through a keyhole, you may prefer the *full-screen zoom* option. That's in **Settings→Accessibility→Zoom→ Zoom Region→Full Screen Zoom**. From now on, whenever you turn on zooming, you enlarge the entire screen.

No matter which zooming region you prefer, the key to the Zoom feature is *three-finger tapping*. For example:

- **Turn zooming on or off** by double-tapping the screen with three fingers. This works in any app at any time.

- **Scroll around the magnified screen** (or inside the lens) by dragging with three fingers.

- **Adjust the magnification level** with a double-tap/drag with three fingers. That is, double-tap with three fingers—but leave them down on the second tap, and drag them up or down the screen.

Fool around with this feature a little bit, and then don't forget you have it available. It's a slick trick that comes in handy.

Set Up Backup

Truth is, you don't encounter as many sobbing lost-data victims these days as you once did. In part that's because so much of our lives are now "in the cloud"—that is, stored online.

You hardly ever hear of people losing all their email, for example. Even if your iPhone dies, all your email is still there on Gmail, Yahoo Mail, Live Mail, AOL Mail, or whatever mail service you use. Hardly anyone loses their music collection, because most of the world now listens to music that streams from services like Apple Music or Spotify.

If you use iCloud synchronization (Chapter 15), then your photos, notes, reminders, address book, passwords, text messages, bookmarks, audio recordings, and documents are automatically duplicated on any other Apple gadgets you have (as well as on iCloud.com), making it unlikely that you'll ever lose your only copy of anything important.

That leaves only a few categories of stuff that you could ever really lose if your iPhone is stolen or destroyed, like your settings (your account settings, home-screen configuration, and any documents you've created on the phone but not synchronized)—along with any data you're *not* synchronizing with other machines.

The iPhone is perfectly happy to back this stuff up in either of two ways:

- **The wireless way: iCloud.** Open Settings→[your name]→iCloud. Scroll down to iCloud Backup and turn it on. From now on, whenever your phone is charging and in a Wi-Fi hotspot, it will back all that stuff up for you.

  Unfortunately, your iCloud account includes only 5 gigabytes of free storage. If your backup needs more space than that, you might have to pay for more iCloud storage (page 408).

- **The free way: your computer.** You can also back up your phone onto your Mac or PC—and this time, there's no charge.

First, connect the iPhone to the computer with the Lightning cable that came with your phone. If the computer is a Mac running macOS Catalina or later, open any Finder window. If it's a Windows PC, or a Mac running an earlier version of macOS, open the iTunes app on the computer.

Click your phone's icon to see what's on it. Now you're ready to back up—or restore.

Mac iPhone backup

Tap whatever **Trust** buttons appear, and supply whatever passcodes the screens ask you for. Apple is just trying to make sure you're not a bad guy extracting information from a stolen phone.

Click the icon of your iPhone in the sidebar at the left side of the window (Finder or iTunes), and then click **Back Up Now**.

Then go away for a while; the process isn't quick.

NOTE: Remember to repeat this process periodically, so your phone is always backed up.

If you drop your iPhone while ice skating and the Zamboni machine pulverizes it, you'll be happy you made this effort.

Once you have your new phone, you can restore all the information and settings to it. During the setup procedure (Appendix A), you encounter the Apps & Data screen. Here you can tap either **Restore from iCloud Backup** or **Restore from Mac or PC**.

Eventually, you'll see a list of your recent backups, so you can choose the one you want. Usually, the most recent one is best.

> **TIP:** You get only one chance to restore from your backup: during this setup process. If you didn't take the opportunity to restore from your backup at setup time, you'll have to erase the phone completely and start over.
>
> To do that, choose **Settings→General→Reset→Erase All Content and Settings**. You'll have to confirm your intention several times, and then enter your phone passcode or your Apple ID password.
>
> When it's all over, the phone will act as though it's brand-new. The setup process will once again invite you to restore your data from the backup.

Prepare the Emergency Screen

Some software engineers at Apple went to a lot of trouble to turn the iPhone into a fantastic emergency beacon. Whenever you're feeling threatened or endangered, it can perform a combination of these safety steps:

- **Make a loud, whooping alarm** to get people's attention and maybe scare away an attacker.

- **Call 911.**

- **Send text messages,** with a map of your current location, to friends or family you've specified in advance.

All you have to do is *tell* the phone you're in trouble, by pressing designated buttons. More on that in a moment.

Setting Up Emergency SOS

First, specify whom to notify when the emergency happens. You do that, weirdly enough, in the Health app.

Tap your icon in the upper right, and then tap Medical ID→Edit→add emergency contact. Your Contacts list opens. Choose the right person, the right cellphone number, and the right relationship to you. Tap Done.

Next, open **Settings→Emergency SOS**. Here you have some decisions to make about how much effort will be required to place the emergency call:

- **Auto Call** makes the phone dial 911 automatically (after a countdown) when you press the trigger keys (described in a moment). If **Auto Call** is turned off, then in the emergency situation, you'll have to confirm the call by swiping your finger across an **Emergency SOS** slider.

- **Call with Side Button,** available only on the iPhone 8 and later, requires some explanation.

The official trigger for the Emergency SOS call on these more recent phones is this: Hold in the side button *and* one of the volume buttons across from it. After two seconds, the standard shutdown screen appears, complete with the **Emergency SOS** slider. At this point, you can place the emergency call by swiping across it.

The Emergency SOS slider is always on the standard Shut Down screen.

But the five-click trick calls 911 with less time and fewer steps.

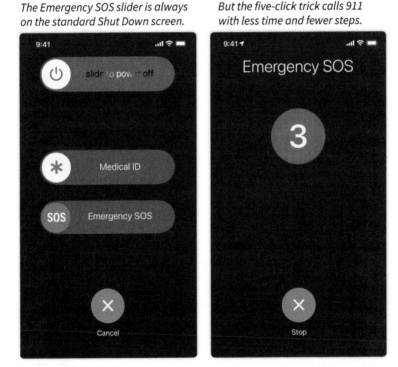

Emergency SOS slider

But if you *continue* holding in the two buttons for *another* couple of seconds, then a big, red countdown appears on the screen: 5, 4, 3, 2, 1. At zero, the phone finally places the emergency calls (911 and texting your loved ones).

All told, that's about eight seconds of holding buttons. If you're in mid-mugging, mid-fire, or mid–home invasion, eight seconds can seem like a very long time.

At last, you can now understand the appeal of the **Call with Side Button** option. It offers a much faster alternative: *Click the side button five times fast.* This is something you can do in your pocket or purse, without even looking at the phone. It produces a 3-2-1 countdown and then places the emergency call.

In other words, **Call with Side Button** and **Auto Call** give you the fewest safety nets, but the least delay, before calling emergency services.

- **Countdown Sound** is a loud, whooping alarm that precedes the emergency call. (It's an option only if **Auto Call** is turned on.)

MEDICAL ID ON THE SHUTDOWN SCREEN

While you're getting your phone ready for emergency situations, you should also enter your own medical information. Open the Health app, tap your icon in the upper right, tap **Medical ID**, tap **Edit**, and then fill out your birthday, medical conditions, medical notes, allergies, medications, blood type, and so on.

Confirm that **Show When Locked** is turned on. You should turn on **Share During Emergency Call**, too. Tap **Done**.

You've just made sure that if you're ever found unconscious or incapacitated, a rescue worker can pick up your phone, wake it, and tap (on the Enter Passcode screen) **Emergency→Medical ID**.

In other words, without having to know your password, first responders will have access to all the important medical details they might need to save your life.

And if you do use the Emergency SOS feature described on these pages, your phone will also share your important medical information with emergency services. ⭐

| 6:34 | .ıl 📶 🔋 |

✳ **Medical ID**   Done

Information

Casey Robin
March 6, 1963 (57 years old)
♥ Organ Donor
⊘ Sharing: Lock Screen

Medical Conditions
SEVERE BEE ALLERGY

Allergies & Reactions
Ragweed

Medications
Methoxalin

Blood Type
A-

Weight
188 lb

Height
6' 2"

Emergency Contacts

spouse

On one hand, it may draw the attention of people nearby who might be able to help you, and it could convince a mugger that maybe you're not worth it. On the other hand, it does alert an attacker that you're doing something with your phone. If you'd prefer to be able to make the emergency call silently, turn off the countdown sound.

Making the Call

If the worst should come to pass, and you find yourself in an emergency, press the side button of your phone five times fast. The phone begins the three-second countdown, produces the audible alarm (unless you've turned it off), and then places a call to 911.

> **NOTE:** Remember that, on an iPhone 8 or later model, the five-quick-presses thing works only if you've turned on **Call with Side Button**. If not, you must hold in the side button *and* one of the volume buttons to begin the emergency calling process.

After another 10-second countdown, the phone texts your designated contacts, and includes a map. The message says: "[Your name] has made an emergency call from this approximate location. You are receiving this message because [your name] has listed you as an emergency contact." If your location changes—maybe you've been kidnapped—the phone updates your loved ones with additional location messages.

> **TIP:** Whenever you do the five-fast-click thing, you do more than bring up the shutdown screen and the emergency countdown. You also *turn off* Touch ID and Face ID. At this point, neither your face nor your fingerprint will unlock the phone— only the passcode.
>
> Why did Apple design it that way? It was thinking of situations where you might be tied up, handcuffed, or unconscious. Now nobody—no mugger, not even law enforcement—can get into your phone without your cooperation.

Hook Up Your 'Buds and Speakers

Bluetooth, of course, is the short-range radio technology that lets phones, tablets, and computers talk to earbuds, headphones, speakers, car sound systems, keyboards, mice, and fitness bands—without wires.

Pairing Audio Devices

The first time you use a Bluetooth thing with your phone, you have to *pair* it—formally introduce it. The first step is making the Bluetooth thing "discoverable," usually by holding down a button until a light starts blinking.

6:26	.ıl 📶 🔋

‹ Settings   **Bluetooth**

Bluetooth   ⬭

Now discoverable as "iPhone dp".

MY DEVICES

Dad's AirPods Pro   Connected ⓘ

Sa7e0d68743a56dbfC   Connected

Sense   Connected ⓘ

Tesla Model 3   Not Connected ⓘ

Westport ✈

Not Your AirPods Pro

AirPods Pro not connected to this iPhone

Connect

Bluetooth pairing

Now, on the iPhone, open Settings→Bluetooth and confirm that Bluetooth is on. After a moment, the iPhone displays the Bluetooth thing's name. Tap it—and that's all there is to it. (Some gadgets may require you to enter a one-time passcode.)

If it's an audio device (headphones, speaker, car), then all sound—music, podcasts, phone calls, chirps—now comes out of those speakers or earbuds instead of the built-in speaker. You can control playback volume with the iPhone's buttons, and the Control Center is especially useful for controlling music or podcast playback (page 54).

Pairing AirPods

Apple's AirPods are much easier to pair. All you have to do is hold the charging case, with its lid open, *near* the phone. On the iPhone's home screen, tap Connect. If you have the AirPods Pro, you get a few additional screens of tips and tricks.

> **TIP:** If the technology gods aren't smiling and the **Connect** button doesn't appear, then holding in the tiny white button on the AirPods case works as an override.

AirPods are supposed to switch automatically from iPhone to iPad to Mac, if they're all running the latest OS versions. But you can always switch them manually on the Control Center. See page 54.

CHAPTER THREE

Cellular, Wi-Fi, and 5G

T he term *mobile phone* communicates the device's most important attribute: You can connect with the world wirelessly, even as you carry it around with you. And because these connections are so important, the modern smartphone has two different ways to get online: Over the cell network and over Wi-Fi.

The Cell Network

The big cellular carriers, like Verizon, AT&T, and T-Mobile/Sprint, have spent billions of dollars building nationwide *cellular* networks. They're cell towers in every populated area that exchange radio waves with our cellphones. These networks were originally intended to carry phone calls, of course; over time, they've evolved to handle internet data, too.

The bars at the top of your screen show how strong your signal is, which is usually related to your distance from a cell tower. (If you see "No service," then you can't make calls or go online over cellular.)

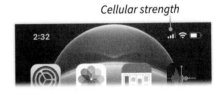

Cellular strength

Signal bars

Every few years, the industry collaborates on a new network standard that's supposed to bring us faster speeds and less network congestion. In 2003, 3G came along ("third generation"); in 2009, it was 4G (also known as LTE); and now, of course, all you hear about is the miracle of 5G. Each new standard requires that you buy—of course!—a new phone to take advantage of it.

The status bar at the top of your iPhone tells you which kind of network—which generation—you're connected to at the moment.

If you're lucky, you'll never encounter the E or 1xRTT° symbols there. Those ancient networks are so slow, they *barely* even count as internet. 3G is more recent but still makes you wait 20 seconds for a typical web page to open.

AT&T developed an enhanced version of 3G and called it 4G, much to the disgust of the rest of the industry, which now had to pick a new name for the actual fourth-generation network. What they came up with is 4G LTE.

LTE was the fastest cellular connection you could hope for—until 5G came along.

5G and You (and Your Expectations)

The biggest change in the 2020 iPhones—the iPhone 12 family and the iPhone Mini—is their ability to use 5G cellular networks. Of course, that means absolutely nothing if you don't know what 5G is all about.

On paper (or online), 5G cellular sounds incredible. Data speeds 25 times faster than 4G! Faster than your Wi-Fi at home or work! Download an entire movie in 30 seconds! Insane capacity, to the point where all cellphone plans are unlimited and full-speed!

In practice, though, there are enough footnotes to fill a podiatry journal. The cellphone companies and cellphone carriers are flooding the airwaves with misinformation, buzzwords, and branding, and it's up to you to sort it all out. The confusion boils down to these elements:

- **5G comes in three flavors.** All cell signals are just radio waves, zooming around at different frequencies. And a 5G network can rely on any of three different bands of frequencies.

  The really fast one, the one Apple and the carriers like to demonstrate, uses high, very short wave frequencies, technically called *millimeter-wave* frequencies. (Naturally, each cell carrier calls this 5G variety something different.)

  These bands can offer ridiculous speeds—4 GB per second, for example, which is over 40 times as fast as the average American's home broadband service. When you're in the presence of this kind of network, you'll see **5G+** (AT&T) or **5G UW** (Verizon) at the top of your screen.

> **NOTE:** T-Mobile/Sprint doesn't have a special indicator for its fledgling millimeter-wave coverage; you just see **5G** all the time. And if you're on AT&T, you may also see **5G E** on your screen. That's not 5G at all; it's AT&T's name for an enhanced LTE.
>
> Told you this was going to be confusing!

Unfortunately, you won't see those icons often. The tradeoff for millimeter wave's impressive speed is its pathetic range—a few hundred *feet*. It's also incredibly finicky. *Everything* blocks millimeter-wave signals: walls, buildings, trees, people, cars, rain, your hand.

Qualcomm, which helped develop the 5G standards (and manufactures the 5G chips inside iPhones), partly solved the blockage problem by intercepting signals that have ricocheted off of buildings and other objects. The iPhone figures out which angles the beams are coming from and returns the signal along the same path.

As for millimeter wave's very short range: The only solution is to put up a *lot* of transmitter boxes on light poles and utility poles. That's why 5G millimeter wave is a luxury that's limited to big cities—in fact, to *parts* of big cities.

You'll most often find it available at people-congregating places like stadiums, concert arenas, and convention centers. If you find it on the streets of a city, stop walking and do your downloading right there, because you probably won't get the same signal half a block away.

The other two kinds of 5G, known as low-band and mid-band, offer lower speeds but better range. A low-band cell tower, for example, can cover hundreds of square miles with signal; a mid-band tower can cover a few miles.

These 5G flavors can still give you better speed than 4G LTE, but it'll be nothing like 40 times faster. Once 5G towers go up in residential and rural America, you can expect a speedup of, for example, five times, and you'll see **5G** at the top of your screen.

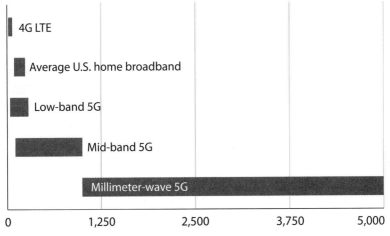

Typical download speeds, in megabits per second

- **Limited coverage.** Installing 5G cell towers is a slow, expensive, manual task that will take years to complete. In these early days, as you move around the country, discovering that you're on millimeter-wave 5G—or any kind of 5G, actually—remains a happy surprise. (Each carrier offers 5G coverage maps on its website, but of course you get different speed results depending on where in a city you're standing, or even how you're holding the phone.)

- **Battery life.** Connecting to a 5G network uses a lot more power than connecting to a 4G network. On 5G, you get 20% less time from each charge—about two hours less.

Apple didn't want its 5G iPhones to become known as the phones with the lousy battery life. So it invented Smart Data mode, which moves the phone between the 4G and 5G networks seamlessly and invisibly. It uses the 5G network only when it *needs* that speed; the rest of the time, it uses 4G or LTE.

In general, Smart Data mode makes its choice depending on whether you're *using* the phone, with the screen on. If you're using video chat, pulling up web pages, and scrolling through social-media posts, you'll be on 5G. But if the screen is dark, and it's just playing music or performing its usual background internet tasks, the phone uses 4G to save your battery. (Apple says you'll also get 5G when you're downloading big files, even if the screen is off.)

You can turn Smart Data mode off, of course; at your option, your phone can use 4G all the time (slower speeds) or 5G all the time (worse battery life). Visit **Settings→Cellular→Cellular Data Options→Voice & Data.**

3:40	..ll 🔋 ▪
‹ Back **Voice & Data**	
5G On ●————————	5G all the time
5G Auto ●———————— ✓	5G only when you need it (Smart Data mode)
LTE ●————————	5G off (best battery life)
5G On uses 5G whenever it is available, even when it may reduce battery life.	
5G Auto uses 5G only when it will not significantly reduce battery life.	

Smart Data settings

- **Who needs that much speed?** The last 5G footnote is this: Do you really *need* a lot more speed? How often do you actually download a movie to your phone?

  For now, the 5G benefits to you are fairly minor. If you have a 5G iPhone in a 5G city, you'll no longer get lower-quality video when you're streaming a movie on cellular. You won't have to wait till you're in Wi-Fi when you want to download a big app. You'll get higher-quality FaceTime video calls.

  In other words, 5G is a marvel not because people do a lot of file downloading on their iPhones. It's because of its potential for machines well beyond phones.

  It turns out that 5G also offers unbelievably low *latency*, which is the split second of lag *before* the data transfer begins. Low latency is a big deal for the future of self-driving cars, which will be able to exchange data with one another *instantaneously* ("Hey, Tesla three cars behind me—I'm slamming on my brakes right now!"). It could open the door for better remote surgery, where a lag in the response time of the surgical robot far away is a matter of life and death. And it could mean that people can, for the first time, make music together over Zoom. (At the moment, the lag is too severe; by the time you hear the other person singing their note, you're already behind.)

  Eventually, all this high speed and low latency will affect the creation of apps, too. Your "pipe" to the internet will be so fast that it will feel like a wired connection—to much bigger, faster computers than you could ever fit into a phone. You'll be able to use apps that require far more powerful processors than your puny iPhone can handle.

Expectations set?

May your days be filled with serendipitous wanderings into 5G+ or 5G UW hotspots, may your city apartment be across the street from a millimeter-wave transmitter, and may you never need remote surgery until high-band 5G is nationwide.

Wi-Fi Hotspots

The beauty of your phone's cellular connection is that it works almost anywhere, even out in the great outdoors. (*Service not available in all locations.)

But there's a second way to get online: Wi-Fi hotspots. These bubbles of internet service are usually limited to areas about 300 feet across. In a school or a company, the network geeks have probably set up Wi-Fi to cover the whole campus, but that won't do you much good when you're in a car zooming down I-95.

Still, Wi-Fi is beautiful. When you *are* in a store, coffee shop, home, school, office, hotel, library, or airport, Wi-Fi gets you online at speeds that can be much faster than most of the cell network. Often, there's no cost; usually, there's no data limit.

When you walk into a hotspot you've used before—your home office, for example, or a hotel you've stayed in—the iPhone auto-joins it. You get no notification or password request; just by walking into the room, you get online.

If you're in a new place, though, the iPhone might discover nearby hotspots and invite you to join one.

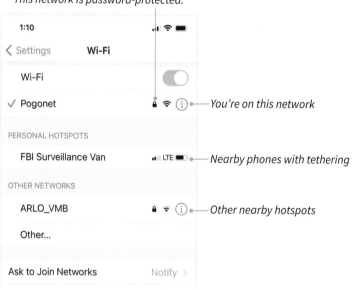

Wi-Fi settings

You can also look over the list of available hotspots on your own, in **Settings→ Wi-Fi**, or using the Control Center (see "Control Center" on page 53). Here, in a tidy menu, is a list of all the hotspots whose signal the phone has picked up. In these lists, a 🔒 indicates a hotspot that requires a password.

Commercial Hotspots

Sometimes, as in a hotel or airport, you'll find yourself within range of a *commercial* hotspot. These are hotspots that require you to pay money, look at an ad, sign in, or all of the above. In these situations, as soon as you choose the network's name, a miniature web page opens so you can complete the sign-up procedure.

Wi-Fi Troubleshooting

There are all kinds of reasons you may seem to be in a Wi-Fi hotspot, but you can't actually get onto the internet. For the sake of your sanity, here's a guide to all the things that can go wrong.

The first one is obvious: Most Wi-Fi hotspots don't let you connect without a password, which you'll have to get from the homeowner, office staff, waiter, or person who pays the bills. You'll know this is the problem because there's a 🔒 next to the hotspot's name in Settings or the Control Center.

Remember, too, that the number of signal-strength bars displayed next to each hotspot's name doesn't indicate how strong your *internet* signal is. It shows only how strong the *Wi-Fi transmitter's* signal is, wherever it sits in the building. That transmitter (base station) might not even be connected to the internet at all. Even if the internet's out in your neighborhood, you may still see a full three-bar 📶 signal. It's just saying you're well-connected to the base station; it says nothing about the internet connection beyond that.

In some situations—data-sensitive offices, for example—hotspots may have been set up to connect only to *specific* computers and phones, and yours isn't one of them.

Finally, you might be connected to a commercial hotspot, but you haven't yet signed in using its Welcome page. Pull out your credit card and get going.

NOTE: The world's most evil internet-access companies have been known to track where you go in the world, and sell this information to advertisers. They do it by grabbing your phone's MAC address (a unique address that's assigned to every single device connected to the internet—nothing to do with Mac computers), and watching which Wi-Fi hotspots it connects to.

In iOS 14, you can slam that privacy door shut. Open **Settings→Wi-Fi**; tap the ⓘ next to the name of the Wi-Fi network you're on right now; turn on **Private Address**. From now on, your iPhone will get a different address every time it hops onto this network, frustrating the companies whose algorithms are trying to watch your movements.

Personal Hotspot

Apple makes four kinds of devices intended to be mobile: watches, phones, tablets, and laptops. You can buy three of them with the built-in ability to get onto the internet over the cellular network—almost anywhere you go—instead of being limited to Wi-Fi hotspots.

But MacBooks are not so lucky. (What does Apple have against its laptops?)

At least there's a solution, called the *Personal Hotspot*. If you do have a MacBook, you can use your nearby iPhone as a glorified wireless internet antenna for it. That way your MacBook will feel as though it, too, has been blessed with the miracle of cellular internet.

Here's how to set it up. On the phone, open **Settings→Cellular→Set Up Personal Hotspot**. Make sure **Allow Others to Join** is turned on, and make a note of the password. Feel free to tap **Wi-Fi Password** to change it to something you prefer.

On the iPhone, in Settings, confirm the password.

On the Mac, choose your phone's name. You're now online.

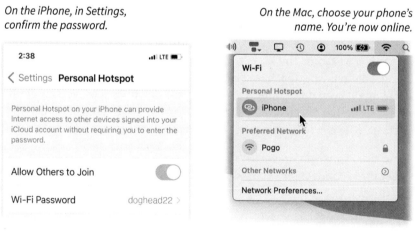

Personal Hotspot

Now, on your Mac, open the 🛜 menu—the list of available Wi-Fi networks—and choose your iPhone's name.

Like magic, the 🛜 symbol on your menu bar changes to ⊘…a blue bar appears at the top of your iPhone to indicate that another device is mooching its internet connection…and your laptop is now online, courtesy of the phone. You probably never even had to enter the password.

> **NOTE:** That no-password business is a feature Apple calls Instant Hotspot, and it's part of the suite of wireless connection features called Continuity (page 396). Continuity requires that each device has Bluetooth turned on, has Wi-Fi turned on, and is signed into the same iCloud account (page 393). If any one of those things isn't true, then you *do* have to enter the iPhone's Personal Hotspot password at this point.
>
> The password is also necessary, of course, for anyone *else* who wants to use your phone's Personal Hotspot.

There is a price to all this magic, however. The first is battery power: Personal Hotspot drains your iPhone's charge right quick. Unless the phone is plugged into power, it's probably wise to save this feature for quick email checks or web articles, not streaming Netflix miniseries.

There may be a financial cost, too. Check your carrier's website to see how much Personal Hotspot use is included with your plan. (They may call it *tethering*.)

On the other hand, Apple's version of Personal Hotspot is especially nice. First, your iPhone's hotspot name always shows up on your Mac's 🛜 menu, ready to use, even if the phone is asleep, and even if Personal Hotspot is turned off. Second, if the phone is running iOS 13 or later, your Mac stays connected to the hotspot even if you close the lid or put it to sleep. That way it can still download new email and messages when you're not actively using it.

The Sequence of Connections

By now, you realize that your iPhone can get onto the internet using various channels. But which one does it use when?

Whenever you or your iPhone wants to check the internet, it tries to get online following this sequence:

1. **Previous hotspots.** First, the phone looks for a Wi-Fi network you've used before. If it strikes gold, it hops on. You're instantly online, without any further red tape.

2. **New hotspot.** If the iPhone detects *new* hotspots nearby, it offers a list to join.

3. **The cell network.** If the iPhone can't find any Wi-Fi hotspots, it connects to the cellular network, like LTE or 5G.

Why is this important information? Because every now and then, you won't be able to get online at all. Safari will keep saying, "You are not connected to the internet," even though you seem to have both Wi-Fi *and* cellular service.

In these situations, the iPhone is trying to use Wi-Fi, but there's some problem with the hotspot: some glitch, some disconnection, some problem with the internet service beyond. It happens.

But as long as the iPhone thinks there's Wi-Fi available, it won't proceed to Plan B, which is trying the cell network!

The solution is to open the Control Center and *turn off Wi-Fi*. Now your phone moves on to the cellular network, and boom: You're actually online.

You may never know what the problem was with the Wi-Fi hotspot, but you won't care.

Wi-Fi on/off

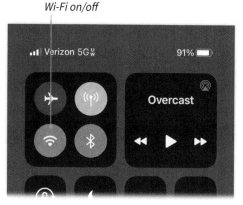

Turning off Wi-Fi

CHAPTER FOUR

Home Screen Extreme Makeover

or the first 13 years of its existence, the iPhone didn't earn much of a reputation for being visually customizable. You could choose your wallpaper. You could turn on dark mode. But that was about it.

With iOS 14, the drought is over. Now you can make your home screens look like almost *anything*. You can change the apps' icons. You can hide their names. You can hide their icons! You can fill the screen with *widgets*—miniature windows containing glanceable information. You can make your phone look utterly unlike any iPhone before it.

Brandon Basara

LilyDesignHaus on Etsy.com

SOSOBranding on Etsy.com

New looks for the home screens

This chapter covers everything you might see, do, and design on your home screens, before you ever open a single app.

Status Icons

As on any self-respecting computer, iOS on the iPhone offers you a little row of status icons—in this case, at the top of the screen. On recent models, where the "notch" creates a gap at the top middle of the screen, there's not a lot of room for these icons, but they're there. Apple really had to do some shoehorning, as shown on page 84.

> **NOTE:** On phones with a notch, some of these indicators don't appear until you expose the Control Center by swiping down from the right "ear" of the phone.

Some of the status icons, like the battery gauge, are self-explanatory. But others are more mysterious—and a few are brand-new in iOS 14:

- **▪▪▪**. Here's your cellular signal strength. Four bars means you're fairly close to a cell tower, and you can expect excellent call quality. (When you're not as lucky, you see the words **No Service** here.)

- **:::!** is your cell-strength indicator when your phone has two SIM cards (two phone lines), as described on page 32.

- **5G, 5G+, 5G UW, LTE, 4G,** etc. These icons indicate which kind of cellular network you're on (see page 72).

- **◥**. You're connected to Wi-Fi (page 76). Once again, the number of lines indicates how strong your connection is.

- **[Carrier name] Wi-Fi.** At some point, you accepted your phone's invitation to turn on *Wi-Fi calling*, which lets you connect your calls to your carrier's network over Wi-Fi—a great help if you don't have good cell service inside your home or workplace. (You can also turn this feature on manually in **Settings→Cellular.**)

- **⁎** appears when there's network activity—that is, your phone is trying to get data from the internet.

- **VPN**. You're connected to a virtual private network—an utterly secure connection to a corporate network (usually your employer) that a network geek set up for you. (When you disconnect, the icon looks like ⊘.)

- ➤ is part of Apple's attempt to make sure you're always aware when apps are collecting data. In this case, the ➤ arrow means some app on your phone is tracking your location. Usually it's obvious which one—it's probably a GPS app like Maps giving you navigation, or a weather app that needs to know where you are. (If the icon is hollow—⟁—then you're using an app that's *allowed* to access your location information, but it's not doing so at the moment.)

- ☎ means you've turned on Call Forwarding (page 239).

- ✈ means you've turned on airplane mode, and your phone can't use Bluetooth or its cellular antenna. (It may still be able to use Wi-Fi.)

- ▦ means you're using the accessibility feature called Teletype, which lets you make phone calls via a human typist who acts as an intermediary.

- ⊕ indicates that you've turned on Rotation Lock, meaning the screen image no longer rotates when you turn the phone in your hands (page 55).

- ☾ means Do Not Disturb is on (page 48). Incoming calls, texts, and notifications won't make the phone chirp, ring, chime, or vibrate.

- ⌷ tells you that your phone is locked. It'll take your face, fingerprint, or passcode to get past the lock screen.

- ⌒ means the phone is paired to Bluetooth wireless headphones or ear-buds (page 69). The icon can be helpful when your phone doesn't seem to be making any sound; you may not be aware that it's directing audio to your AirPods or some other wireless headset.

- ▬ shows your battery charge level. If there's a lightning bolt on it, the phone is charging. If it's red, you've got 20% charge or less. If it's yellow, Low Power Mode has kicked in (page 25).

- ▯ shows the battery charge of your Bluetooth earbuds or headset. Handy!

- ↻ means your phone has generously offered its cellular internet connection to a nearby laptop, tablet, or other Wi-Fi device. That is, you're using Personal Hotspot (page 78).

- ▷ means your phone is connected to your car, courtesy of CarPlay (page 433).

You may see patches of color on that little icon bar, too. On a Face ID phone, it's a little colored oval behind the clock; on home-button phones, it's a stripe across the entire top.

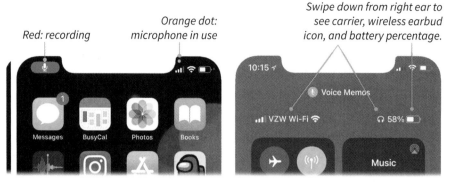

Red: recording

Orange dot: microphone in use

Swipe down from right ear to see carrier, wireless earbud icon, and battery percentage.

5 *Status icons*

In any case, the color is meant to make you aware that your phone is transmitting sound, video, its location, or its internet connection:

- **Blue** is a general privacy heads-up color. It means you're projecting your phone's image to an Apple TV (or another AirPlay receiver; see page 427), that an app is using your location, or that Personal Hotspot is on. (The latter two features also have their own icons. Apple just wants to make *really* sure you know what's happening.)

- **Red** tells you that your phone is recording. Either you're using the Voice Memos app or you're recording the screen (page 437).

- **Green** means you're on a call.

In iOS 14, Apple has introduced yet another color indicator: what appears to be a tiny LED dot. It appears in the top-right corner, just above the signal-strength bars (see "Status icons").

This dot appears when any app or iOS function is using your camera (green) or microphone (orange). For example, the orange dot appears when you're using Siri (to remind you that you're transmitting a voice recording to Apple's computers), and the green dot appears whenever you're using FaceTime, Zoom, or another video chat app (to remind you that you're on camera).

> **TIP:** The top of the Control Center (page 52) reveals the name of the app that's using your camera or microphone (or the app that did most recently). It might say, for example, "Zoom" or "Camera, recently." It's intended to answer the question: "OK, that little screen dot is on—what app is it talking about?"

Pick Out Your Wallpaper

You paid good money for your phone, and a decent chunk of it paid for your glorious, vivid color screen. You may as well show it off by choosing a great picture to use as its *wallpaper*.

On a computer, wallpaper is your desktop background—the tablecloth behind all your files, folders, and windows. But on the iPhone, wallpaper can appear in two places: on the lock screen and on your home screens. Either way, performing this small feat of interior design is an important step toward making the phone your own.

To look over your options, open **Settings→Wallpaper**. Here, for your reference, are your *current* lock-screen and home-screen images. Tap **Choose a New Wallpaper** to proceed.

The Choose screen offers you three prominent categories of options:

- **Dynamic** wallpapers incorporate movement: animated colorful bubbles, slowly pulsing and swimming through a black void. Fortunately, they're not fast enough to be distracting.

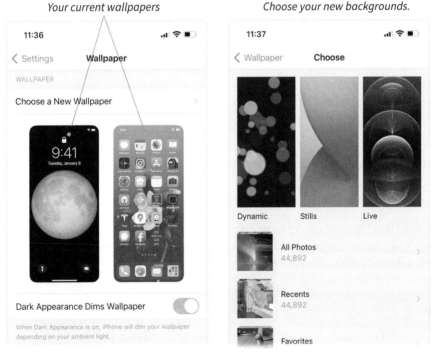

Your current wallpapers

Choose your new backgrounds.

Wallpaper options

- **Stills.** Many of these abstract paintings and nature images appear to have split personalities—and they bear the **◑** logo. If you install these, they look *different* in light mode and in dark mode. (And as you know from page 87, you can turn on dark mode either manually or on a schedule—like when the sun goes down.)

- **Live** wallpapers are a lot like Live Photos (page 181): They're not so much still images as three-second movies. When you install one onto the lock screen, and then long-press it with your finger, you see that three-second animation. (Nobody's quite sure why this is useful.)

On your home screen, these wallpapers don't do anything special.

> **TIP:** If you install one of your *own* Live Photos as a lock-screen wallpaper, it also plays its three-second loop when you long-press.

Below those three categories, you see something that should look familiar: all your pictures from the Photos app, in the same categories and albums. Any of them can serve as your wallpaper, although complex ones can make it hard to read the names of your icons on the home screen.

Once you've found a promising wallpaper image, tap it. It immediately appears on a sample lock screen.

> **TIP:** This sample lock screen always seems to be stuck at 9:41 a.m. on Tuesday, January 9. Anyone know why? Anyone? Anyone?
>
> Because that was the precise moment in 2007 when Steve Jobs unveiled the very first iPhone. (You may notice that whenever the iPhone appears in an Apple advertising photo, that's the time and date on it, too.)

If you've chosen one of your own photos, you can pinch or spread two fingers to adjust its size or drag with one finger to adjust its position within the screen frame, as shown on the facing page.

The little ⦚ is the on/off switch for Perspective mode, which makes the photo shift slightly when you tip the phone, as though it's a few inches away. At one time, people got really excited about this stuff.

> **NOTE:** If you've chosen a Live Photo, another icon appears here: ◎. That's the on/off switch for "Play the three-second video when I long-press the glass on the lock screen."

If it all looks good, tap Set.

Spread two fingers to zoom in.

Prepping the wallpaper

At this point, you can specify where you want this wallpaper to appear. Tap Set Lock Screen, Set Home Screen, or Set Both.

Dark Mode

In dark mode, the backgrounds of your windows turn dark gray, and the text is white.

Dark mode doesn't make you more productive. It doesn't save any battery power. *Maybe* it's less disruptive for somebody trying to sleep next to you in bed. Mainly it just looks cool.

To try it out, open Settings→Display & Brightness. At the top, tap Light, Dark, or Auto—which uses light mode during the day and dark mode at night.

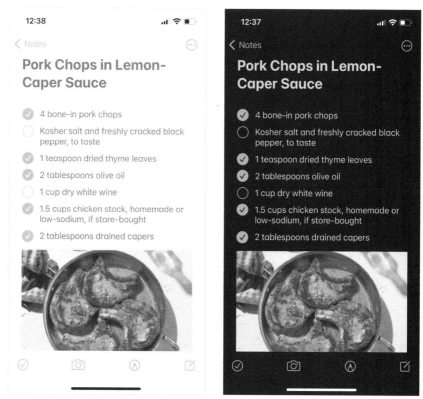

Light mode   *Dark mode*

Light vs. dark

> **TIP:** And what, exactly, does the iPhone define as night? That's up to you. When you turn on **Automatic**, you can tap the **Options** line and choose either **Sunset to Sunrise** or **Custom Schedule**, which lets you specify exactly which hours you want dark mode.

Of course, you can always turn dark mode on or off whenever the mood strikes you, using your Control Center (page 52).

You'll discover that dark mode mostly affects Apple's built-in programs, like the home screen, Calendar, Photos, Messages, Notes, and Mail. Dark mode doesn't interfere with color—for example, it doesn't invert any colors in your photos, the pages of your Books, or the web pages in Safari. And it may not affect apps from other software companies at all.

Give it a try. If you hate the look, restoring light mode is only one tap away.

Manage Your Apps

In the beginning, there was one home screen, and 16 app icons upon it. And lo, the people were sore amazed.

Nowadays, you can install all the apps you want. Each appears as an icon on your home screen—or, rather, home *screens*, because iOS sprouts more side-by-side screens as necessary to hold them all, up to 15 of them. Swipe your finger right or left to move among the screens.

At the bottom, a special set of four app icons doesn't move when you change home screens. This is the Dock: a place to park the four apps you think you'll use most. You never have to worry that you're on the wrong home screen to find them; they are, in effect, on every home screen.

The Dock and home screen

As apps became the dominant software form of our time, more apps filling more screens eventually meant more hassle trying to find a certain app quickly.

The obvious solution is to use Siri, at least when it's not rude to speak aloud. Say, "Open Notes" or "Open Camera." No matter how buried that app's icon, it opens instantly.

Another way to find and open an app: Use the search feature; see page 106.

But maybe the best tool you have against app chaos is to organize them, just as you would your closet or your desk. When you know where you can expect to find an app, you'll spend less time swiping around searching for it.

Some people organize the apps on their home screens by function, or in order of frequency of use, or even by color: All the red ones are on the first screen, all the blue ones on the second screen, and so on.

Deleting and Rearranging Apps

The key to commencing any app-organizing binge is to long-press anywhere on any home screen. (Doesn't have to be on an icon, as it was before iOS 14.)

You've just entered home-screen-editing mode, better known to iPhone fans as *wiggle mode*. If you've done it right, you'll see why: All the icons and widgets on your home screen are jiggling in place, like a 3-year-old who needs a potty break.

Another clue: Every app icon sprouts a ● button.

> **Tip:** If you long-press an app icon, a shortcut menu appears first—but after another couple of seconds of your pressing, it vanishes and wiggle mode begins.

Wiggle mode is the key to performing a couple of important tasks:

- **Remove an app** by tapping the ●. The phone offers you two choices: **Delete App** (you still own the app and can re-download it at any time from the App Store) and **Remove from Home Screen** (the app stays on your phone but is now visible only in the App Library described on page 103). You're not allowed to delete the apps that came installed on the iPhone.

> **TIP:** For the first time in iOS history, you're now allowed to leave a home screen completely blank. Minimalists and absurdists, this one's for you.

Long-press for the shortcut menu.

Ready to move or delete things

Wiggle mode

- **Rearrange your app icons** by dragging them into new positions. The other icons scoot out of the way.

  If you drag an app to the edge of the screen, the next home screen slides into view, as though you're flipping pages or changing channels. If you keep your finger down and continue the drag, you move the icon onto that second home screen. Or third, or fourth.

  If you drag an icon to the edge of the *last* home screen, you create a *blank* home screen. That's how you create new ones.

  Wiggle mode is also how you organize the apps on the Dock. Drag one away to open up a new slot on the Dock; now you can drag a different app into its place.

> **TIP:** You can even drag multiple icons at once—if you have the dexterity of a master coin magician.
>
> Here's the trick: Start dragging the first app. Move it just a little bit. At this point, you can tap other icons with another finger. Each flies magnetically to your first fingertip, and a number (❷) appears to show how many apps you've captured.
>
> When you've rounded up all the icons you want to move, drag your original finger, still down, to a new place: onto a folder, onto another home screen, or whatever. The icons land there in reverse order of your tapping.

Long-Pressing Icons

A handy menu opens when you long-press something on your home screen, whether it's an app, a folder, or a widget (page 95).

Most of these commands are unique to the icon you're tapping. For Messages, you get the names of your most frequent (or pinned) correspondence. For Photos, you can jump directly to Favorites, Most Recent, or Search. For Settings, you get direct access to your Battery, Cellular Data, Wi-Fi, and Bluetooth settings pages.

But these long-press menus always include Edit Home Screen (another way to enter wiggle mode) and Remove App (another way to delete an app).

Apps in Folders

To provide some assistance for your organizing binge, Apple offers you app *folders*.

There's no menu bar on an iPhone, and therefore there's no New Folder command. Instead, you create a folder by dragging one app directly on top of another one. The iPhone creates a new miniature window containing both of those apps. If they're both the same type—music, photos, games, or whatever—it even tries to name the folder for you. (You're welcome to type whatever name you want, though.)

To create a new folder, drag one app onto another.

When a folder gets full enough, it sprouts its own pages and page dots.

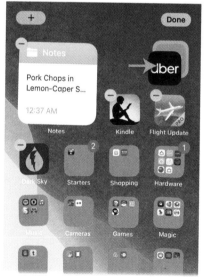 

App folders

To close a folder, tap anywhere outside it; at this point, you can drag more apps onto it to install them inside.

In wiggle mode, you can move a folder around just as you would an app icon.

When a folder is open, it's like a miniature home-screen network of its own. You can rearrange the apps by dragging them (in wiggle mode). You can remove an app from a folder by dragging it outside the folder window's borders. If you add enough apps to a folder, it sprouts additional "pages" that scroll sideways, making a folder's app capacity nearly infinite.

> **TIP:** You can even install a folder onto the Dock, where it becomes something like a pop-up menu of favorite apps.

To rename a folder, long-press its icon and then tap **Rename**. Or open the folder and then long-press its name.

To get rid of a folder, long-press its icon and then tap **Remove Folder**. If you confirm by tapping **Remove from Home Screen**, iOS gets rid of the folder but dumps all the apps that were inside it into your App Library (page 103). In other words, you never lose apps by deleting a folder.

Exiting Wiggle Mode

When you've finished fooling around with your apps and folders, tap **Done**, tap a blank area of the screen, swipe up from the bottom edge of the screen, or press the home button, if you have one.

> **NOTE:** If, at some point, you're feeling that you've made a real mess of your home screens, you can use the nuclear option: **Settings→General→Reset→Reset Home Screen Layout**. That function eliminates blank pages, consolidates all your apps into the smallest possible number of home screens, and puts all your downloaded apps in alphabetical order.

Secrets of the Home-Screen Dots

Beneath your app icons but above the Dock, a small row of dots appears. Most people grasp quickly that they represent your various home screens. If the far-left dot is lit up, you're looking at the first home screen. If the last dot is lit up, you're looking at the last home screen.

But there's more to this display. First, you can tap it repeatedly to cycle through your home screens—first screen-by-screen to the right, then screen-by-screen to the left.

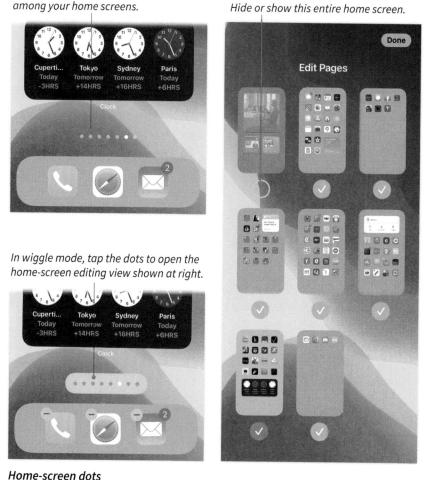

Tap or swipe the dots to move quickly among your home screens.

Hide or show this entire home screen.

In wiggle mode, tap the dots to open the home-screen editing view shown at right.

Home-screen dots

Second, in iOS 14, you can now *drag your finger across it.* That lets you zoom through your home screens much faster than you'd be able to by swiping. It's a useful technique if you have a huge number of home screens.

Hiding Home Screens

But the biggest deal with the home-screen dots is that they let you *hide home screens completely.* And yet they're hidden only temporarily—you can bring them back as needed.

And why would you want to do this? Apple discovered that millions of iPhone fans don't bother organizing the apps beyond the first home screen (and maybe not even on that one). A lot of people let their apps fall, as they're downloaded, pell-mell onto additional home screens, in an appalling state of disorganization. At that point, those people either use Siri ("Open Messages") to open the app they want, or they do a quick search for it (page 106). In iOS 14, they can exclusively use the App Library (page 103).

If that's how you open your apps, what's even the point of having home screens showing their icons? You may as well hide the screens you're not using.

> **TIP:** Remember, you're just *hiding* these home screens. At any point, you can return them to visibility, a quirk that opens up some clever possibilities.
>
> For example, you might fill one particular home screen with apps that you use only when you're on the road. Or only on the weekends. Or only when you're on kid duty. Or you might put all your time-killing apps on one page, which you hide during the workday so you're not tempted to get distracted.
>
> And then, when you can, in fact, justify playing a game or wasting time—the Department of Motor Vehicles comes to mind—you can unhide it and dive in.

Begin by entering wiggle mode (long-press anywhere on any home screen). Once everything is wiggling, tap the strip containing the home-screen dots. You enter something new in iOS 14: the Edit Pages screen (shown at right in the "Home-screen dots" figure.)

You're now facing miniatures of all your home screens. At this point, you can tap one to jump there, which is useful in itself.

But if you turn off a checkmark and then tap any blank area, then that home screen, and all the app icons on it, are hidden.

Now, don't panic. You still have access to all your apps—in the App Library (page 103). And, of course, you can also find them with a search, or by giving a command to Siri.

But in the meantime, you've streamlined your home-screen environment. And should you wish to bring one of these pages back, you can just repeat the process: Enter wiggle mode, tap the dots below the screen, turn the checkboxes on, and tap a blank spot. They're back.

Widgets

When you get right down to it, the iPhone and its archrivals, Android phones, are awfully similar. But one difference used to hit you between the eyes: Android phones have *widgets*.

These are little floating miniwindows that you can position directly on your home screens, nestled among your app icons. They reveal at-a-glance information, like today's weather, the value of your stock portfolio, breaking news, a photo, the next slice of your calendar for the day, and so on.

For many years, Apple sniffed at this obscene invention. Clearly, it violated the purity of the original apps-on-home-screen design.

But with iOS 14, the sniffing is over. Widgets have come to the iPhone.

Widgets at different sizes   The widgets catalog

Widgets on the iPhone

The Widgets Catalog

To add a widget, begin in the home screen where you'll want to put it. Enter wiggle mode (long-press anywhere).

Tap + at the top-left corner. And presto: a scrolling catalog of widgets for your inspection (see "Widgets on the iPhone").

If you scroll down past Apple's homemade widgets, you come to a more complete list—not only of the Apple ones, but of all the apps you have that offer companion widgets.

But choosing a widget to install is only the first step. The next step is choosing what *size* you want.

Most widgets come in small, medium, and large, which you can examine by swiping. (They occupy the screen space of two, four, and eight app icons, respectively.) Bigger ones show more information but, of course, eat up more of your screen.

Swipe to view the different sizes and designs for this widget.

Widget sizes

When you find the size you like, tap **Add Widget**. It now appears on your home screen—still in wiggle mode, so you can drag it to move it. As with app icons, you can drag a widget onto a different home screen, too. When the new widget is where you want it to be, tap an empty spot on the screen.

Once you've installed a widget, you may need to change some of its settings. For example, you'll need to tell the Weather widget what city you want to show, and the Notes widget which notes folder to show you. Once you're in wiggle mode, just tap the widget. It "flips around" so you can make your settings changes on the "back panel." Tap outside the widget to flip it back around and return to "work."

Here are Apple's starter widgets:

- **Batteries** shows your current battery level for the iPhone and—in the medium and large sizes—your AirPods and other connected Bluetooth gadgets.

- **Calendar** shows a slice of your calendar, so you can inspect your upcoming appointments.

- **Clock** shows the current time in one or four cities simultaneously. You specify which cities you want on the "back" of the widget. (The large size has room for the names of the cities.)

- **Files** gives you one-tap access to files you've recently opened in the Files app (page 327).

- **Maps.** This widget is a slice of the Maps app, complete with icons that let you quickly search for restaurants, gas stations, coffee shops, or any address. The idea is that if Maps is helping you navigate somewhere, this little window shows when you'll get there.

- **Music** gives you quick access to albums or playlists you've played recently. The small, medium, and large sizes offer one, four, or eight album covers to tap.

- **News** shows one, two, or four headlines from the News app, based on the topics you've indicated you'd like to follow.

- **Notes** lets you install, as a widget, a single note from your Notes app—something you might want for quick reference. Great uses for this idea:

frequent phone numbers, a Brainstorms note, an Account Logins list, or a packing list you've created in the app's checklist style (page 354). The medium and large sizes show the first couple of lines of several notes in the same folder.

Once you've installed this widget, you can "flip it around" to specify which note or folder it displays.

- **Photos.** Awwww, what a sweet idea: a floating panel that shows different photos from your Photos collection throughout the day; tap it to open the Photos app.

- **Podcasts** lets you pick up where you left off in one of your episodes in the Podcasts app. The small version of the widget shows the most recent episode you've played; the larger ones show the last few episodes. Tapping either opens the Podcasts app itself so you can begin listening.

HOW TO BUILD YOUR OWN WIDGETS

For the first months after the debut of iOS 14, the number-one most downloaded free app on the App Store was Widgetsmith. It's an app that lets you create your own widgets. Sort of.

Widgetsmith and similar apps, like Color Widgets and Widgeridoo, offer templates for widgets in various categories: calendar, battery, reminders, weather, health tracking, tides, astronomy—and then let you choose the font, colors, borders, and other aspects of your design. In other words, the widget will display canned information, but the design is yours.

(The resulting widget, once installed on your home screen, always bears the name of the creation app. You can't call it *Fancy Weather* or *My Clock*—it's called *Widgetsmith*.)

These apps have become very popular among people who use the design trick described starting on page 108 to overhaul the entire look and feel of the

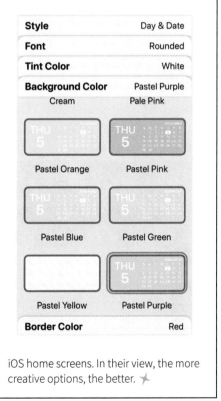

Style	Day & Date
Font	Rounded
Tint Color	White
Background Color	Pastel Purple
Cream	Pale Pink
Pastel Orange	Pastel Pink
Pastel Blue	Pastel Green
Pastel Yellow	Pastel Purple
Border Color	Red

iOS home screens. In their view, the more creative options, the better. ✶

- **Reminders** displays the single most urgent to-do item in your Reminders app (page 361)—or, in the medium and large versions, a longer list of them. To mark them as done, you have to tap the widget to open the Reminders app itself. (Flip the widget around to choose which reminders list you're seeing.)

- **Screen Time** gives you a quick and horrifying glance at how much time you've wasted on your phone today (page 439).

- **Shortcuts** gives you one-tap access to one, four, or eight of the self-running software robots you've created (page 451).

- **Siri Suggestions** (page 106) are the icons of apps or websites that Siri thinks you might enjoy right now, based on the kinds of things you usually do at this time of day and in this location. You might be surprised at how often the app you want is right here waiting for you.

- **Smart stack.** Here's a weird one: It's a widget that *changes* through the day. It might be the Music widget when you wake up, the Weather widget as you get ready for work, and then Notes as you dig into your workday. All of them sit in the same spot on the same home screen; Apple is trying to save you screen space by showing you only the widget you need at the time you need it.

  You can also flip through them manually, by flicking *upward or downward* on the widget.

In wiggle mode, drag one widget onto another of the same size.

Swipe up or down to move through the widgets of a stack.

Building a stack

To pull this off, Apple uses the same kind of intelligence—its discovery of your location, time, and activity patterns—that it uses to calculate your Siri suggestions.

> **TIP:** You can also build a widget stack of your own. It's the same idea—multiple widgets occupying the same space—except that this time, *you* get to choose the component widgets.
>
> To create a stack, begin in wiggle mode. Drag one widget on top of another. (They must be the same size.) You can add up to 10 widgets to a stack.
>
> Once they're stacked this way, you can long-press and then tap **Edit Stack** to view your options. You can drag the ≡ handles up and down to specify the order of the combined widgets. You can swipe left to remove one of the widgets from the stack. You can even turn on **Smart Rotate**, which attempts to give your homemade stack the same kind of anticipatory intelligence that Apple's own Smart Stacks have. Tap ⊗ to exit the editing screen.

- **Stocks** shows the current share prices and trend lines of three stocks on the small or medium widget, or six on the large one. You can't choose these stocks on the widget. Instead, the widget displays the contents of the "watchlist" you've created in the Stocks app (page 366).

- **Tips** offers periodic "helpful hints and hidden gems" (Apple's words) pertaining to iOS 14.

- **TV** is a quick way to pick up where you left off watching shows or movies in the TV app. The three widget sizes offer one, two, or three thumbnails of things you've been watching.

- **Weather.** The small size shows today's current, high, and low temperatures; medium adds a five-day forecast; and large shows a five-day forecast and hourly conditions for today. You specify which city's weather appears here by choosing on the back of the widget.

Set up the Widgets display just right, and you'll discover that these things save you a lot of web searches and app-opening when all you need is a quick lookup.

> **TIP:** Widgets are largely intended for looking, not touching. To make a change—to check off something on a to-do list, for example, or to choose different music to play—*tap the widget* to open the corresponding app.

Widgets Beyond Apple's

Apple's starter widgets are...a start. But widgets were such an instantaneous hit that thousands of them quickly flooded the App Store, from software companies as small as someone in their basement and as big as Google.

Usually, widgets are associated with full-blown iPhone apps. If you inspect your widgets catalog as described above, you see a long list of widgets that go with apps you already use. Each one provides a home-screen glance of information without your having to open the app.

But you can also search the App Store for "widgets" to find thousands more. Many of them offer better-looking or more flexible weather, stocks, sports, calendar, checklist, or workout widgets than the ones Apple provides.

There are widgets for all the big-name apps—Spotify, *The New York Times*, TripIt, Twitter, Dropbox, airlines, Waze, Yelp, NPR, Google Maps, Kindle, Evernote, Chrome, Amazon, and many others.

Most of them are free, but a few offer in-app purchases, meaning you can pay for more flexibility or designs. Some of them offer only a couple of sizes; others offer not only multiple sizes, but various designs and layouts of the same information. The Fantastical widget, for example, offers 11 different designs.

Home screens. They're not just for apps anymore.

The Today Screen

Technically, widgets aren't actually new to the iPhone. It's just *putting them on the home screen* that's new.

They're still available in their original spot: what Apple calls the Today screen. It's a screen full of widgets, scrolling vertically, that hangs out at the left end of your home screens. (It's also to the left of the Notification Center screen and the lock screen. In other words, you can look at these widgets even before you've unlocked the phone.)

Let's say, for example, that you want to check your calendar to see what's next. Just wake the phone and swipe to the right on the lock screen. There's your calendar widget.

Editing Modern Widgets

You choose which widgets you want here much the way you do on the home screens: Long-press an empty spot to enter wiggle mode (or tap Edit at the very bottom). Tap the **+** at the top to view the widget catalog. (It's exactly the same list you see when putting widgets on the home screen.)

Tap a widget, choose a size for it, and tap **Add Widget**.

> **TIP:** If you long-press the Today screen until the widgets are wiggling, you can drag a widget to the right—directly onto one of your home screens.

Today screen   Home screens   App Library

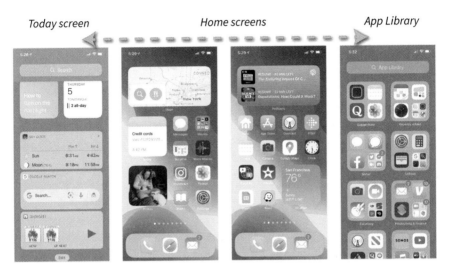

The Today and App Library screens

Editing Other Widgets

As noted above, the Today screen has always offered widgets. You didn't have a choice of size or shape, but they were there.

Those older widgets are still available, but they rely on a completely different installation routine: Long-press an empty spot to enter wiggle mode; this time, though, tap **Customize** at the very bottom. Now you're looking at a different kind of catalog: a list of the old-style widgets.

It has two parts: the widgets you've already installed on the Today screen at the top and, below those, the ones you haven't. See page 4 for instructions on using this standard iOS list editor.

The App Library

The App Library is a game-changing new iOS 14 feature. It's a problem to a solution that many had, but few could articulate: home-screen hell.

It's a truism, and a stock joke of any standup comic, that when you use Google to do an internet search, you never look past the first page of results. The iPhone is pretty much the same. You probably know what icons are on your first home screen, and *maybe* your second. But without looking, can you say what's on your third, fourth, or fifth home screen? It's just chaos. You download apps, and they just land wherever. Apple noticed that more and more people were having to use the search command just to find apps amid the landfill of downloads.

The App Library is a final screen to the right of all your other home screens (see "The Today and App Library screens" on the previous page). *Every app you own* is on this single screen.

Tap the search box...   ...to view an alphabetical list of all apps.

App Library

Tap to see the rest of the apps in this category.

You have four ways to locate an app on the App Library screen:

- **Tap the search box** at the top. Instantly, you're looking at an alphabetical list of all your apps. You can search, you can scroll, or you can tap a letter of the alphabet in the index down the right side to jump to, for example, the P's.

- **Suggestions.** Statistically speaking, the four apps in **Suggestions** at the top left are the ones you're most likely to want at this time and in this location. The iPhone has been watching you, noticing what apps you open when (and where).

- **Recently Added** are the apps you've most recently downloaded.

- **Social, Creativity, Utilities, Entertainment...** All the other foursomes on the App Library screen are app *categories*. The App Library tries to put every one of your apps into one of these categories. Now you don't have to remember what home screen a certain app is on—you only have to remember what the app *does*.

> **NOTE:** In every group of four icons, the *lower-right* icon is usually a *folder*. (You can tell because its icon looks like four even tinier icons; see the "App Library" figure on the facing page) Tap it to open a screen that displays all the apps within.

The App Library has all the tools you need to find any app quickly. Perhaps not as quickly as you can find one of the apps on your first home screen—but much faster than it would take you to scroll through all your home screens, trying to read the names of all the little apps.

Searching the Phone

Apple figures you'll probably need to search a lot. So it's thoughtfully built a search bar on every home screen—just *above* the home screen. To pull it into view, put your finger in the middle of any home screen and tug downward.

> **NOTE:** The search bar is also sitting on the Today screen described on page 102.

You can, of course, use the search box to find an *app*; that's what most people use it for. But this is a global search. It can find information inside any of your Apple apps (Notes, Mail, Reminders, and so on), *plus* information from the web. You can use it as a quick Google searcher when you want information about movies, sports, restaurants, news, and so on.

Ordinarily, when you want to search for an email, you probably start by opening Mail; to search for a note, you probably start by opening Notes. Then you use the search function within those apps.

But you could instead *always* start here. This search *incorporates* the search functions in all those other apps.

How to Use Search

When you tap into the search box, two things appear:

- **Siri Suggestions.** Here's a set of eight icons. In theory, they're the apps *most likely* to be the ones you want, based on the time of day, where you are, and whatever patterns of behavior iOS has observed in you.

  It sounds a little improbable that iOS could anticipate what app you want right now. But you might be amazed at how often it guesses correctly.

> **NOTE:** The term "Siri Suggestions" may imply that this has something to do with Siri, the voice assistant described in Chapter 6. It doesn't. Apple has taken to adding the word "Siri" to any artificial-intelligence feature on the phone, whether it involves voice recognition or not.

- **The keyboard.** Just start typing (or dictating) to specify what you want to find. With each letter you type, the iPhone refines the results.

The results list is usually too tall for your screen; keep scrolling.

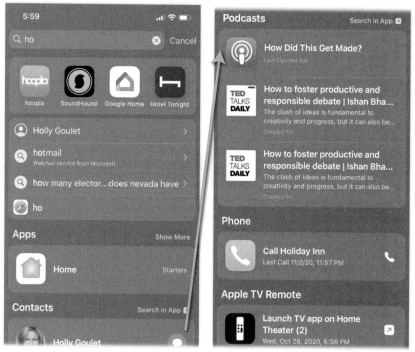

Global search

The results list arrives in sections, by category. For example, if you've typed *em*, the suggestions may include **Emily Martin** (someone in your Contacts),

Emmy Awards 2020 (a list from your Notes app), **Emergency SOS** (one of the Settings pages), and so on.

> **TIP:** To get a better view of your results, drag your finger to scroll the list—and hide the keyboard.

Here are the categories you'll see:

- **Apps.** At the very top: apps whose names match what you've typed. If you've typed *not,* you might see the icons for **Notes, JotNot Pro, Sticky Notes,** and **RetailMeNot,** for example. Tap to open the app.

- **Autocomplete suggestions.** If you've typed *not,* the screen might offer *notes, jotnot pro,* and *notre dame football.* These aren't search results; they're just guesses at what you might be trying to enter, to save you typing time.

  This section may also include web-search suggestions (marked by Safari's compass icon) and logical guesses at what you want from within your apps.

- **Siri Knowledge.** This heading offers snippets of information from the kinds of things Siri "knows" about (Chapter 6): movies, sports, weather, restaurants, and so on. If you've typed in part of a current movie's name, here's where you'll see the results.

- **Information inside Apple apps.** Next come clearly labeled sections for all the iPhone's standard apps: Books, Contacts, Settings, Podcasts, Reminders, Maps, Notes, TV, Mail, Voice Memos, News, Calendar, Clock, and so on.

  Search is finding text within all of them. In Calendar, it searches the names of your calendar events, including meeting invitees and locations. In Voice Memos, it searches the names and descriptions of your audio recordings. In Music, it can hunt through song, performer, and album names, and so on.

- **Other companies' apps.** Search can't see into the guts of many non-Apple apps, but you may find headings for Venmo, Uber, and a few other big-name apps.

Tapping anything in the results list opens it into the corresponding app.

At the very bottom of the results list, a section called Search in Apps appears. These are apps whose databases are too massive to search right from the global search bar. (If you type *mc,* do you really want Maps to clutter your search results with every McDonald's in America?)

Instead, tap one of these (**Search Maps** or **Search App Store**, for example) to pass along what you've typed so far directly into those apps.

> **TIP:** In **Settings→Siri & Search**, you'll find a master list of apps that Search can "see into." Tap one and turn off its switch to exclude its contents from the search results.

Radical Makeovers

In the months after iOS 14 arrived, TikTok and YouTube exploded with videos that showed people performing radical plastic surgery on their home screens. People were redesigning the look, the feel, the *emotion* of their phones' screens in ways Apple never dreamed of. They were replacing the Apple app icons with stunning new ones, uniformly and boldly designed as a set, in harmony both with one another and with the amazing wallpapers behind them.

Custom icons   *Coordinated wallpaper*   *Blank Dock*

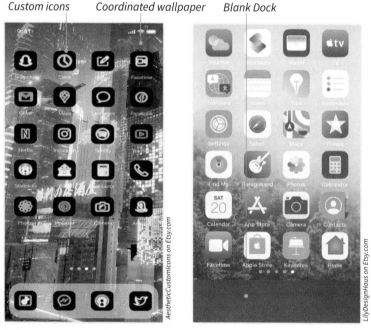

iPhone extreme makeovers

Creating these looks is a hack. It takes time and patience. And you should be aware that when it's all over, on the resulting home screen, your apps' icons can no longer display the little badges (❷) that show, for example, how many emails or messages you've missed.

But you might consider that a small price to pay for the ability to change iOS's look completely. Here's how to do it:

1. **Round up the replacement icons.**

   If you have infinite artistic talent and patience, you can draw them yourself and save them as images in your Photos app. If not, visit a website like icons8.com/icons/ios or www.flaticon.com/packs, where you can find gorgeous sets of ready-to-use app icons.

   Or you can buy complete sets—icons and matching wallpaper—on Etsy.com for a few bucks. (Do a search for *ios aesthetic.*) After all, what your friends care about is how cool your phone looks—not how long it took you, right?

2. **Create a shortcut.**

   As you'll learn on page 451, Shortcuts is an iPhone app that lets you create software robots—automated sequences of steps that you can trigger with only one tap. Once you've created a shortcut, you can save it to your home screen as an app—and in the process, *you get to choose its icon.* (You can probably see where this is going.)

   You're going to create a shortcut that does nothing but open an app. For this example, let's say it's going to open Mail.

   Open Shortcuts. Tap +, then **Add Action**→**Scripting**→**Open App**→**Choose**. Find and select **Mail**; tap **Next**.

   On the next screen, you can name the shortcut. This isn't the name you'll see on the home screen—this is the name Siri will listen for if you want to open it by voice—so type anything you want for now.

   Tap **Done**.

3. **Choose a new icon.**

   At this point, you're sitting on the All Shortcuts screen, with your new app at top left. Tap the ● icon, tap the next ●, and tap **Add to Home Screen**. Where it says **Home Screen Name and Icon**, tap your shortcut's icon, and from the shortcut menu, tap **Choose Photo** or **Choose File**.

   Here's your chance to choose the specialized icon you downloaded in step 1. You get the chance to frame it inside a square—drag it to shift its position, spread or pinch two fingers to change the size within the square—and then choose **Choose**.

Begin in the Shortcuts app; tap +.
Create a **Scripting** shortcut.

Choose a new icon and name...
...and marvel at the results.

Building an app icon

4. **Choose a new name.**

    And *here's* where you specify the name the icon will have on your home screen. It can be *Mail*, it can be *Epistles*, or—if you delete the text here—it can have no name at all. It will just be a naked icon on your home screen.

5. **Tap Add, and then Done.**

If you return to your home screen at this point, you'll see the happy result: the Mail shortcut with the icon and name you designed.

Of course, this isn't the *real* Mail app. It's a shortcut that *opens* the Mail app. That's almost the same thing, but not quite:

-   **The *real* Mail app's icon** is still sitting on your home screen. You may want to hide it. To do that, long-press its icon; from the shortcut menu, choose Remove App→Move to App Library.

- **When you tap your custom icon,** Mail doesn't open instantly; you're first forced to make a brief stopover in the Shortcuts app—before *then* arriving in Mail. That's just how shortcuts work.

  Yes, total creative freedom costs you one second every time you open an app.

You can now repeat these steps for all your other icons. You can combine your app artistry with widgets that work well.

The App Switcher

Once you've sorted, redesigned, and organized the icons of all your apps, there's one more bit of business: *using* them. That's right: At some point, you'll want to tap an app and actually open it.

Mostly, you'll use one app at a time. But what happens when you want to duck into a different app to look something up? Maybe you're on the phone and need to check your calendar. Maybe you've copied a recipe from the web and want to paste it into Notes. Maybe you're blasting away in a game during work hours just as your supervisor walks by.

In all of these situations, it would be nice to be able to hop between open apps, without having to return to the home screen in between. That's why Apple invented the app switcher. Here's how to summon it:

- **Face ID phones:** Swipe up from the bottom of the screen—only half an inch or so.

- **Home-button phones:** Double-click the home button.

> **TIP:** If you're using Voice Control (page 147), you can just say, "Open app switcher."

And voilà: The app switcher appears. It looks like a horizontal row of playing cards. Each one represents an open app, the older ones scrolling off to the left.

Sometimes, just peeking at another app's card right here is enough; you can tap your original app's card and get right back to what you were doing. Of course, you can tap any card to leap fully into that app.

Because the cards are chronological, it's easy to bounce back and forth between the two most recent apps you've used.

Swipe to review your open apps.   *Swipe up to force-quit one.*

App switcher

Force-Quitting an App

When a Mac or Windows program glitches up, you force-quit or force-exit it. Believe it or not, the iPhone is a computer, too—and sometimes, its apps lock up or start behaving badly, too.

There's no Force Quit menu command or Task Manager dialog box on the phone, though. So when trouble arises, you jettison a flaky app by opening the app switcher—and then *flicking the app's card upward*, right up off the top of the screen.

This is, remember, a troubleshooting technique; you won't need to use it often. There's no need to quit an app when you're finished with it. No matter what you've read online, having dozens of open apps doesn't slow the phone down or bother it one bit.

NOTE: One more thing may show up on the app switcher: a strip at the very bottom bearing the name of an app like Mail or Safari. That's Handoff at work—a half-finished bit of business you started on your Mac. See page 396.

Face ID Phones: Bypass the App Switcher

If your phone has Face ID (that is, no home button), get a load of this trick: You can switch apps just by sliding the black bottom bar horizontally! No app switcher needed!

Try it now: Swipe that bar to the right. The next-most-recent app you've used slides into view, ready to go. Swipe that bar again, and the second-oldest app appears. And so on.

Swipe the bar to the right...   *...to bring the previous app into view.*

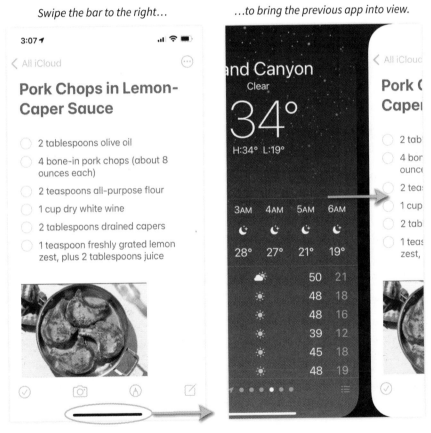

Sliding between apps

TIP: In general, you always swipe right first, because whatever app you're using is always at the right end of the app lineup. But when you're flicking back and forth between two apps—for example, comparing an email draft with an outline in Notes— you can swipe the bar in either direction. You can swipe right, left, right, left, as long as you're never in one app for more than six seconds.

Apple does intend for the app you're in to be at the far right of all the app screens—but it's willing to make an exception when you're very obviously just hopping between two apps.

PART TWO

Info In, Info Out

CHAPTER FIVE

Typing and Dictating

I t was one of the iPhone's most dramatic and revolutionary design concepts: all screen, no plastic keys. In most ways, that's a blessing; maps, photos, videos, ebooks, spreadsheets, and even emails do a lot better when there's more screen area.

But without physical keys, how do you type? Directly on the glass, using an onscreen keyboard that appears whenever you might need to enter text.

At first, that was every bit as slow and clumsy as it sounds. But Apple's engineers have spent years trying to make it smarter and more efficient. Today, features like dictation, swipe typing, third-party keyboards, autocorrect, and autocomplete suggestions make the iPhone typing life a lot more pleasant.

The iOS Keyboard

There's no "open keyboard" button in iOS. The keyboard just appears whenever you might need it: when you're entering or editing text in any app, when you're entering an email or website address, when you're renaming a folder, and so on.

Eventually, you'll find a typing style that suits you. You can hold the phone in one hand and type with the other; your fingers can support the phone from the back as you type with both thumbs; or you can type with the thumb of one hand (with your other fingers holding the phone) and the index finger of the other hand.

As you type each letter, a little "speech balloon" pops up to show the letter you're typing. The iPhone has always offered that balloon as a courtesy, on the assumption that your finger is blocking the actual key you're typing.

> **TIP:** You can turn off the balloon in **Settings→General→Keyboard→Character Preview**. You know—if you think a spy with a telescope can figure out what you're typing by looking at those balloons.

The bubble peeking from under your finger

The onscreen keyboard

The primary keyboard offers the letter keys A through Z, plus these control keys, in gray:

- ⇧ **is your Shift key.** If you tap this key, the next one you type will be a capital letter.

> **TIP:** If you double-tap the ⇧ key, it changes to look like ⬆. You've just engaged Caps Lock. Everything you type now appears in ALL CAPITALS until you close the keyboard or tap ⬆ again. (There's an on/off switch for this feature in **Settings→ General→Keyboard**.)

- ⌫ **is your backspace key.** You can tap it repeatedly to delete what you just typed, or you can *hold* it down to delete the preceding text rapidly. In fact, if you *keep* holding it down, it starts deleting entire words at a time.

- ☺ **opens the standard iOS emoji picker,** a horizontally scrolling array of thousands of emoji symbols. They're organized by category. The ones you've used recently appear first, followed by **Smileys & People, Animals &**

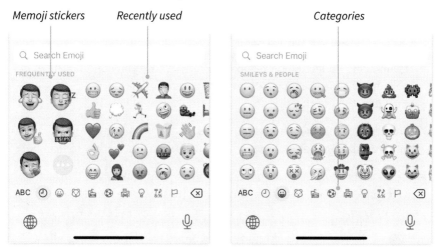

Memoji stickers   *Recently used*   *Categories*

Emoji palette

Nature, Food & Drink, Activity, Travel & Places, and so on.

If you're not sure what category contains a certain symbol (would "fish" be under **Animals** or **Food?**), you can use iOS 14's new **Search Emoji** box. (You can also type a word; if iOS has an emoji to match, it appears as a QuickType suggestion above the keyboard.)

Tap the ABC key to restore the original alphabetic keyboard.

> **NOTE:** The first chunk of the emoji picker is reserved for *Memoji stickers*—cartoon versions of your own face, frozen into various expressions you might find useful in an online chat. Details on page 264.

- ⊕ **appears only if you've installed additional language layouts** (page 61). You can hold it down for a menu of languages, or tap it to hop back and forth between your most recent two.

- ⬚ **is the Dictation button.** Tap it when you want to speak your thoughts instead of typing them. It's much, much, much faster than typing—but produces occasional mistranscriptions. Details on page 132.

- **return is like the Return key** on a regular keyboard; it advances you to the next line. (The iPhone keyboard has no Tab key or Enter key.)

So far, so good. But clearly, there's not enough room on this tiny keyboard for number or punctuation keys. How are you supposed to find those?

It turns out you've been looking at only one of *three* keyboards that rotate in a cycle:

- **The alphabet keyboard** is the one you use most of the time. At its lower-left corner, you can tap the 123 to switch to:

- **The numbers + punctuation keyboard.** Here's where you can find the digits 1 through 0, plus the most common punctuation marks, like ?, @, $, and &. You can return to the alphabet keyboard by tapping ABC.

  But even on this special keyboard, there's *still* not enough room for all the punctuation marks. From here, you can tap the #+= key to bring up:

- **The brackets + currency keyboard.** This one's got symbols you'll probably use even less often, including [brackets], {braces}, #, =, and the bullet (·).

  From here, you can backtrack to either of the other keyboards. You can go back to the numbers/punctuation keyboard by tapping 123, or all the way back to the alphabet layout by tapping ABC.

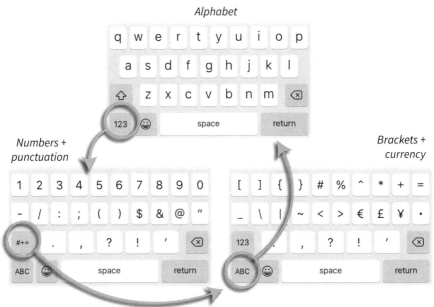

Keyboard layouts

Typing Faster by Typing Less

The iPhone keyboard does its best to make your life easier—by handling the fussier aspects of punctuation automatically. For example:

- **Automatic capitalization.** You don't have to capitalize the first word of a new sentence. The iPhone does that for you.

- **Automatic periods.** When you reach the end of a sentence, *tap the space bar twice.* You get a period and a space—and a capitalized next word—automatically.

- **Automatic apostrophes.** You don't have to type the apostrophes in common contractions, either. You can just type *dont, cant, im, youre, wouldnt, werent, theyll,* and so on; the iPhone adds the apostrophe automatically as soon as you end the word.

- **Autocorrect.** The iPhone does its best to fix typos and misspellings automatically as you go. For example, if you type *I'm puting a pikle on my sqndwich,* the iPhone realizes you meant *I'm putting a pickle on my sandwich* and replaces the erroneous words automatically.

ONE-HANDED TYPING

The huge majority of people need two hands to type on an iPhone. But if you have some kind of freaky six-jointed thumb, it's possible to slide the entire keyboard over to one edge of the screen so you can reach every letter with one hand—and use the other for carrying a cup of coffee.

To set that up, long-press the 😊 or 🌐 button; in the pop-up menu, tap one of the keyboard buttons at the bottom. The outer ones make the entire alphabet keyboard huddle, with miniature keys, against the left or right side of the screen.

And the center one restores the original centered keyboard. ✦

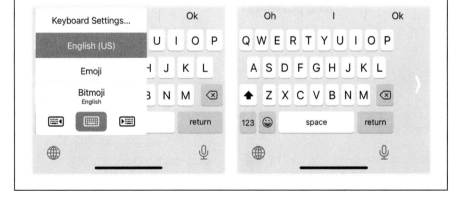

Of course, autocorrect has become the butt of a thousand jokes and inspired millions of swear words, because it doesn't always correctly "fix" what you've typed. (Just try to type *gnight!* to someone. You'll discover that the iPhone insists that you mean *night,* every time.) It's a pretty sheepish feeling to realize that you just texted utter nonsense to somebody, thanks to autocorrect gone wrong.

If you pay attention, though, you never have to suffer that shame. The QuickType bar (the three word buttons above the keyboard) always

1. Hints that autocorrect is about to step in: a lit-up button here, and highlighting here.

2. Sure enough: Your word got nuked.

3. Backspace once; tap your original word here.

4. You've got your intended word back.

Overriding autocorrect

shows you the replacement it wants to make—in advance of making it. (The phone also highlights the word in color before replacing it.)

If you just keep on typing, you're allowing autocorrect to do its thing. If you don't want the replacement, tap the *first* QuickType word—the one in quotes—as shown in "Overriding autocorrect."

If it's too late for that, and autocorrect has just replaced a word incorrectly, tap your ⊗ key. iOS instantly offers the original word in a tappable bubble above the correction.

> **TIP:** If you discover that there's a word autocorrect continues to "correct" against your will, add it as an entry in **Settings→General→Keyboard→Text Replacement**, as described on page 41. (Leave the Shortcut box empty.) Once you've done so, autocorrect will leave that word alone. In fact, it will even use this word as a proposed suggestion if you type something similar.
>
> This is a handy trick for "teaching" the phone to capitalize abbreviations like JFK and PTA when you type *jfk* or *pta*. You can also use this feature to teach the phone that, in your world, certain naughty words are allowed to exist.

- **Spelling checker.** A red, dotted underline identifies a word that the iPhone thinks you've misspelled. If you think so, too, double-tap the word, and then tap **Replace**, to see some suggested spellings. Tap one to fix the mistake.

You can turn all these features off individually in **Settings→General→ Keyboard**. But in general, learning to embrace them saves you a lot of time and fussing.

> **TIP:** Some people prefer to use a stylus on the iPhone—a specialized inkless pen that offers more precision than your big, fat fingers. You can find dozens of these inexpensive styli on Amazon.com.

The QuickType Bar

While you're typing, three tappable words appear just above the keyboard. This is the QuickType bar, known to the geeks as *predictive typing*. In short, the iPhone uses artificial intelligence to guess the next word you're going to type. (Or, if you've begun typing a word, it tries to guess how you're going to *finish* the word).

Suppose, for example, you begin a sentence by typing *I can't*. QuickType offers the three most common verbs that would follow that phrase: **believe, wait,** and **get**.

But maybe what you really wanted to type is *I can't understand,* and those three suggestions don't help you. So you continue to type. As soon as you've got *un,* sure enough, the QuickType bar offers *understand*. Tap it to save yourself eight keypresses.

QuickType hasn't guessed your word yet.

You type two more letters— ah, there it is! Tap that word…

…to drop it in, correctly spelled.

QuickType suggestions

This is yet another feature you can turn off in Settings→General→Keyboard. You'll save yourself a little strip of screen space, but you'll lose what can be a helpful time-saver (and spelling helper).

Swipe Typing

Apple didn't invent swipe typing—it was available on other phones for years, quietly amassing a huge following of fans. But the iPhone version of it works very well. (Apple's marketing calls it QuickPath; in Settings→General→Keyboard, it's called Slide to Type.)

The concept here is that instead of tapping individual keys, you're going to *slide your finger across them*, quickly and sloppily. You're tracing a path that hits all the keys you want along the way. When you lift your finger, the iPhone's software will analyze your fingertip motion and somehow figure out which word you were going for.

Just hit the letters of your word in the right order during your swipe (here, "tomato"). Somehow, the phone knows to ignore any other keys you hit along the way.

Slide to Type

This is definitely not what you learned in typing class in seventh grade. But it's very satisfying—your finger leaves a cool animated trail—and, with a little practice, you can "type" incredibly fast this way.

What's especially cool is that you can *swipe* some words and *type* others, without ever switching modes or settings. You can swipe common words and type uncommon ones, for example.

> **TIP:** You can even switch freely between words in English, Simplified Chinese, Spanish, German, French, Italian, and Portuguese as you go. iOS recognizes the words in all the languages you've installed (page 61) as you swipe.

Don't worry about double letters; the software adds them automatically. To enter the sentence *Sleep is good for cattle,* just swipe *Slep is god for catle.* The iPhone knows what you mean, even if nobody else would.

Clearly, swipe typing can never be perfect. Some swiping patterns cross over identical sets of letters, and the iPhone can't read your mind. For example, if you swipe to the left across the top row of keys, the iPhone doesn't know if you mean *pot, pit,* or *put;* the swipe for all three words is exactly the same.

Similarly, the iPhone doesn't always guess right about double letters. There's no way to begin a sentence with *god* or *beet,* for example; the phone always presents *good* and *bet,* because those words are more common. (It reserves the right to change those guesses based on context when you've typed a little more, though.)

For that reason, when you're swiping to type, the first tap of the backspace key (⌫) deletes the *entire word* you just typed, rather than deleting letter by

PUNCTUATION WITH ONE TOUCH

That business about having to switch between keyboards every time you want to type a punctuation mark gets old fast. Even creating the common comma means hiding the alphabet keyboard, tapping a key on the punctuation keyboard, and then switching back again.

True iPhone aficionados, however, have mastered the art of the *punctuation slide*. It's a way to get a punctuation mark with only a single finger motion.

To perform this advanced stunt, touch 123 to bring up the punctuation keys— and *without lifting your finger*, slide over to the symbol you want, and *then* lift your finger.

The alphabet layout returns automatically. Instead of three keystrokes, you've just entered punctuation with only one.

You can use the same shortcut, by the way, to produce capital letters. Drag from the Shift key (⇧) directly onto the letter you want. Presto: a capital letter with a single, quick drag. ✦

letter as usual. The phone realizes that it's imperfect, and it doesn't want you to get frustrated having to delete an erroneous word one letter at a time.

Swipe typing is freaky and alien at first, and some people never cotton to it at all. It doesn't get in your way if you never use it, but there is nonetheless an off switch for **Slide to Type** in **Settings→General→Keyboard**. (There's also an off switch for the whole-word backspace feature.)

> **TIP:** You don't have to use Apple's onscreen keyboard. You can download and install other virtual keyboards just as freely as you can install other apps: SwiftKey, Gboard, Fleksy, TouchPal, and so on. Some of them have smarter swipe typing; some of them have different key shapes; each has its fans. Once you've downloaded a keyboard, install it at **Settings→General→Keyboard→Keyboards→Add New Keyboard**. Then, when you're typing along and want to choose one of your new keyboards, hold down the ⊕ button to see your keyboards menu.

Editing Text

Getting your words onto the iPhone screen without the benefit of a physical keyboard is only half the battle. The rest of the battle is *editing* your text without the benefit of a mouse.

Moving the Insertion Point

On a laptop or desktop, it's easy to fix a typo you spot a few sentences back: Click with your mouse or trackpad to position the little blinking bar (the insertion point).

On the phone, you have three options for moving the insertion point:

- **Tap.** Your fingertip isn't the world's finest-tipped pointing device, but it's usually precise enough to tap before or after a word, for example.

- **Drag.** Using your fingertip, you can "grab" the blinking insertion point and drag it directly to a new spot. It gets jumbo-sized while it's in transit, to help you see it around the flesh of your finger.

This, of course, was not the first time. Alex's parents had, by this time, grown used to her frequent departures. She had run away from home so many times that lately, her parents just rolled their eyes at each other and went back to their TV show.

Dragging the insertion point

- **Use the space-bar remote control.** This might sound crazy, but it's by far the most precise way to move the insertion point: Long-press the space bar until all the keys go blank (about half a second)—and leave your finger down. You've just turned the entire keyboard into a trackpad. Without lifting your finger, you can now slide anywhere on the screen, moving the insertion point as you go. If you reach the edge of the window, it scrolls automatically.

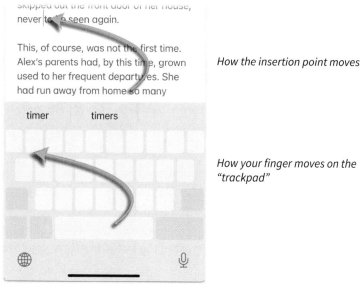

How the insertion point moves

How your finger moves on the "trackpad"

The space-bar remote control

The beauty of this trick is that the insertion point *isn't* underneath your finger. Because you can see it perfectly, you gain much better precision in placing it (by releasing your finger from the glass).

> **TIP:** On models with pressure-sensitive screens (3D Touch)—the 6s, 6s Plus, 7, 7 Plus, 8, 8 Plus, X, XS, and XS Max—you don't have to *long*-press the space bar to enter trackpad mode. You can just hard-press it.

Once you've positioned the insertion point, you can begin typing to add text, or tap the ⌫ key to start deleting it.

Selecting Text

Sometimes you need to operate on bigger chunks of text. You might need to cut, copy, or delete entire sentences or paragraphs. As the first step, you have to *select* the text—to highlight it. How do you do it without a mouse? A few ways:

- **Drag the selection handles.** Double-tap the first or last word you want to highlight. Not only is that word highlighted, but colored dots now appear at diagonal corners. Drag these handles to expand the selection.

Double-tap the first word.

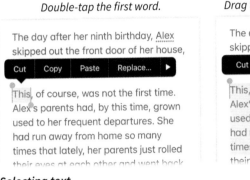

Drag the lower handle to the end point.

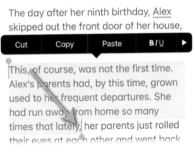

Selecting text

- **Use selection mode.** Tap the insertion point itself to produce the command bar. When you tap **Select**, handles appear at the corners of the nearest word. You can drag these handles to highlight a block of text.

> **TIP:** If you've mastered the space-bar remote control trick, you can also use it to enter selection mode.
>
> Long-press the space bar; when all the keys go blank, slide to the beginning of the text you want to select. Now—still without lifting your finger—tap the blank keyboard with a *second* finger to enter selection mode. At this point, drag your original finger to highlight text, as far and as fast as you want. (It's less tricky than it sounds.)

- **Select all.** The command bar that appears when you tap the insertion point also contains a **Select All** button. It highlights everything on the screen, or in whatever text box you're editing.

Cut, Copy, Paste (Basic Method)

The key to cutting and pasting on the iPhone is the command bar. The whole process goes like this:

1. **Select the text you want to cut or copy.**

   Use the techniques described already to highlight some text and open the command bar, as shown in "Selecting text."

2. **Tap Cut or Copy.**

You've now got the text on your invisible Clipboard. (Of course, if you hit Cut, the text is also now gone from its original location.)

3. **Navigate to the spot where you want to paste the text.**

It might be a part of the same screen, or a different app entirely. In any case, position the insertion point where you want the text to reappear.

4. **Tap the insertion point, and then tap Paste.**

This is the only hard part—remembering to tap the insertion point itself to make the command bar appear.

In any case, whatever you cut or copied now appears.

At the top of the screen, you may notice the brief appearance of a strange little message: "Mail pasted from Notes" (for example). That's a new iOS 14 security feature, designed to alert you when apps have accessed what's on your Clipboard—a rare tactic of evil apps.

TYPING ACCENTED CHARACTERS

Accented characters don't appear on any of the phone's keyboard layouts. You'll have a hard time requesting a résumé, or swatting at a piñata, without those diacritical marks.

Here's how to get them: *Leave your finger down* on the letter key that will have the marking. You get a cool pop-up palette of possible diacriticals to choose from.

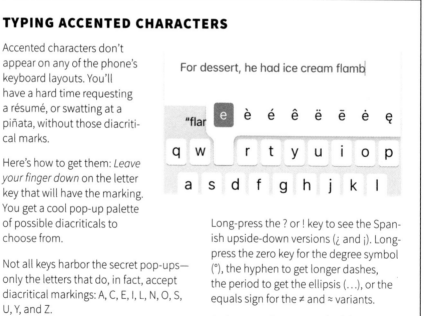

Not all keys harbor the secret pop-ups—only the letters that do, in fact, accept diacritical markings: A, C, E, I, L, N, O, S, U, Y, and Z.

Some of the punctuation marks offer these menus of alternatives, too.

Long-press the ? or ! key to see the Spanish upside-down versions (¿ and ¡). Long-press the zero key for the degree symbol (°), the hyphen to get longer dashes, the period to get the ellipsis (…), or the equals sign for the ≠ and ≈ variants.

And you can long-press the $ for a menu of other currency symbols, like ¢ and €. ✦

Cut, Copy, Paste (Slick Method)

When you write a phone operating system, you're constantly balancing the needs of power users, who want shortcuts and advanced techniques, and beginners, who just want to get through the day.

Here's a copying/cutting technique in the first category that you've probably never seen anybody use; it's that obscure. But in true power-user fashion, it truly is quicker than the traditional way.

Once you've highlighted some text, *pinch inward with three fingers* anywhere on the screen. Believe it or not, that's the Copy command. (The word "Copy" appears briefly at the top of the screen.)

If you do that three-finger pinch twice in a row, you get Cut.

Now navigate to the spot where you want to paste the text. Once the insertion point is in place, *spread outward* with three fingers. That's your Paste command.

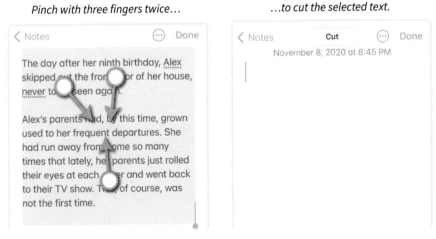

Pinch with three fingers twice...   *...to cut the selected text.*

Three-finger cut

Those three-finger pinches and unpinches might seem like a pretty weird sequence of finger gestures. But if you squint a little, they make sense. It's as though you're plucking the selected material right off the screen and then splatting it down again later.

Undo, Redo

When you cut, paste, or delete something by mistake, you can use the Undo command. And where is that? Hiding:

- **Shake the phone.** That's right—give the entire phone a quick shake. Tap Undo to confirm. (If it doesn't seem to work, make sure **Shake to Undo** is turned on in **Settings→Accessibility→Touch**.)

> **TIP:** If you shake again, you're offered the **Redo** button. That's useful when you want to undo the Undo.

- **Swipe left with three fingers.** Strange but true: That's your Undo.

> **TIP:** Swipe *right* with three fingers for Redo.

Dictation

Dictating to type, using the iPhone's speech-to-text feature, is always faster than typing it out by hand. You can literally enter text as fast as you can say it.

The only problem: Dictation usually isn't as accurate as typing. And when there are people around, dictation isn't as polite—or as private.

CONNECTING A REAL KEYBOARD

If you're not out and about, there's no reason you can't use a regular, full-size, Bluetooth wireless computer keyboard to type into your phone.

Put the keyboard in pairing mode (check the manual—it's usually a matter of pushing some button or switch). After a moment, the keyboard's name shows up in **Settings→Bluetooth**. Once you select it, that's your new keyboard. The onscreen keyboard no longer appears when you're preparing to type.

Not only can you type with better speed and precision using a full-size keyboard, but—if you're using Apple's Bluetooth keyboard—the keys for brightness, volume, and playback control actually control the iPhone now. ✦

Eventually, you'll weigh these considerations and decide whether dictation is right for you.

Dictation Basics

You can dictate wherever you would ordinarily type. In fact, the dictation button—$\underline{\mathbb{Q}}$—is right there on the typing keyboard.

When you tap that button, iOS replaces the keyboard with dancing sound waves that respond to your voice.

Say whatever you would type. You don't have to speak any slower or louder than you ordinarily do. The only new skill required is remembering to speak your own punctuation, like this: "Hi Alex (dash)—I'm running about two hours late (period). Please (comma), please don't start without me (period)."

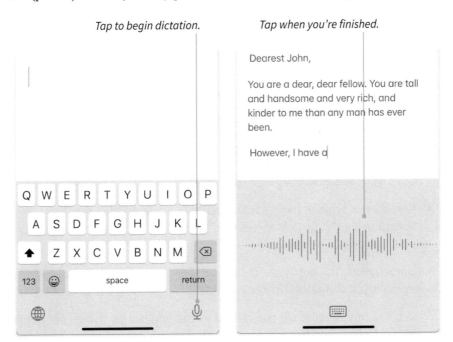

Tap to begin dictation. *Tap when you're finished.*

Dictating text

As you speak, your transcribed words fly onto the screen. Tap the dancing sound waves when you're finished. The keyboard returns, ready for you to make whatever corrections you need.

> **TIP:** Don't get too hung up on the words that first appear while you're speaking. Often, they *change* as you dictate more of the sentence. That's the iPhone using the context of your overall thought to correct its initial guesses.

You'll probably discover that the accuracy of the dictation feature is very good, especially for everyday sentences. If you use imaginative words ("I'll come over to have a lookie-loo!") or phrases from other languages ("You and the boss need to have a tête-à-tête"), well, some manual corrections may be in your future.

Advanced Dictation

Behind the scenes, the iPhone uses the same technology as the Dragon transcription software for desktop computers. If you've ever used those products, you'll discover that most of the commands are the same. They include these:

- **Punctuation** like "question mark," "exclamation point," "ellipsis," "asterisk," "ampersand," "space bar," "open paren" and "close paren," "quote" and "unquote," and so on.

> **TIP:** Conversely, you can say "no space" after a word if you wanted to run it together into the next word. To "getthis," you'd say, "get (no space) this."

- **Line break commands** like "new line" and "new paragraph."

- **Emoticons** like "smiley," "frowny," and "winky." These aren't emoji; they're the old-style emoticons like :-).

- **Currency symbols** like "dollar sign" ($), "cent sign" (¢), "euro sign" (€), "yen sign" (¥), "pounds sterling sign" (£).

- **Mathematical symbols** like "percent sign" (%), "greater-than sign" or "less-than sign" (> or <), "degrees sign" (°), "tilde" (~), "vertical bar" (|), and "pound sign" (#).

- **Capitalization.** Say "cap" right before any word you want capitalized. For example, you can say, "I'll meet you outside (cap) The (cap) Café."

  Or, if you have a bunch of capitalized words, you can say, "Caps on" Before Saying Them, And Then "Caps off" to stop capitalizing them.

  When you want one word to appear in all capital letters, you can make "all caps" HAPPEN.

  You can also say "All caps on" IF YOU WANT EVERYTHING IN CAPS UNTIL YOU SAY, "All caps off."

- **Letters of the alphabet.** Most of the time, the iPhone understands what you want if you just speak the letters. You can say, for example, "I'm turning you over to the FBI."

The phone is also smart enough to autoformat phone numbers, email and web addresses, and money amounts. If you say "eighteen dollars and forty-one cents," the phone writes "$18.41."

> **NOTE:** Dictation becomes a whole new ball game when you add Voice Control to it. That's the iPhone's hands-off voice-control-over-everything feature, described in Chapter 6. Among other virtues, it lets you fix Dictation's mistakes just by speaking the corrections you want.

The iPhone Speaks

Not only can the iPhone convert your voice into text—it can also convert your text into voice. That is, it can read whatever's on the screen to you. That's handy when, for example, you want to listen to some web article or long-winded email without having to look at the screen.

To set this up, open **Settings→Accessibility→Spoken Content** and turn on **Speak Selection** and/or **Speak Screen**. While you're here, you can tap **Voices** (and then your language) to choose a speaking voice for your phone. Some of them, including Ava and Alex, sound incredibly lifelike. (Most of them require downloading before use. Tap ⬇.)

> **TIP:** There's an option here called **Highlight Content**, which highlights each word in color as the phone speaks it. That's a useful effect for beginning readers, anyone who's dyslexic, or anyone who just wants to follow along.

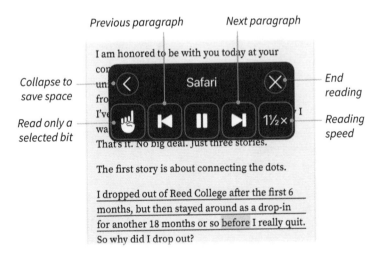

Speaking palette

Once you set all this up, you're ready to command the phone to read aloud.

- **If you turned on** Speak Selection: Select some text as described on page 128. On the command bar that appears, tap Speak. (You may have to scroll the command bar to see this button.)

- **If you turned on** Speak Screen: Beginning above the top edge of the screen, swipe down with two fingers. The phone begins to read all the text on the screen.

  Fortunately for your sanity, that two-finger swipe also produces the Speaking palette. It offers a speaking-rate control, previous/next paragraph buttons, and, maybe most usefully of all, a ❚❚ button.

CHAPTER SIX

Siri and Voice Control

After living with the voice-activated assistants in 3.5 billion smartphones, 150 million smart speakers (like the Amazon Echo), and a growing number of talking cars, refrigerators, wristwatches, and eyeglasses, we get it: We can speak to our gadgets and get spoken answers. We can ask about the weather, find out the score of last night's game, or get help remembering who won the 1995 Best Picture Oscar.

It all began, though, with Siri.

> **NOTE:** And Siri began as a U.S. Department of Defense project called CALO (Cognitive Assistant that Learns and Observes). That technology was spun off to the Stanford Research Institute (SRI) and then became an app called Siri, which Apple bought in 2010. (No, Siri wasn't named for SRI. "Siri" is a Norwegian name meaning "beautiful woman who leads you to victory.")

But using your voice to ask for information and open apps—both tasks Siri has become very good at—is only the beginning of the phone's ability to understand your vocal requests. Voice Control is a more recent feature that's designed to let you control *the entire phone* by voice: opening apps, tapping buttons, changing settings, and so on. This chapter covers both of these mega-features.

Siri

Siri, Apple's voice assistant, is built into every machine Apple makes— including, of course, the iPhone.

You can ask her all kinds of questions without ever worrying about the phrasing. If you want to know what the weather's going to be, you can say,

"What's the weather outlook?" "Will I need a sweater tomorrow?" "Give me the Boston weather," "Will I need an umbrella in Cleveland?" or just about any variation. She answers you with a spoken voice, typed text, and often relevant images.

You can do more than ask her for information, though; you can also ask her to *do* things. You can say, "Remind me about the muffins at 7:30 p.m." or "Make an appointment with Chris for Monday at noon" or "Open Safari." She can even find files for you: "Show me all the emails from Alex" or "Find all my pictures from the Grand Canyon last year."

If you've set up home-automation gear that bears Apple's HomeKit logo, you can even control your house: "Turn off all the lights," "Close the curtains," "Who's at the front door?" and so on.

Two Ways to Ask Siri

Not everybody wants Siri. That's why, the first time you fired up iOS 14, it invited you to turn the feature on. If you're now changing your mind, her master switch is in Settings→Siri & Search→Press Side Button for Siri (or Press Home for Siri).

Even when Siri is turned on, though, she's not listening all the time. For that, you'd need a human assistant, which costs a lot more.

Instead, she listens for a command or a question only when you do one of these things:

- **Hold in the side button (or home button)** until the swirling Siri orb appears at the bottom of the screen.

> **TIP:** If you use CarPlay (page 433), you can hold down the **Talk** button on your steering wheel; if you're wearing wired earbuds, you can hold in the clicker; if you're wearing AirPods, you can set things up so double-tapping them awakens Siri; if you're using some other brand of Bluetooth earpiece, you can hold down the **Call** button. In these cases, you hear a double-beep when Siri is ready for your request.

The phone doesn't have to be unlocked or awake. Just pick it up and hold down the button.

Once the Siri orb appears, you can let go of the side or home button. On the other hand, if you continue holding it while you're speaking, you won't have to worry that Siri might execute your command before you've finished saying it.

- **Say, "Hey Siri."** This feature, too, is something you were supposed to activate during the phone's initial setup. If it's too late for that, visit Settings→Siri & Search and turn on **Listen for "Hey Siri."** You're asked to speak a few sample Siri queries to teach her what you sound like.

> **NOTE:** In Settings→Siri & Search, you can change Siri's language, gender, and regional accent. Here, too, you can delete your dictation history so Apple's computers no longer have any recordings of you issuing commands.

Once the animated Siri orb appears at the bottom of the screen, you can speak your question or command.

Within seconds (depending on your internet connection), Siri speaks her response and displays an answer panel at the top of your screen. Often, the name of the app responsible for producing the answer appears at the top of the answer panel (**Maps**, **Stocks**, or **Weather**, for example); you can tap the panel to open the app for more details. (In iOS 14, Siri tries to keep her initial answer panels compact instead of taking over your screen.)

Siri is listening!

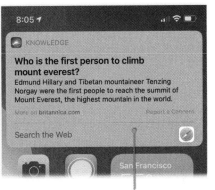

Her reply

Siri sequence

When you've fully absorbed the answer, you can flick the panel up and away, or press the side or home button, or speak a Siri command like "Goodbye," "See you later," or "Go away."

> **TIP:** If you're in a public library, it may be more polite to *type* your Siri requests instead of speaking them. To set this up, open **Settings→Accessibility→Siri**. Turn on **Type to Siri**. From now on, triggering Siri with the side or home button opens—instead of the swirling orb—a panel containing a **Type to Siri** box.
>
> (You can still operate Siri by voice, but only by using the "Hey Siri" command.)

Questions to Ask

Siri is one with the internet; in fact, she can't do much without it. But with it, she's ready to help you with answers to all kinds of questions:

- **Search the web.** "Search the web for a used Tesla Model 3." "Search for banana bread recipes." "Search for news about the next *Die Hard* sequel." "Search the web for pictures of pineapples."

  "What movie won the Best Picture Oscar in 1999?" "What's the tallest mountain in the world?" "When was Barack Obama born?" "How many days until Thanksgiving?" "How many inches in a mile?" "What's the exchange rate between dollars and yen?" "What's the capital of Germany?" "How many calories are in an Almond Joy bar?" "What's a 20% tip on $71 for three people?" "What's the definition of 'erudite'?" "How much is $142 in francs?"

  "How old is Robert Downey Jr.?" "When's the next solar eclipse?" "Give me a random number." "Graph x equals 2y plus 7." "What flights are overhead?" "Where was Gwyneth Paltrow born?" "How much caffeine is in a cup of tea?" "How much fat is in a Big Mac?"

  "Search Wikipedia for John Lennon." "Tell me about Fiorello La Guardia." "Show me the Wikipedia page about bioluminescence."

- **Consult the clock.** "Time." "What's the time?" "What time is it in Dallas?" "Today's date." "What's the date a week from Friday?"

- **Find restaurants.** "Good Indian restaurants around here." "Find a good pizza joint in Boston." "What are the reviews for Red Lobster in Cincinnati?"

  "Table for two in Virginia Beach tonight." "Find a reservation at an inexpensive Italian restaurant Tuesday night at six."

- **Identify a song.** "What's this song?" "Who's singing this?" "What's playing right now?" "Name that tune!"

  Siri listens to whatever music is playing around you, and—thanks to her built-in Shazam skills—displays the name of the song, the album, and the performer.

- **Ask for directions.** "Give me directions to the airport." "How do I get to 200 West 70th Street, New York City?" "Take me home." "Navigate to the nearest gas station."

"What's a 17% tip on $80?"

9:13	
CALCULATOR	
Tip: $13.60 Total: $93.60	
Amount	$80.00
Tip (17%)	$13.60
Total	$93.60

"How tall was Abe Lincoln?"

9:14	
KNOWLEDGE	
6 ft 4 in Abraham Lincoln · Height Wikipedia	
Messages Movies Instagram Books	
San Francisco	

"What's this song?"

9:15	
MUSIC	
Kitty's Rumba by Benedetti & Svoboda	
SHAZAM	

"Good pizza places in Boston."

9:15	
MAPS	
Ciao Pizza And Pasta ★5.0 (722) on Yelp · Closed Now 59 Williams St Chelsea $$	
Carmine's Cafe ★5.0 (20) on Yelp · Closed Now 69 Huckins Ave Quincy	

Siri responses

- **Consult the Yellow Pages.** "Find pizza near me." "Where's the closest CVS?" "Find a hospital in Houston." "Search for gas stations." "What are the hours of the Apple Store?"

- **Check the weather.** "What's the weather today?" "What's the forecast for next week?" "What's the temperature in Austin this weekend?" "Check the forecast for Prescott, Arizona, on Friday." "Tell me the windchill in Juneau." "What's the humidity right now?" "How's the weather looking in Paris?" "How windy is it?" "Should I wear a jacket?"

  "When's the moonrise?" "When will the sun set today?" "When will Mars rise tomorrow?"

- **Ask about stocks.** "What's Nike's stock price?" "What did GM close at today?" "How's the S&P 500 doing?" "How are the markets doing?" "What's Verizon's P/E ratio?" "What's Best Buy's average volume?"

- **Request movie trivia.** "Who was the star of *101 Dalmatians*?" "Who directed *Moonlight*?" "What is *Three Amigos* rated?" "What movies are opening this week?" "What's playing at the AMC theater on 42nd Street?" "Give me the reviews for *Tenet*." "What are today's showtimes for *Star Wars XVIII: The Rise of the Dawn of the Force*?"

- **Request sports stats.** "How did the Mystics do last night?" "What was the score of the last Marlins game?" "When's the next Cavs game?" "Are there any baseball games on today?" "Who has the best batting average in the American League?" "Who has scored the most goals in German soccer?" "Show me the roster for the Oilers." "Who is pitching for San Francisco this season?" "Is anyone on the Braves injured right now?"

- **Consult your address book.** "What's Ann's work number?" "Give me John Jacoby's office phone." "Show Hannah's home email address." "What's Sarah Cooper's home address?" "When is my wife's birthday?" "Show Avery Smalling." "Who is Payton Phoenix?" "Show my boss's work number." "Give me directions to my girlfriend's house."

> **TIP:** If you request information on people with relationships to you—mother, father, parent, brother, sister, son, daughter, child, friend, spouse, partner, assistant, manager, wife, husband, fiancé, boss—Siri can't comply until she knows who that is. So the first time you mention your brother (for example) in a query, Siri asks you for his name and then asks if you'd like her to remember that for next time. That's how she learns.

If Siri mispronounces somebody's name, you can say: "Learn to pronounce Siobhan Hbzrazny's name" (or whatever it is). With tremendous deference, Siri invites you to speak the correct pronunciation and then displays a set of ⊙ buttons so you can pick the closest match from her offerings.

- **Look up your passwords.** "What's my Netflix password?" "Look up my Amazon password." "I need my Bank of America password." (She requires your password, fingerprint, or faceprint before displaying your Passwords screen.)

> **NOTE:** This feature is awesome! Admit it: You never knew it existed.

Commands to Issue

Now we're not talking about questions you can ask Siri; we're moving on to commands you can give. Really, the only hard part of Siri is *remembering to use her*—which entails remembering all the kinds of things she can understand.

The complete cheat sheet begins here:

- **Open apps.** "Open Notes." "Play Chess." "Launch Instagram."

- **Camera.** "Take a selfie" (or "panorama" or "photo" or "slo-mo video").

- **Flashlight.** "Turn on the flashlight." "Turn it off."

- **Change settings directly.** "Turn on Do Not Disturb." "Make the screen brighter." "Dim the screen." "Turn on Bluetooth." "Turn off Wi-Fi." "Make the volume softer." "Mute the sound." "Louder."

- **Open Settings panels.** "Open Wi-Fi settings." "Open Battery in Settings." "Open Sounds settings." "Open Wallpaper settings." "Open Personal Hotspot settings." "Open the settings for Notifications." "Open Netflix settings." "Open Twitter settings."

- **Place calls and check voicemail.** "Call Janet." "Call the office." "Start a FaceTime call with Alton DeVries." "FaceTime Todd." "Do I have any new voicemails?" "Play my voicemails."

- **Set or cancel alarms.** "Wake me at 8 a.m." "Change my eight o'clock alarm to 8:15." "Wake me up in two hours." "Delete my alarm." "Cancel my 3 a.m. alarm." "Turn off all my alarms."

- **Start or stop the timer.** "Set the timer for 15 minutes." "Show the timer." "Pause the timer." "Resume." "Reset the timer." "Stop it."

- **Find photos.** "Show me the Yellowstone Park album." "Show me the videos from Christmas last year." "Get me the videos from New York." "Show me the slo-mo videos from Middlebury." "Give me the pictures from last summer." "Show me my cat photos." "Show pictures of Tia last year in Arizona."

- **Find people.** "Where's Robin?" "Is my mom home?" "Where are my friends?" "Who's here?" "Who is nearby?" "Is Janet at work?"

Thanks to the miracle of the Find My app (page 330), Siri can show you on a map exactly where these people are—provided, of course, that they've given permission for you to track them.

- **Listen to your email in your inbox.** "Read my latest email." "Read my new email." "Any new mail from Blake today?" "Show new mail about the office party." "Show yesterday's email from Taylor."

- **Dictate mail.** "Email Charlie about the reunion." "Email my dad about the dance on Friday." "New email to Emery Holstein." "Mail Mom about Saturday's flight." "Email Richard Panton and Cynthia Powell about the picnic." "Reply, 'Dear Finley (comma), thank you for your concern (period). My goldfish is just fine (period).'" "Email Carter and Tatum about their work on the Jenkins project and say, 'I couldn't have done it without you.'"

- **Check your calendar.** "What's on my calendar today?" "What's my day tomorrow?" "When's my next appointment?" "What's the rest of my day look like?" "What's on my calendar for April 4?" "When is my meeting with Rowan?" "Where is my next meeting?"

- **Make calendar appointments.** "Make an appointment with Jeri for Tuesday at 3 p.m." "Set up a haircut at 11." "Set up a meeting with Harold this Monday at noon." "Meet Bobby Cooper at 8." "New appointment with Dexter, next Wednesday at 7." "Schedule a Zoom call at 7:30 tonight in my office."

- **Reschedule appointments.** "Reschedule my meeting with Harold to a week from Monday at noon." "Move my 3 o'clock meeting to 3:30." "Add Juanita to my meeting with Jeri." "Cancel the Zoom call on Sunday."

- **Listen and reply to your texts.** "Read my new messages." "Read that again." "Read my last message from Frankie." "Reply, 'Oh wow (period). Why didn't I see that coming (question mark)?'" "Tell him I won't be there because I have to rearrange my sock drawer." "Call her." "FaceTime him."

> **TIP:** If you're using earbuds, headphones, CarPlay, or a Bluetooth speaker, Siri's thinking is that you're probably not looking at your phone. So when you dictate a text, Siri reads the message back to you and then asks if you want to send it.

- **Send texts.** "Send a text to Eden London." "Send a message to Lennox saying, 'You're so fired.'" "Tell Noel I'm running late." "Send a message to Drew's mobile asking her to pick me up at the bus station." "Send a text message to (800) 922-0200." "Text Terrell and Amy: Did you pick up the pizza?"

- **Get directions.** "Show me 1520 Amsterdam Avenue, New York City." "How do I get to the airport?" "Directions to my dad's house." "Navigate to Mateo's work." "Take me home." "What's my next turn?" "Are we there yet?" "Stop navigation."

- **Operate the Notes app.** "Make a note that the back door key is under the mat." "Note: Do not put as much jalapeño into the recipe as it calls for." "Create a 'Things to do this weekend' note." "Find my Master Passwords note." "Show all my notes." "Start a note called 'Good TV.'" "Add *Breaking Bad* to my Good TV note."

- **Set a reminder.** "Remind me about Maya's birthday on August 25." "Remind me to bring the muffins to the meeting." "Remind me to take my hormone pill tomorrow at 7 a.m." "Set a reminder to change the air filter when I get to Mom's house." (When you do, in fact, get near your mom's house, the reminder will pop up. Thank you, iOS location sensors!)

> **TIP:** When you're looking at a note, an email message, or a web page, Siri understands the command "Remind me about this."
>
> In other words, you can say, "Remind me about this tomorrow," "Remind me about this at 8 p.m.," or "Remind me about this when I get home." Siri will display a reminder notification at the appropriate time or place—with a link that takes you right back to the original note, email, or web page!

- **Play music or podcasts.** "Play some Bangles." "Play some blues." "Play my workout playlist." "Shuffle my 'Romantical Evening' playlist." "Play 'Bohemian Rhapsody.'" "Play." "Skip." "Play podcasts." "Jump back 15 seconds." "Skip ahead four minutes." "Find the *Unsung Science* podcast in the Podcasts app."

  If you're an Apple Music subscriber, you can *really* go nuts. Just ask for any song, album, or performer. "Play 'Rocket Man.'" "Show me some B.B. King albums." "Play 'Eleanor Rigby' next" (or "…after this song"). "Shuffle Beyoncé." "Play more like this." "Skip this song." "Play 'You and I' by Jacob Collier." "Play that song from the movie *Coco*." "Play the top song from 2007." "Play the top 20 songs of the 1980s."

  And in midsong: "Like this song." "Rate this song two stars."

- **Start playing the radio.** "Play WKYC." "Play BBC Radio 1." "Play 96.1 The Snooze." The phone can stream any of 100,000 radio stations.

- **Translate a phrase.** "How do you say, 'Where's the exit?' in French?" "Say, 'That's too expensive' in Mandarin." "Translate 'That's a gorgeous building' into German." (Siri can translate phrases to and from English, Mandarin Chinese, French, German, Italian, Spanish, Arabic, Japanese, Portuguese, and Russian.)

Other Apps

Siri can extract information from a few non-Apple apps, too, which can save you all kinds of tapping and fumbling.

- **Lyft and Uber.** "Order an Uber." "Call me a Lyft." Once you indicate which kind of car you want, it takes only one more tap to confirm the ride request.

- **Apple Cash, Venmo, Cash App.** "Pay Frances Doyle $30 in Venmo for babysitting." "Request 10 bucks from Mark in Apple Cash." Siri displays what she believes you've ordered her to do; confirm with a tap.

- **WhatsApp, WeChat, Skype.** "Tell Madison, 'I think I love you' with WhatsApp." "Send a WeChat to Shanice saying, 'So sorry I missed you at the pig roast!'" "Let Samson know, 'I may have accidentally scratched the car' in Skype."

- **Pinterest.** "Find sundress design pins on Pinterest."

- **LinkedIn.** "Send a LinkedIn message to Starr that says, 'Looking for a good accountant?'"

Siri's Personality

Believe it or not, Apple actually employs professional comedy writers whose sole job is writing wisecracks for Siri. (Don't worry—she'll never answer you sillily unless you start it by *asking* a silly question.)

For example, she's got a couple of dozen different answers for "What is the meaning of life?" (Example: "I Kant answer that. Ha ha.")

The web teems with collections of Siri's wit and wisdom. But for a small sampling, ask her:

"Are you married?"

"How old are you?"

"Can I borrow some money?"

"Do you like Android phones?"

"How much wood would a woodchuck chuck if a woodchuck could chuck wood?"

If the conversation stalls, try, "Tell me a joke." Or "Read me a haiku." Or "Tell me a story."

In general, it's safe to say these things about Siri:

- **She's gotten much better** and smarter over the past few years.

- **She still doesn't always** have the answer.

- **You're probably not** using her enough.

Voice Control

In Chapter 5, you can read about Dictation, which lets you speak to type. In this chapter, you've now read about Siri, which lets you speak commands.

Now imagine that those two features fell in love and had a superbaby. Its name: Voice Control.

Apple designed Voice Control as an accessibility feature, to help people who can't use their hands and fingers to manipulate the iPhone. But you may find situations where it's handy no matter who you are.

Dictation by Voice Control

That's correct: The iPhone has two different dictation features. Dictation, described in Chapter 5, is the older feature; Voice Control is more recent. Each has its own personality:

Classic Dictation	Voice Control
Types what you say	Types what you say *and* lets you control the iPhone: Taps buttons, scrolls lists, opens apps, and so on
You have to correct errors manually	You can correct errors by voice
Requires an internet connection	Does not require an internet connection
Maximum utterance length: 60 seconds at a time	No length limit

The actual dictation techniques, and the resulting accuracy, are the same for both technologies. But Voice Control adds a powerful edit-by-voice feature to the cruder act of dumping text into an app.

To set up the Voice Control feature for the first time, open Settings→ Accessibility→Voice Control and turn on Voice Control. Tap Learn more to see a list of starter commands. A tiny 🔵 appears on the status bar when the iPhone is listening.

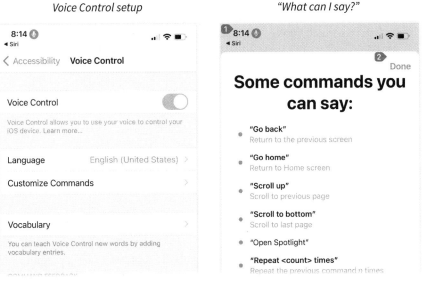

Voice Control setup   *"What can I say?"*

Voice Control

> **TIP:** From now on, tell Siri, "Turn Voice Control on" or "Turn Voice Control off." That's many fewer steps than burrowing into Settings 30 times a day.

Ready to try it out? Open an app that accepts typed text, like Mail, Notes, or Messages. Tap as though you're about to type–but instead, just start talking.

Speak normally; as you speak, the transcribed text appears on the screen, exactly as when you're using the Dictation feature. All the same commands for creating capitals, punctuation, and new paragraphs work here, too.

But when you're using Voice Control, you have another world of voice power at hand: correcting mistakes by voice.

Editing with Voice Control

Maybe you're correcting mistakes it made, or maybe you've just decided to *rewrite* something. Either way, you can say things like these:

- **"Delete that"** or **"Scratch that"** to erase the last word or phrase you said. "Undo that" is pretty great, too.

- **"Move after 'Dear Mom** (or whatever)' "** to move the insertion point to a certain place in the text.

- **"Move to end"** or **"Move to beginning"** of whatever you've been writing.

- **"Move forward** (or **backward**)" and specify any number of lines, sentences, or paragraphs.

- **"Select all"** to highlight all the text in the document. But you can also highlight only a single word or phrase ("Select 'Regarding last week's meeting'").

- **"Select next** (or **previous**) **word"** or character, sentence, line, or paragraph. Or specify how many: "Select next four words." You can also "Deselect that."

- **"Bold** (or **italicize**, or **capitalize**) **that"** to format text you've selected. Or specify the specific words you want to format.

- **"Correct 'oxymoron'"** (or **"Correct that"**) to view a list of alternative interpretations, numbered for your convenience.

- **"Replace 'you had the gall to' with 'you chose to.'"** That right there is an amazing feature: You can speak to replace any word or phrase, without fiddling around with commands to move the selection first.

Say, "Correct 'hero.'"
Say the number of the result you prefer.

Speak a correction

All of this works in any app.

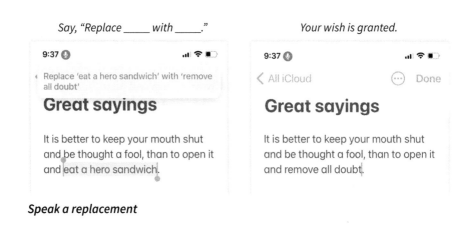

Say, "Replace _____ with _____."   *Your wish is granted.*

Speak a replacement

> **TIP:** You can teach Voice Control new words. In **Settings→Accessibility→Voice Control→Vocabulary**, you can add new terms, jargon, or names that Voice Control doesn't recognize on its own.

Control the iPhone by Voice

But dictation is only half of Voice Control's purpose in life. It also lets you control *the entire phone* by voice: opening apps, tapping buttons, changing settings, and so on.

Here are some of the most useful verbs you can use:

- **"Tap."** "Tap OK." "Tap Reply." "Tap Done." "Tap and hold Next Keyboard." "Double-tap." "Long-press Notes."

- **"Open."** "Open Weather." "Launch Photos." "Open app switcher." "Open Control Center." "Open Notification Center."

- **"Go."** "Go home." "Go back."

- **"Turn off"** and **"Turn on."** You can use Voice Control as the on/off switch for any accessibility feature: Zoom, VoiceOver, Color Filters, Switch Control, or even Voice Control itself.

- **"Scroll"** and **"Swipe."** "Scroll up" (or down, left, or right). "Swipe up" (or down, left, or right). "Swipe right three" (or any number of items, such as photos). "Zoom in" (or out). "Rotate right" (or left). "Pan up" (or down, left, or right). Panning is what you'd do to move around on an enlarged photo, map, or PDF document. You see a little finger dot actually sliding across the screen.

> **TIP:** You can say things like, "Repeat five times." That's useful when you've just used a Scroll or Zoom command.

Finally, you can control the physical buttons on your phone. "Lock screen." "Take screenshot." "Turn volume up" (or down). "Mute sound" (or unmute). "Rotate to landscape" (or portrait). "Reboot device."

> **TIP:** Don't miss the subtle text balloons at the top of the screen; every time you say a Voice Control command that doesn't work, it gently suggests a wording that *will* work.

How to Quiet Voice Control

Voice Control is smart enough to ignore any talking it hears that *isn't* one of the commands it knows. If you say, "Open Notes," Notes opens. But if you say, "Where were you last night? I was worried sick!" Voice Control sits tight and shuts up.

That's your first line of defense against triggering accidental Voice Control commands. But you have two others:

- **Turn on** Settings→Accessibility→Voice Control→Attention Aware. Now Voice Control listens only when you're actually *looking* at the phone. (This option isn't available on home-button phones, because they have no idea when you're looking at them.)

- **Command Voice Control to stop listening altogether** by saying, "Go to sleep." Later, you can say, "Wake up" to use it again.

> **TIP:** If you wind up turning Voice Control on and off often, consider defining a triple-press of the side button or home button to be its on/off switch; see page 451. Or maybe dedicate your double-tap-on-the-back-of-the-phone to turning Voice Control on and off; see page 421.

How to Tap Things Without Names

Telling Voice Control to "Tap OK" or "Tap Send" works great—when there's a clearly labeled button to tap. But what if you want to tap a spot that's *not* labeled? What if you're trying to tap an unlabeled toolbar icon? Or if you're trying to tap a particular spot in Maps?

Apple came up with a couple of ingenious workarounds:

- **Show names.** In some cases, you imagine that iOS knows what something is called, but you don't. Consider the ⊕ button that you can install below the keyboard. What's that called?

  In these cases, say, "Show names." iOS instantly adds clean little tags to every single doodad on the screen so you can learn what they're called.

- **Show numbers.** In other cases, there are things to tap, but even iOS doesn't have names for them—like the photo thumbnails in Photos. In that case,

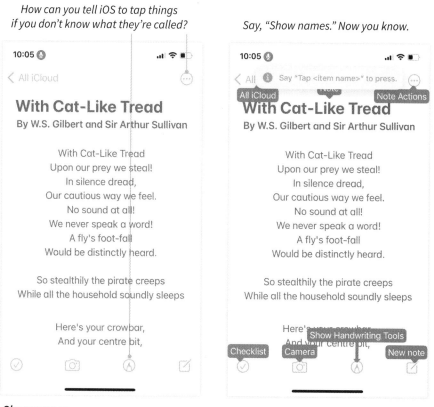

How can you tell iOS to tap things if you don't know what they're called?

Say, "Show names." Now you know.

Show names

say, "Show numbers."

iOS superimposes tiny *number* tags on every tappable item in the window. Now you can just say, for example, "Nine" to tap whatever is identified by the 9 tag.

- **Show grid.** There's one more situation that might seem truly impossible: trying to tap a spot that has no name *and* no numberable shape on the screen—like a Maps map or a zoomed-in photo.

In these last-resort situations, you can say, "Show grid." This time, the phone overlays a grid of numbered squares on the entire screen.

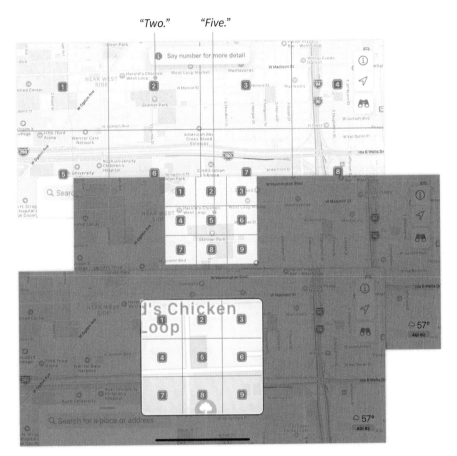

Show grid

TIP: You can even specify the dimensions of the grid. Say, for example, "Show grid with four columns by six rows."

Each time you speak the number of a grid square, the grid squishes down into just that square, making your number-calling increasingly more precise. Eventually, you'll have shrunk the grid enough that you can say, "Tap seven" (or whatever the square's number is), even if it's as small as a single pixel. You can even say, "Drag two to nine" to drag from one spot to another. When you're finished with the gridding, say, "Hide grid."

The Master Commands List

In Settings→Accessibility→Voice Control→Customize Commands, you can define new commands. You might want to do that for any of five reasons: to replace spoken commands that strike you as clumsy or too long; to insert a block of canned text; to reproduce a finger gesture that's more complex than a simple "scroll" or "pan"; to run a shortcut (a macro you've created in advance; see page 451); or—most ambitiously of all—to play back an entire, complex, multistep *chain* of spoken commands.

Here's how it works:

1. **Tap Create New Command.**

   Once you've created your first command, by the way, there'll be an extra tap here. You'll tap Custom and *then* Create New Command.

   In any case, you're now looking at the New Command dialog box.

2. **In the Phrase box, type what you'll want to say to trigger this command.**

   For best results, choose a phrase that you'd never utter in any other context. Don't make "Open Calendar" your trigger phrase for opening up a scandalous website (unless you're setting up the ultimate prank on someone else's phone).

3. **From the Action menu, specify what you want this command to do.**

   You have four options here:

   Insert Text makes a canned chunk of text pop into place, instantly and perfectly. You might use this one to enter a password—which is great, because what you say doesn't have any relationship to what gets typed. You say "Enter Citibank password," and iOS types out *iLOV3cOOki3Z*.

   Run Custom Gesture flawlessly reproduces a finger gesture—a pattern you've tapped or drawn on the screen, maybe something more complex than a simple drag or swipe. This could be something that helps out when you're playing a video game, for example.

   As soon as you tap this option, the New Gesture screen appears. You're supposed to perform the gesture here—any sequence of drags, taps, or swipes, with as many fingers as you want. Later, iOS will replay them flawlessly, with precisely the same timing you use here.

Run Shortcut assumes that you know what a shortcut is. It's a series of steps that the iPhone memorizes for you, so you can trigger them later with a single command or tap. Instructions start on page 451.

Playback Recorded Commands is a catch-all, meaning "Anything else you can dream of!" It allows your phone to reproduce, with precise timing, *anything at all* that you can do on the phone by issuing Voice Control commands. Maybe that's opening Photos, selecting the most recent picture, and AirDropping it to yourself. Maybe it's the multistep login process for some really dumb corporate network system. If you can achieve it with a series of spoken commands, you can record and play back the entire chain.

To teach the phone what you want it to do, say, "Start recording commands." Walk through the process you'll want it to reproduce, using any Voice Control commands you like. When you're finished, say, "Stop recording commands."

> **NOTE:** This option is very powerful—and very picky. To play back correctly, the setup of the screen must be exactly as it was when you recorded the commands; otherwise, the playback will bomb out with a little error message.

Here's what you'll say...   *...and here's what Voice Control can do.*

10:18 🎙   .ııl 🗢 ▮
Cancel   **New Command**   Save
PHRASE
Enter my password
A command must have a speakable phrase that is unique. Examples: • Insert my name • Insert my home address
Action   Insert "Fjuy5?;$8aswdf" >
Application   Safari >
A command must have an action.

10:18 🎙   .ııl 🗢 ▮
‹ Back   **Action**
Insert Text
Run Custom Gesture
Run Shortcut
Playback Recorded Commands
To use Recorded Commands, say "Start recording commands," speak a sequence of commands, and end with the phrase "Stop recording commands."

Creating a custom command

4. **From the Application menu, choose the name of the app where you want this custom command to work.**

By limiting this custom command's *scope* (as they call it) to a single app, you reduce the chances you'll trigger it accidentally. Meanwhile, you open the possibility of using the same spoken command to trigger a different custom command in other apps.

Or you can leave this menu set to Any, meaning you want this command to work no matter what app you're in.

5. **Tap Save.**

Your custom voice command is ready to test (and debug)!

> **NOTE:** To delete a command, tap its name in **Settings→Accessibility→Voice Control→Custom**. Tap **Edit** and then **Delete Command**.

Using this feature, some Googling, and a good deal of patience, you can make Voice Control do things Apple never dreamed of.

CHAPTER SEVEN

The Camera

For the first couple hundred years of photography, "thin camera" was an oxymoron. Real cameras were bulky. Good cameras were huge. How could there be a camera capable of quality like this—the winner of the 2020 iPhone Photography Awards, taken by Kristian Cruz—in a slab a quarter of an inch thin?

Photo by Kristian Cruz. Instagram: @kris.from.brooklyn

Every year, Apple introduces even better cameras. The latest ones take insanely clear photos in near darkness, zoom 2.5 times, employ *computational photography* to adjust their color and lighting to best effect, and use artificial intelligence to combine multiple exposures for the most flawless result. They shoot video in stabilized 4K. As a result, the iPhone is the most-used camera of the human race.

Starting the Camera

You already know about the hardware elements of the iPhone's cameras—one, two, or three dark, round holes on the back, and a much less noticeable lens on the front.

The software element, though, is the Camera app. You can't take a picture without it.

But photo ops don't wait around, especially when animals or children are involved. As a result, Apple has tried to make it easy to get to the Camera app quickly:

- **Wake the phone and swipe to the left.** That's right: The Camera app is waiting off to the right of the lock screen. (Swipe on the background, not on one of the notification banners.)

- **Wake the phone and long-press the 🔘 button.** Same thing: The Camera app opens right from the lock screen.

> **NOTE:** On the Force Touch iPhone models—the iPhone 6S, 7, and 8 families, plus the iPhone X, iPhone XS, and iPhone XS Max—you can *hard*-press the 🔘 button instead. That's about half a second quicker than long-pressing it.

- **Open the Control Center (page 52) and tap the 🔘.** This method assumes, of course, that you've installed the Camera icon onto the Control Center, as described on page 56.

- **Say, "Hey Siri, open Camera."** Hands-free!

Each of these techniques fires up the camera app without your having to unlock the phone first.

That's great for your photographic response time, but it does mean that a passing evildoer could, in theory, pick up your phone and take a picture. They won't be able to do *much* evil, though—they can't transmit or share any pictures they take, and they have no access to the pictures you've already taken.

> **NOTE:** Whenever the Camera app is open, a tiny green "LED" dot appears at the top of the screen. As noted on page 84, that's the standard security warning that iOS 14 displays whenever an app is using your camera.

Snapping the Shot

Once the Camera app is open, you can take a picture by tapping the white ◎ button.

Or, because taking pictures is one of the things people do a lot, Apple wants to make sure you have other options for snapping:

- **Press either volume button on the edge of the phone.** If you're holding the phone horizontally, using those physical keys makes the whole thing feel more like a traditional camera. Often, using them keeps the camera steadier, too, than tapping a button on the screen.

- **Press a volume button on your earbuds clicker.** If you've got your wired earbuds connected to the phone, this technique has some advantages. First, it means the phone doesn't get jiggled at all. Second, it lets you take a selfie from a little farther away than arm's length.

- **Say, "Hey Siri, say cheese."** This is photography by voice control—a hands-free method that lets you set down the phone wherever you want and then run back to be in the shot.

  This works only if you've installed the "Say Cheese" shortcut as described on page 454.

If the sound is turned on, you hear a nice shutter-click noise, and the photo appears as a thumbnail on the lower-left corner of the screen. You can tap that thumbnail to look at the photo full-screen—or you can take another picture right away.

Camera Modes

Open the Camera app; tap the shutter button: That's about as simple as picture-taking gets.

How *complex* it can get, though, is another story.

The Camera app has different photographic modes for taking different kinds of pictures and videos: **Time-Lapse, Slo-Mo, Video, Photo, Portrait, Square**, and **Pano.**

To switch modes, you can drag horizontally on these labels. But you can also drag horizontally *anywhere on the screen*, which gives you a much bigger target.

Change modes…by swiping across the viewfinder.

The mode buttons

Here's a secret, though: There are other modes hiding in the Camera app. You've also got Selfie mode, Night mode, Flash mode, Zoom mode, and so on.

> **TIP:** Ordinarily, the Camera app opens into Photos mode every time you start it up. If you use one of the other modes most of the time, visit **Settings→Camera→Preserve Settings→Camera Mode**. Now the Camera will remember whatever mode you were using the last time you used the app.

All these modes are described on the following pages.

Photo Mode

You can, of course, just tap the white shutter button and be done with it. Your $1,200 phone serves perfectly well as a Polaroid camera (although the pictures look a lot better and develop a lot faster).

But there can be much, much more to it, depending on how much control you want—and how much patience you have. Here are eight settings you might want to adjust before snapping the shot.

Orientation

For the first couple of centuries of photography, almost all photos were taken in *landscape* orientation—with the camera held horizontally. But the rise of the smartphone made upright, *portrait*-orientation photographs a thing. (Portrait orientation is not to be confused with Portrait *mode*, described on page 171.) That's mostly because the phone is easier to hold with one hand when it's upright, both for taking pictures and for viewing them.

In any case, you make the choice. The picture you take will match the orientation of the phone at the moment you take it.

> **TIP:** Sometimes, when you're shooting straight down or straight up, the phone gets confused. It's not sure whether it should record a portrait or landscape photo. Fortunately, you can always rotate a picture later if it guesses wrong. See page 213.

Focus, Exposure, and White Balance

Unless you intervene, the Camera app makes all photographic calculations automatically. Tap ◎ and be glad you got the shot.

But when you have the time and inclination to take manual control, the Camera app is ready for you.

The first step is to tap the subject of the scene—the part of the photo you consider most important. Tapping adjusts three settings:

- **Focus.** Ordinarily, the phone focuses on whatever object is closest. Or, if there are people in the picture, it's smart enough to focus on their faces— up to 10 of them.

    But you can tap anywhere in the scene to set the focus on something else. A yellow square appears to show the phone's new focus target.

> **TIP:** Tapping to set focus is especially useful when you're taking a super-close-up (macro) shot, where the subject is between 4 to 8 inches from the lens. In those situations, the iPhone automatically focuses on the subject and *blurs everything else*, just as in professional magazine-type photos.

- **Exposure** means overall brightness level. The phone ordinarily examines the light in the scene and sets the exposure automatically—but that process may leave your *favorite* part of the scene too light or too dark.

    In that situation, tap the subject. The iPhone recalculates to make *that part* correctly exposed, even if it sends other parts of the photo into shadow or overblown brights. That's really useful when the phone, on its own, makes the important part of the scene look too bright or too dark.

    At this point, a tiny yellow sun slider appears next to the focus square. Swipe repeatedly up or down on it to change the exposure of the entire scene.

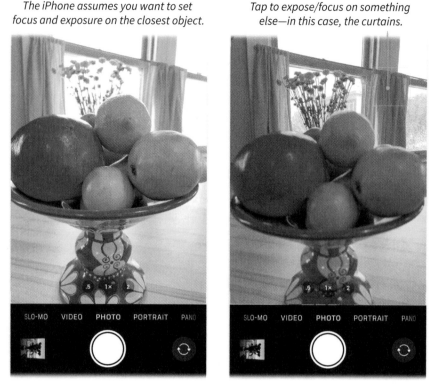

The iPhone assumes you want to set focus and exposure on the closest object.

Tap to expose/focus on something else—in this case, the curtains.

Setting focus and exposure

> **NOTE:** In iOS 14, you have the new exposure control described starting on the next page, which lets you tweak exposure independently of focus and white balance.

- **White balance** is the color cast of a scene. Occasionally, for example, you'll see that the shot is going to be a little too bluish or too yellowish. Here again, the iPhone was doing its best by looking at the entire scene—but your tap tells it to worry mostly about the look of the *subject*.

To reset these settings, tap somewhere else, or aim the phone somewhere else.

Focus Lock/Exposure Lock

Now that you appreciate the majesty of tap-to-focus, tap-to-expose, and tap-to-white-balance, you're ready to understand the appeal of *locking* these attributes. That's what happens when, instead of tapping something in the scene, you long-press it.

In that situation, the iPhone adjusts those aspects of the scene and *freezes* them from shot to shot. That's what you want when you're taking multiple pictures of some fast-moving subject, like animals, divers, or motorcycle racers. There's just not time to do that tap-to-focus thing over and over again.

Using this technique, you can *prefocus*. You can aim the camera at something that's at the same distance away—and has the same lighting—as the fast-moving subject you're hoping to capture. If your kid is going to go down the slide, long-press the bottom of the slide. If somebody's going to do a cannonball into the pool, long-press the empty water where it's going to happen.

The yellow square blinks twice, and the phrase "AE/AF Lock" appears. It means, "You have locked the automatic exposure and automatic focus." (To unlock it, tap somewhere else.)

> **TIP:** In AE/AF Lock, you can once again drag the yellow sun slider to adjust the exposure.

Long-press to lock...and then drag the slider to change exposure.

AE/AF Lock

Now you can take a bunch of rapid-fire pictures without pausing to recalculate focus, exposure, and white balance.

Exposure, Part 2

You now know the traditional way to adjust the exposure of a shot: Tap the subject, and then, if necessary, use the little sun slider.

HIGH DYNAMIC RANGE (HDR) AND DEEP FUSION

This sidebar is here for those interested in extra credit. It won't be on the test.

It does, however, help explain why iPhone pictures look so darned good.

Dynamic range is the spectrum of darks to lights in a scene, and it's always been a weak spot of digital cameras. The human eye detects a much broader dynamic range than any computer sensor. That's why somebody in front of a window who looks fine in real life comes out as a silhouette in the photograph.

If you brighten up the exposure, your subject may now be properly exposed, but the window behind them will be bleached nuclear white.

Years ago, photographers learned a trick called *high dynamic range* photography (HDR). They set the camera to take at least three photos at different exposures, and then, in software, superimposed them in such a way that the details are preserved in both the darkest and the brightest parts of the image.

The iPhone can do all this HDR stuff automatically. In fact, it *does* do all this HDR stuff automatically unless you turn it off in **Settings→Camera→Smart HDR** (or **Auto HDR**).

Smart HDR is the newer feature, available on the iPhone XS and later models. It performs additional calculations on each shot, like recognizing common elements—faces, skies, hair—and studying each frame of its multiple shots to find the most detailed capture of each one.

If HDR is turned off in Settings, you can now turn HDR on or off for each shot individually in the Camera app, using the **HDR** button that appears. (In the example here, you can see that without HDR, the sun creates a bleached-white ball, rather than gradations of color.)

Apple's marketing materials sometimes mention "Deep Fusion" (available on the iPhone 11 and 12 families). It's yet *another* bucket of computational photography tricks. It replaces Smart HDR when the light is low and you're using the standard or zoom lens.

This time, the camera combines the best parts of nine shots—not just based on their brightness or darkness, but on their detail and clarity. The results are subtle—you can't even compare Deep Fusion shots and regular ones, because there's no on/off switch or indicator when Deep Fusion is turned on—but every little bit of help to make your photos look great is welcome. ✦

HDR off   HDR on

That method has never been ideal. It requires that you repeat the process for every photo. And you can't adjust the exposure *independently* of focus and white balance.

That's why, in iOS 14, on the iPhone 11 and later models, Apple has blessed you with a dedicated exposure control.

To see it, tug upward on the screen to reveal the hidden photo controls, including the ⊕.

When you tap it, a horizontal slider appears. You can drag right or left to make the scene darker or brighter. As you go, keep your eye on the tiny light meter at the corner of the screen. It lets you know how far off you are from what the phone considers to be perfect exposure (the center of the meter).

Ordinarily, the Camera app does a good job of exposing the subject.   *But for more artistic control, tap Exposure and then drag the slider.*

Exposure control

To turn off this exposure override, reset the slider to zero.

It's quick, it's smooth, and it's available in both Photo and Video modes.

Filters

After investing millions of dollars developing technologies intended to make your photos' color as clear and true as possible, Apple then went the other direction—and gave you the option to mess them all up.

The ⊛ button opens a row of nine color mutations for the picture you're about to take: black-and-white, vivid, warmer, cooler, and so on. (On the iPhone 11 and 12 families, swipe up on the screen to reveal the adjustments row. If you're holding the phone upright, you may have to scroll this row of icons to see the ⊛ button at the far right.)

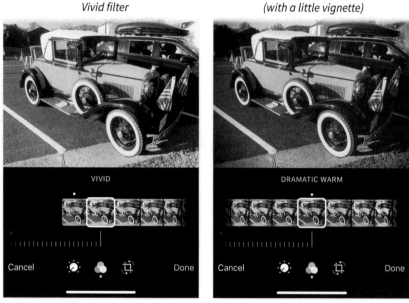

Vivid filter

Dramatic Warm filter (with a little vignette)

Filters

You can play with these filters either before you take the shot or—in the Photos app—afterward (page 213).

Zooming and Framing

Fiddling with your camera's settings is important—but as any pro will tell you, not nearly as important as the composition of the shot. How does it fill the frame? What's the angle? How does the subject sit in the frame?

The iPhone offers a surprisingly long list of ways to answer those questions—beginning with its ability to zoom in.

Unfortunately, the laws of physics and mechanics impose one limitation on smartphone cameras that even Apple's clever engineers can't seem to work around: You can't put a telephoto (zoom) lens into a slab a fraction of an inch thick.

The zoom lens on a real camera is made of telescoping cylinders that slide in or out as you zoom; there's just no room for that on an iPhone. (At least not until the rumor comes true—that Apple is working on a periscope design inside the body of the phone.)

For now, therefore, Apple can offer only two options for getting closer to a subject that's far away.

Digital Zoom

Put two fingers on the screen and spread them apart. Sure enough, you magnify the image. Pinch inward to shrink it again.

If your iPhone has only one lens on the back, then what you're getting is a *digital* zoom. In essence, the phone is just blowing up the image. Everything gets bigger, but everything also loses resolution and clarity.

Optical Zoom

If, on the other hand, your iPhone model has at least two lenses on the back, then you have at least two levels of zoom at your disposal—and everything in between.

On these phones, you can zoom in or out using any of three methods:

- **Pinch or spread two fingers.**

- **Tap a zoom button** near the shutter button. On the Pro models (three lenses), for example, you can tap 2 to zoom in 200% (you're now using the telephoto lens) or .5 to get the widest possible shot (using the wide-angle lens). On two-lens iPhones, tap 0.5x for the wide shot or 1x to return to the standard lens.

- **Put your finger anywhere on the zoom button or buttons** (.5, 1, 2x) and drag horizontally. You get a cool zoom dial that lets you choose any intermediate amount of zoom. The iPhone ingeniously combines the image from the two lenses to provide a photo with exactly the magnification you've dialed up—0.6x, 1.7x, or whatever.

Keep in mind that anything over 1x (on two-lens phones), 2x (on Pro models), or 2.5x (on the iPhone 12 Pro Max) is a fake zoom—digital zoom—that degrades the picture quality.

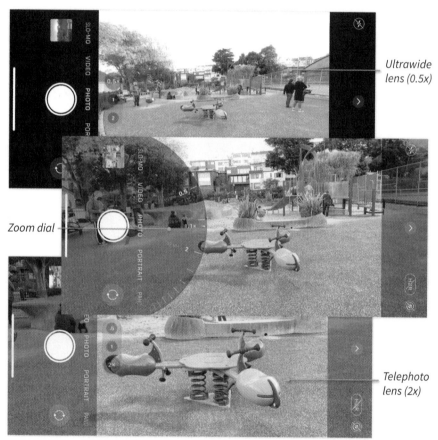

Ultrawide lens (0.5x)

Zoom dial

Telephoto lens (2x)

Zooming in

The Ultrawide Lens

If your phone has two lenses on the back—an iPhone 11 or 12, for example—then one of them is standard and the other is an "ultrawide-angle" lens, great for capturing a group of friends, a tall building, or a huge hotel interior.

The iPhone 11 Pro and 12 Pro models add the third lens—the telephoto lens.

On those models, you may notice something intriguing whenever you're framing up a shot with the normal or telephoto lens: The parts of the Camera app screen that are usually black now show you translucent *previews* of a wider shot. The app is reminding you that you have a wider view available—that you could zoom *out* a little more (tap .5, or use the zoom dial) and capture an even broader area.

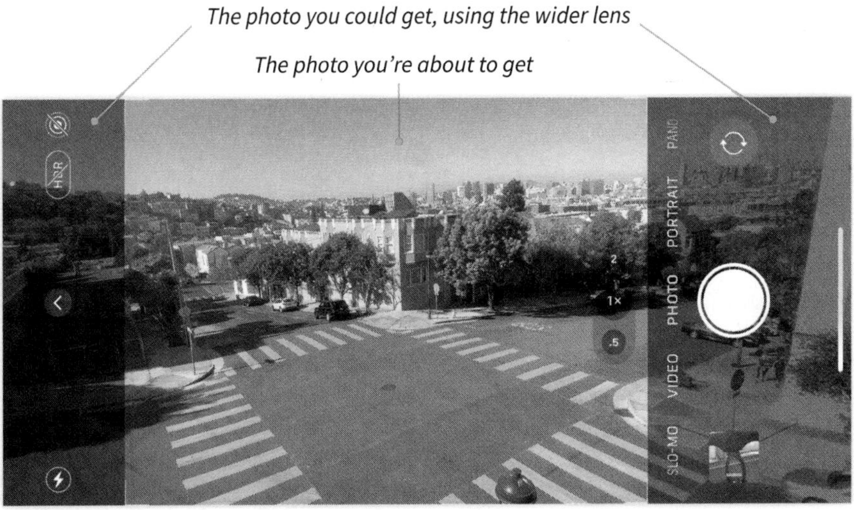

The photo you could get, using the wider lens

The photo you're about to get

Larger area preview

> **NOTE:** On the iPhone 12 models, Apple fixed the fisheye effect that used to distort pictures taken with the ultrawide lens. (You fisheye fans can turn this lens correction off in **Settings→Camera**.)

Unfortunately, the ultrawide lens isn't quite the supercamera its siblings are. It's not stabilized like the others, and it doesn't do as well in low light.

Changing the Proportions (iPhone 11 and 12)

The iPhone 11 and 12 families have a secret: a row of advanced photographic controls that doesn't appear until you *drag upward* on the screen.

Among other controls, one of them manages the *aspect ratio* of the shot you're about to take—its proportions. You have three choices: **Square** (this one's for you, Instagram fans); **4:3** (standard iPhone dimensions, and TV dimensions until HDTVs came along); and **16:9**, which matches the shape of high-definition videos and modern TVs.

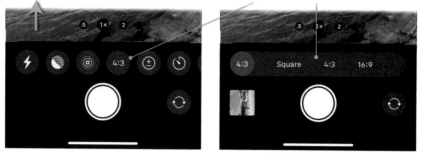

Tug up to reveal the advanced controls.   Choose an aspect ratio.

4:3   Square   16:9

Aspect ratios

Even if you never change the setting, you can now impress people at parties by knowing what an aspect ratio is.

The "Rule of Thirds" Grid

If you've ever been to film school or taken a photography class, you've probably heard about the rule of thirds. It's really more of a *guideline* of thirds: a suggestion that in any photographic composition, you should position the most important parts of the scene at the intersections of the lines of an imaginary tic-tac-toe grid. Supposedly that's a stronger composition than having stuff in the center of the frame.

Truth is, you can probably *imagine* the tic-tac-toe grid just fine. But if you'd like to see it superimposed on your viewfinder, you can turn it on in Settings→Camera→Grid.

> **TIP:** Once you've turned on the grid, you get another handy bit of visual guidance: a carpenter's level. It appears when you're shooting straight down—onto a table, for example—and it resembles two crosshairs, one each yellow and white. When you've got them aligned, you're holding the phone perfectly flat.

Portrait Mode

For many decades, one of the key advantages of big, heavy, bulky SLR cameras was their ability to get blurry-background portraits—where the subject is sharp and clear and the background has a gorgeous, soft-focus look. Getting that kind of shot required a lens with a wide-open aperture, like f/1.4 or f/1.8, and a camera with a big sensor inside.

Impressively enough, the iPhone can now create precisely that look, even though it's not in the least big, heavy, or bulky. Apple calls this Portrait mode, and it pulls off this effect in one of two different ways:

- **iPhones with two or more lenses on the back.** On these models, iOS compares the incoming light from two lenses. Even though they're less than an inch apart, they can analyze the parallax well enough to determine what's foreground and what's background.

  The phone then creates the blurry background—not optically, like an SLR camera would, but with a special effect in software. Still, the two lenses provides a crisp enough separation that the effect is incredibly convincing.

- **Single-lens iPhones.** Portrait mode is available on these phones, too. But without the benefit of the two offset lenses, the results usually aren't as crisp or foolproof. Furthermore, these cameras attempt to isolate the subject from the background only if it's a person; two-lens iPhones can also isolate pets, objects, bowls of fruit, or whatever.

> **TIP:** On Face ID phones, you can use Portrait mode for selfies, too. The same TrueDepth camera that identifies the contours of your face (page 38) also helps to distinguish the foreground from the background.

To take a Portrait shot, swipe sideways across the Camera app viewfinder until you've selected the **Portrait** mode.

Your subject must be between 15 inches and 8 feet away from you, or the trick won't work. If you've got the distance right, the words **NATURAL LIGHT** (or whatever lighting effect you've chosen) light up in yellow. (If the distance isn't right, a message on the screen tells you "Move farther away," for example.)

You can make a few other adjustments before you snap the shot:

- **Zoom in.** On Pro models, you can tap or slide the zoom buttons.

- **The flash, the self-timer, and filters** are all available in Portrait mode.

Portrait mode

- **Adjust the blur.** If your iPhone is a member of the XS, 11, or 12 families, you can control the *degree* of the blur before you take the picture. Tap the ⓕ in the corner of the screen to view a slider above the shutter button, which you can drag right or left.

  Of course, you can also adjust the blur *after* taking a shot, as described on page 217. Try *that* on an SLR!

- **Set up Studio Lighting.** On the iPhone 8 Plus and all Face ID models, don't miss the "wheel" of six enhanced-lighting effects that hug the bottom edge of the viewfinder.

  Apple calls them *studio lighting*, because they're meant to simulate fancy lighting setups in a photography studio. For example, the one called Stage Light literally cuts out the background entirely, making it black, and the one called High-Key Light Mono makes the photo black and white but gives it a pure-white background.

  Once you've selected one of these effects, you can adjust how much of it you want to apply. Tap the ⬡ to make a slider appear that controls the intensity of the effect.

Once you've taken the shot, feel free to open it, admire it, and edit it.

Original

Stage Light

High-Key Light Mono

Studio Lighting

Portrait Mode + Night Mode

On the iPhone 12 Pro models, you can take Portrait-mode photos without a flash, in very low light, courtesy of Night mode (page 177).

The results are amazing, especially when you consider that Portrait mode has a reputation for needing *lots* of light. (On lesser phones, Portrait mode may not work at all if the light isn't good.)

Selfies (the Front Camera)

When you use the front camera to take a picture of yourself, you can use the screen itself as a mirror. You can try different expressions, experiment with your pose, and fix your hair.

> **NOTE:** In fact, the front camera is *exactly* like a mirror. It reverses your image right-for-left, as a mirror does. If there's writing on your T-shirt, it looks backward as you frame the shot.
>
> For the first 12 years of the iPhone's existence, the Camera app corrected that horizontal swapping when taking the actual picture. The writing on your T-shirt always looked correct in the photo. But in iOS 14, for the first time, you can make the finished selfie match the preview: reversed. If that's what floats your boat, duck into **Settings→Camera** and turn on **Mirror Front Camera**.

All you have to do is tap the ⌖ button to activate the front-facing camera. Tap ◎ or press a volume key as usual.

The resolution, quality, and low-light sensitivity of the front camera are not, alas, anywhere near as good as the back camera's. But sometimes, just getting a good, clear shot of yourself is enough.

NIGHT-MODE SELFIES

In the iPhone 12 models, Apple has blessed the *front* camera with fancy light- and color-detection features like Deep Fusion, Smart HDR 3, and Dolby Vision, making it just as capable of soaking up light over time as the back camera.

You take a Night-mode selfie the same way you'd take any selfie: Tap the ⌖ button. The only difference is that you do it in dark situations.

The 🌙 icon appears exactly as it does when you're using the back camera. The iPhone may assist by using its screen flash for better clarity and color, too (unless you've turned the flash off).

If you wear glasses, you might especially appreciate this feature—because when you rely on the screen-flash feature alone, you look like you have bright-white rectangles superimposed on your eyes. It's not a good look. ⭐

The Self-Timer

The Camera app has a self-timer, too. You can give yourself either a three- or ten-second countdown before the camera takes the shot by itself. (It's available for both the front and back cameras.)

That's a blessing when you want to be in a group picture with your friends, because you can prop up the phone and give yourself time to run back and join them. The self-timer is also a fantastic way to avoid jostling the phone when you take the picture, because you have time to put it down. Here's how to find it:

- **iPhone 11 and 12 families:** Swipe up on the screen to reveal the ⏱ button among the advanced controls at the bottom. Tap it.

- **Other models:** Tap the ⏱ button.

Choose the countdown length you prefer.

Self-timer options

Set up the phone where it will have the right view, and then tap the ◎ or a volume key. If you're using the front camera, enormous countdown digits appear on the screen; if you're using the back camera, the flash blinks brightly three or ten times as the countdown. When the counter reaches zero, the phone takes the picture.

Actually, it takes a burst of 10 pictures, because you know very well that *some-body's* eyes will be closed in half of them. This way, you have the opportunity to scroll through multiple shots later and find one that's usable.

Low-Light Shooting

Digital cameras need a lot more light to "see" than your eyeballs do. That's why camera makers spend billions of dollars on techniques for boosting the illumination when the scene is dim—and Apple is no exception.

The LED Flash

On this "camera," the "flash" is a very bright LED on the back of the phone. It has a range of about 6 feet, so don't bother using it in a sports arena or at a concert.

Ordinarily, the phone uses the flash whenever it's necessary—that is, in low light. That's what the iPhone calls Auto mode, and it's indicated by the standard ⚡ button at the top of the screen.

Tug upward on the screen to reveal the advanced controls.

Here, you can force the flash to fire (or prevent it from firing).

Advanced flash controls

But you can control it:

- **Prevent it from firing.** On Face ID phones, tap the ⚡ at the top of the screen, so it changes to a ⚡̸. Now the flash won't fire. (On other models, tap ⚡ and then tap Off.)

- **Force it to fire.** Professional photographers often force the flash to fire—even if the camera itself doesn't believe it's necessary, even in broad daylight—to provide some illumination on the subject's face when it's in shadow, or when there is strong light coming from behind.

  On iPhones with home buttons, tap ⚡ and then tap On.

On Face ID phones, there's a little more to it. *Swipe upward* on the screen (or tap the ∧ at the top of the screen) to reveal the little drawer of photographic controls at the bottom. The first button is the flash control. Tap to view all three options: Auto, On, or Off. Tap On.

No matter what settings you fiddled with, you'll always know in advance when the flash is going to fire: The ⚡ icon turns yellow.

The Screen Flash

The LED flash does a great job of shining light on the subject—when you're using the *back* cameras. But there's no LED on the front of the iPhone. How are you supposed to illuminate selfies?

Fortunately, the iPhone comes to the rescue with a bright *screen flash* when you're taking a selfie in low light. (A *very* bright screen flash—three times the usual maximum brightness of your screen, for a fraction of a second.) The phone even adjusts the color of the screen flash to adjust for the white balance of the scene (see page 162).

It works incredibly well.

Night Mode (iPhone 11 and Later)

Camera flashes have made many a nighttime photo possible, but they're really not ideal. A flash picture tends to bleach out people's skin and turn the background into a cavelike blackness. It would be much better if the camera could capture the scene the way you see it, with natural lighting and shadows.

That's now possible, at least if you have an iPhone in the 11 or 12 family. As long as the subject isn't moving, these phones are capable of *soaking in* feeble light over several seconds—without ever firing the flash. The resulting photos look like they were taken in much brighter light, with color and detail that even your eyes couldn't detect at the time.

It's called Night mode, and it's represented by the ☾ button in the corner of the Camera screen. It may take on any of three looks:

- **Missing.** In bright light, there's no point to Night mode, so the ☾ button doesn't even appear.

- **White.** The lighting is…OK. The iPhone believes it can pull off the shot without using Night mode, but you can tap the ☾ to turn it on if you'd like to try it.

iPhone 8   iPhone 12 with Night mode

Night-mode comparison

- **Yellow.** It's really dark where you are, so the phone wants to use Night mode. The yellow ☾ also shows a time, like **3s**, that indicates how long you'll have to hold the phone steady.

At this point, you can tap ☾ to view a yellow time slider next to the shutter button. This is your opportunity to *override* the phone's estimate of how long it will need to soak in light.

Increase the time for a brighter picture; decrease it for a dimmer one. When you finally tap the shutter button to begin the shot, the slider acts like a countdown to let you know when the process is over.

> **TIP:** If you drag this slider all the way to zero, you turn off Night mode altogether. If you *also* turn off the flash, it's possible to get some really awful pictures.

You can use Night mode while holding the phone, as long as you're not moving much. But putting it on a tripod, or even propping it on something, gives you even better results. In fact, if the phone detects that it's propped up, you get much longer options, like 30 seconds—long enough to take pictures of stars.

> **NOTE:** Night mode doesn't work when you're using the ultrawide lens (page 168).

In iOS 14, Night mode tries to alert you if it detects that the subject is moving. It displays a pair of crosshairs—one white, one yellow. You're encouraged to keep them superimposed for the clearest shot.

Night mode on, set for 3 seconds

Night mode off (slider at zero)

Night-mode slider

Burst Mode

Some moments in life happen so fast, it's almost guaranteed that you won't hit the shutter at exactly the right moment. Great moments in sports, backyard nature, and children come to mind.

That's the beauty of burst mode on the iPhone, which captures 10 shots a second for as long as you're pressing the button. Later, you can scroll through them to find the best ones.

The method of capturing a burst depends on the phone model:

- **iPhone 11 or 12 family.** Drag the ◎ button to the left and hold it there for as long as you want it to fire.

> **TIP:** Or, if you prefer, you can set things up so pressing the *volume-up* key acts as a burst-mode button. Open **Settings→Camera→Use Volume Up for Burst**.
>
> Just remember that you've turned that option on. At this point, your volume-*up* key triggers Burst mode, and your volume-*down* key takes a regular still photo. If you're used to using *either* volume key as the still-photo button, you may be in for quite a surprise.

- **Earlier models.** Hold down the ◎ button or either volume key.

While you're holding down the burst-mode button, the "shots fired" counter rapidly increments to show how many pictures you've captured.

> **NOTE:** You can also capture bursts with the front-facing camera.

It's perfectly possible to capture hundreds of pictures in a burst before you even know what's happening. Fortunately, iOS makes it easy to inspect them, find the perfect frames worth keeping, and delete all the rest.

Tap the thumbnail in the lower left to open the burst, which the iPhone represents as one photo. The phrase "Burst (35 photos)" clues you in that you're looking at a whole batch.

If you tap Select, you'll find that you can now scroll through all the pictures the burst captured. Below them, a tiny "filmstrip" shows where you are in the stream. The filmstrip frames marked with dots are the ones the iPhone recommends keeping, based on their clarity (and whether people's eyes are open).

Those dots are only suggestions, however. Your job is to tap the selection circle in the lower-right corner of each photo *you* want to keep. A blue checkmark appears on each one. Finally, tap Done.

At this point, the phone invites you either to Keep Everything or Keep Only 2 Favorites (or whatever number you've checked). The iPhone deletes all the unchecked rejects.

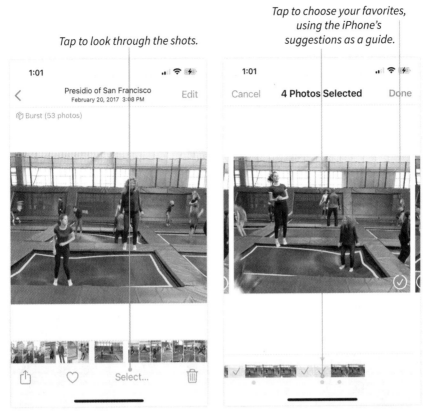

Tap to look through the shots.

Tap to choose your favorites, using the iPhone's suggestions as a guide.

Burst-mode selection

Live Photos

Nobody's quite sure who pounded on Apple's door asking for this feature, but here it is: the Live Photo. It's a still photo with a three-second video attached, with sound—not quite as lifeless as a still photo, not quite as big and bulky as a video. A Live Photo is perfect for capturing...anything that can happen within three seconds.

Technically, you're getting 1.5 seconds before you tap the shutter, plus 1.5 seconds afterward. Therefore, when using Live Photo, remember to keep the phone steady for longer than you normally would.

Tap ◉ in the corner of your screen to turn Live Photo on or off for the next shot you take. If it looks like ◈, then you'll take a still photo.

During the three seconds of the shot, the word LIVE appears in yellow. That's supposed to warn you to hold the phone still.

TIP: A Live Photo takes up about twice as much space as a still photo, so be judicious about leaving it turned on for everything.

You're about to take a Live Photo.   Now recording! Hold still!

Live Photo indicators

What To Do with a Live Photo

As you scroll through the images in the Photos app, each one plays, as a silent video, for half a second. You can long-press one to view the entire three-second video, with sound.

TIP: You can find all your Live Photos in one place by opening the Photos app. Tap **Albums**, and then scroll down to **Live Photos**. They're all inside.

Live Photos are easily viewable on any Apple machine. On the iPhone or iPad, long-press one to play it. On the Mac, click the ⊙ button to play it.

But things get trickier when you try to send a Live Photo to somebody outside the Apple cult.

When you try to share a Live Photo, a ⊚ LIVE badge appears to remind you that the recipient might not see what you see—and to give you the chance to turn *off* the video portion of the photo, by tapping the badge.

Here's what you can and can't do with these weird hybrid photos:

- **You can't email a Live Photo** or send it as a standard text message (to, for example, an Android phone). It arrives on the other end as a still photo.

- **You can't open it in an editing program** like Photoshop. Again, the video portion gets stripped away.

- **You *can* post a Live Photo to Facebook;** it plays just fine there.

- **You can convert it into a three-second video.** When you're looking at it in Photos, tap ⫟, scroll down, and tap **Save As Video**.

- **You can convert it into an animated GIF,** which is a far more universal short video format. To do that, open the Shortcuts app (page 451) and find the ready-made shortcut called **Live Photo to GIF**.

Pano Mode

It's very cool that some iPhones have an ultrawide lens. But it can't compete with the ultra-*ultra*wide lens on every iPhone: Panorama mode.

The idea here is that you swing your arm around you, capturing an enormous swath of the view—240 degrees of it. The iPhone creates a very wide strip of photograph, in full resolution. Panoramas are fantastic for capturing grand scenes of nature, conveying the feeling of certain streetscapes, or photographing the inside of Grand Central Terminal.

This isn't the old trick of taking a bunch of side-by-side photos and then stitching them together on your computer. The iPhone creates the panorama in real time as you move your arm, gracefully merging changes in exposure throughout the scene.

Once you've slid over to choose the **PANO** mode, an arrow shows you which way to move the phone through space. If you would like to begin the panorama at the other end, tap the arrow to reverse it.

Panorama so far

Move the phone through space.

Try to keep the arrow on the line.

Stop whenever you've got enough.

Panorama in progress

> **TIP:** Most people think of a panorama as a wide horizontal strip. But this mode is also useful for capturing very tall things—buildings, giant redwoods. Just turn the phone 90 degrees before you begin.

Tap the shutter button (or press a volume key), and then swing the phone in an arc around you, smoothly and slowly. You can see a miniature of the panorama growing as you swing.

Try to keep the big white arrow on the center line, so your panorama is level. As you go, messages on the screen may tell you to slow down, or to keep the arrow on the center line.

Now, 240 degrees is a *really* wide slice of life, and capturing it may force you to twist like a Twizzler. Fortunately, you're under no obligation to make every panorama that wide. At any point, you can tap ⏺ to end the panorama at its present width.

In fact, for all the iPhone cares, you can use the Pano feature to take photos that are just a *little* bit wider than your current lenses can handle. (In any case, you can also crop a panorama later to any width.)

Pano shots usually come out looking amazing. They're seamless and high-resolution, and they do a much better job of recreating the feeling of being in a place than a single photo can.

You can even print them out, courtesy of local printing shops or online services like mpix.com. A printed panorama looks pretty great when hung over a couch.

Finished pano

Capturing Video

The iPhone's video capture is so good that Hollywood movies have been made using nothing but the phone. We're talking high-definition quality or even 4K (which is four times the resolution of high definition), stabilized, with incredible color and clarity.

> **NOTE:** You can capture video in either traditional widescreen format (hold the phone horizontally) or the newfangled, phone-friendly vertical format (hold the phone upright). Keep in mind that if you shoot vertically, your video may fit people's phone screens nicely—but on computer screens and TV screens, it will look tiny and ridiculous and be flanked by massive empty bars.

When you're in the Camera app, swipe horizontally on the screen until you hit **VIDEO** mode.

You can tap ◉, or either one of the volume keys, to begin shooting video. You know you're rolling because a time counter, in red, tracks how long your shot has been.

Tap ◉ to end the shot.

To review the newly captured video, tap the thumbnail in the corner. The video opens and plays silently; tap ◥ to turn on the sound. (For more on playing videos you've captured, see page 193.)

Settings Before You Shoot

Before you shoot the video, you can also set up many of the same parameters that you would when shooting a still photo (page 160):

- **Zoom level.** You can zoom in exactly as described on page 167.

- **Front camera.** Tap ⊡ to shoot a selfie video.

- **Focus, exposure, white balance.** Tap the screen to show the phone where you want to calculate focus, exposure, and white balance.

- **AE/AF lock.** Long-press the subject to lock the focus, exposure, and white balance. Drag the tiny yellow sun slider to adjust the exposure manually.

- **Exposure control.** If you have an iPhone 11 or 12, tug upward to reveal iOS 14's new ⊕ button, which you can tap to adjust the overall exposure of the scene (page 163).

But some controls are unique to shooting video. For example:

- **Video light.** If you could use a little more light in a dark scene, turn on the LED on the back of the phone by tapping 𝟃 and then **On**. (On Face ID phones, swipe upward on the screen first to reveal the 𝟃 button.)

> **NOTE:** This is one of the few options you can't change *during* filming. You have to turn the LED on or off before you begin.

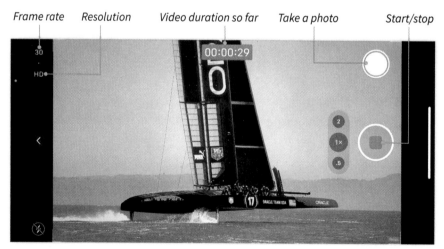

Frame rate   *Resolution*   *Video duration so far*   *Take a photo*   *Start/stop*

Shooting video

- **Resolution.** In iOS 14, you no longer have to fiddle around in Settings to change between high definition and 4K shooting. Just tap the HD/4K indicator in the corner of the screen to choose the resolution you want.

> **NOTE:** And what does **HD** mean in this case? Is that 720p or 1080p (the two common high-definition formats)? That's a choice you make in advance, in **Settings→ Camera→Record Video.**

4K, of course, is the latest standard in picture clarity. Every frame of 4K video has four times as many pixels as a frame of high-definition video. It looks amazing when it plays on a 4K television.

WHY VIDEO ZOOMS IN

Maybe you've noticed: When you switch from **PHOTO** to **VIDEO** mode, the viewfinder suddenly *zooms in*, making everything bigger than it looked in Photo mode.

This peculiar effect has to do with the iPhone's *stabilizer*. When you shoot a video, or take a still photo in low light, the slightest movement of your hand can create blurriness.

Therefore, the iPhone employs some astounding technology to counteract your hand jiggles. Inside, electromagnets literally *move the camera lenses* relative to the sensor (the "film")—rapidly and in real time, to offset any bumps of your hand. (On the iPhone 12 Pro Max, the *sensor* moves, not the lenses. It's a fancier technology with even better results, but the general principle is the same.)

So why does the video image zoom in?

Imagine that your hand suddenly jerks upward by a fraction of an inch. Inside the phone, the lenses must move down to compensate. But now the subject is no longer centered in the frame—it's now a little bit lower than center.

Fortunately, the iPhone has a buffer—a margin of additional capturable area around the frame. Without it, the lens shift would move your subject outside the frame momentarily, chopping it off.

And that's why Video mode is slightly more zoomed in than Photo mode. The phone is enlarging the image enough to give it some extra buffer area in case of bumps. It's all in the name of giving you video that's smooth and stable, even if your hand wasn't. ✦

But be careful: 4K video also consumes a huge amount of storage space on your phone—at least 190 MB a minute. You have to ask yourself, too, where you're going to *play back* 4K video that you recorded. No iPhone has enough pixels to play 4K; the videos will play, but without the glorious clarity. Your only options are a 4K television, a 4K computer screen, or YouTube.

In other words, it doesn't always make sense to shoot a video in 4K, even if you have the option.

- **Frame rate.** Tap the tiny indicator at the top right to switch among **24**, **30**, or **60** frames a second.

You can spend many happy hours reading articles online about the aesthetic and psychological effects of videos shot with different frame

THE INSTANT-SHOOTING TRICK

You probably assume that to shoot a video, you need to swipe the Camera screen over to **VIDEO** mode. And that's what most people do.

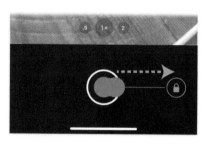

But on the iPhone XR, XS, 11, and 12 families, there's an even quicker way to start shooting. Apple calls it QuickTake.

When time is of the essence—you catch your dog barking in a way that sounds human, your kid is suddenly doing a handstand—open the Camera app and immediately *hold down* the ◎ with your finger. The phone begins rolling video for as long as you keep your finger down. You've saved one precious second of switching modes.

If it looks as though the clip is going to continue long enough to make your finger cramp up, slide it to the right (onto the 🔒) and let go. Now the phone will continue recording until you tap ⦿. (As a convenience, in this record-lock mode,

you get a ◎ button that snaps a still photo *during* the video recording.)

There's just one fly in this very convenient ointment, and it has to do with aspect ratio (page 214). It turns out that when you use QuickTake, your video has whatever aspect ratio is set in *Photo* mode. Usually, that's 4:3, which is more square than traditional 16:9 videos.

Of course, you can change the aspect-ratio control to 16:9 before you shoot (page 169)—but at that point, you've lost whatever time advantage you had by using QuickTake in the first place. ✦

rates. As a general guide, though, **24** frames per second is what you're used to seeing in movies, **30** is what TV uses, and **60** frames per second is sometimes useful when there's a lot of motion (or commotion) in a scene. It keeps everything perfectly smooth and crisp—and also gives you the option to create buttery-smooth slow motion when you edit the footage on a computer.

You can even make most of those adjustments *while* you're rolling. You can change focus, exposure, zoom, and so on. You can even snap a still photo while you're capturing video, thanks to the ◎ button that appears.

In general, the iPhone does a spectacular job capturing video—and you never have to worry about changing tapes. A 512-gigabyte iPhone can capture 272 hours' worth of video, which should just barely make it through your kid's birthday party.

Slo-Mo Videos

Slow motion never gets old. It's amazing for analyzing golf strokes and baseball swings, dogs shaking off water, and videos of anything crashing into anything else.

Slo-Mo mode works just like Video mode, except that the frame-rate readout in the corner of the screen offers a choice of **120** or **240** frames a second instead of the usual **24, 30,** or **60**.

The basic idea: The more frames in a second you capture upon *recording*, the slower the action when you play it back at the standard 30 frames a second.

> **TIP:** On the iPhone 11 and 12 families, you can shoot slo-mo using the front camera, too. Apple calls these clips "slowfies."
>
> Nobody else ever has.

When you play back a slo-mo video, iOS does something really cool: It begins the playback at full speed, gracefully slows down to one-quarter or one-eighth speed for the middle part, and then accelerates to regular speed for the last second of video.

The points where it slows down and speeds up are up to you. When the video is open for playback, tap **Edit** and then drag the little vertical handles on the timeline.

Shorten the video.   *Adjust where the slo-mo part begins and ends.*

Adjusting slow motion

When you send this video to anybody else, it becomes a frozen video with the speedups and slowdowns exactly as you set them. But when this video shows up on your own iPhones, iPads, and Macs, you can continue to adjust the starting and stopping points of the slow-motion part in the Photos app.

Time-Lapse Videos

If you're the kind of person who gets a thrill from watching sped-up videos of flowers blooming, snow sculptures melting, or buildings going up, time-lapse is for you. It massively speeds up scenes that are ordinarily too slow-moving to be interesting.

The Time-Lapse mode compresses everything down to 20 or 30 seconds, no matter how long the scene took in real time. In other words, the longer you film, the greater the acceleration.

> **TIP:** On the iPhone 12 models, you can even shoot time-lapse videos in Night mode (page 177). This feature opens up a world of time-lapse videos of starscapes wheeling through the nighttime sky, red taillight trails on city streets, and the northern lights blowing your mind overhead.

CHAPTER EIGHT

The Photos App

f you're like most people with an iPhone, you take a lot of pictures. And if you take a lot of pictures, you need some way to organize them.

That's what Photos is all about. It's a massive digital shoebox for collecting, editing, organizing, and sharing the photos and videos of your life.

It's a big, sprawling, sophisticated app with a lot of power. Fortunately, if you've ever used Photos on the Mac, you'll feel right at home; the iPhone version is almost identical. And if you haven't, you'll catch on soon enough.

Opening a Photo

On many of the Photos screens, you look at your pictures in the form of thumbnails, small enough that you can view a whole bunch simultaneously.

TIP: You can pinch or spread two fingers to make the thumbnails smaller or bigger.

But when you tap a photo, it opens up at full size, or as big as your screen will allow. You get to see when and where you took it, and you can try tricks like these:

- **Zoom into a photo** by spreading two fingers against the glass. At that point, you can pan around the enlarged shot by dragging with one finger.

- **Rotate the photo**—because you're looking at a horizontal photo while holding the phone vertically, and you'd like to see the shot at full size—by

A thumbnail view. Tap one…   *…to open it.*

The Photos app

rotating the phone. (On the other hand, if you want to rotate a photo because the phone took it sideways, see page 213.)

- **Share the photo** by tapping ⬆ (page 203).

- **"Favorite" this photo** by tapping ♡. You've just added the photo to your Favorites album, which is on your Albums screen (page 197).

- **Delete the photo** with a tap on, of course, the 🗑. Tap **Delete Photo** to confirm.

- **Edit the photo** by tapping **Edit**. You'll probably be amazed at how much power you have over the color, brightness, and other aspects of your photos and videos. It may not quite be Photoshop, but it's at least Phonoshop. See page 210.

- **Hide the controls** by tapping anywhere on the picture. Now you're showing off the photo as handsomely as possible: uncluttered, full-screen, framed in black (if it doesn't, in fact, fill the screen).

  Tap again to bring the controls back.

- **Close the photo**—to go back to your thumbnail farm—by pinching with two fingers or by tapping the ‹ at top left.

Playing a Video

You wouldn't think there would be much involved in playing a video—but, in fact, two important tools are hiding in plain sight.

First, every video begins playing silently. Apple is trying to save you embarrassment in public places. Tap 🔇 to hear it.

Second, you may notice that there are no ◀◀ or ▶▶ buttons for moving quickly through a video—not even a scroll bar. But if you tap the video's thumbnail at the bottom of the screen, it expands to a full-length "filmstrip." Tap inside it, or drag through it, to zip around in the video.

Tap to expand the "filmstrip." Drag to skim through the video.

Unmute the video.

Playing a video

Navigating Photos

It's pretty hard to get lost in the Photos app. If you pinch enough times (or tap ‹ at top left enough times), you eventually wind up at a screen with four tabs across the bottom: **Library**, **For You**, **Albums**, and **Search**. Here's what they do.

Library

Tap here when you want to see all your photos at once, represented as scrolling thumbnails. The oldest are at the top.

Apple's big challenge here was to design an app that shows you every photo you've ever taken without *losing* you in them. How does an app display 12 years' worth of photos—thousands of them—while still making it easy for you to find a certain one?

For starters, you can look at them in groupings by Years, Months, or Days; tap the buttons at bottom.

Tap (or spread two fingers) to go Years→Months→Days. ⟶

Pinch to go back. ⟵

Years, Months, and Days

Within each of these chronological displays, Photos shows you representative photos, using artificial intelligence to choose *good* shots for each time period:

- **Years** represents each year with a single photo.

> **TIP:** Photos represents each year with a big, bright thumbnail image from the current month (of that year).
>
> But you can swipe horizontally across a Year thumbnail to view shots from other months of the same year.

- **Month** view *isn't* grouped by month—gotcha! Instead, each thumbnail represents what Apple calls a Moment: a set of photos taken at one time and place. A Moment might contain all the shots from a wedding, one day of your vacation trip, or a beach walk. Each is identified both by date and by location.

- **Days** view offers a gorgeous grid of photos. Scroll through it for a warm wash of nostalgia, or tap any one of the thumbnails to open it full-screen. At that point, you can swipe horizontally to browse through the adjacent shots.

> **NOTE:** Whenever you swipe through the photos like this, videos and Live Photos play, but with the sound off. That's Photos' clever way of letting you know that they *are* videos and Live Photos; you can always tap 🔇 to hear them.

- **All Photos** is all your photos and videos, on one massive, chronological, endlessly scrolling page. As usual, tap one to view it at full size.

> **TIP:** Long-press any thumbnail to see a larger version of it—along with a shortcut menu containing **Copy, Share, Favorite**, and **Delete from Library** buttons.

For You

This screen is where Photos flexes its artificial-intelligence muscle. It shows thumbnails of photos it has assembled according to what it thinks are your interests. Here's what you might find:

- **Shared Album Activity.** Here's where you see, for example, who's accepted your invitation to see some pictures you shared. See page 207 for more on Shared Albums. (OK, this category's not so artificially intelligent.)

- **Memories.** Apple calls them Memories; you might call them "musical slideshows, created by AI, based on commonality of time, place, or subject."

And sure enough: This is a row of miniature billboards of ready-to-play slideshows with names like "San Francisco Trip," "On This Day Last Year," "Grand Canyon," "Dad Over the Years," and so on.

Photos automatically creates about three new Memories every day, which do a great job of resurfacing shots that you would otherwise never look at again.

When you tap a memory, its Details page opens. What you'll usually do here is just play the darned thing by tapping ▶; you get a beautiful, animated, sweetly crossfading slideshow, accompanied by (and actually *timed to*) glorious music. But you can also scroll up to look at the photos *in* this slideshow, and the people and the places related to it.

> **TIP:** Once the video is playing, tap it for some creative modification buttons (see "Playing Memories"). You can change the background music (**HAPPY**, **UPLIFTING**, **EPIC**, and so on) and the slideshow length (**SHORT**, **MEDIUM**, **LONG**).
>
> The 🔼 button appears here, too. If the Memory is really good, you can share it with other people as a standalone video.

Tap during playback to view the music and length controls.

Playing Memories

- **Featured Photos.** There's that nosy AI doing its thing again! This time, it's displaying what Apple claims are "your best photos." And sure enough: They tend to be the best lit, best photographed shots from the past few years.

- **Effects Suggestions** are pictures that Photos humbly thinks would look better if you applied one of its effects. Usually, it's either a Portrait-mode shot (Photos suggests applying one of the studio-lighting effects—see page 172) or a Live Photo (Photos proposes **Loop**, **Bounce**, or **Long Exposure**—see page 216).

- **Sharing Suggestions** are like Moments, in that they're groups of shots that you took at about the same time and place. The difference is that this time, Photos recognizes who's *in* the photos, using face recognition, and suggests that you share the shots with those people. ("Share with Alex?") Impressive.

- **Recently Shared** rounds up pictures you've shared—or that have been shared with you.

Albums

You may know what people 50 years ago called "photo albums": books of photos, fastened to paper pages in a certain order, that someone handpicked for some purpose.

The electronic versions have certain advantages. For example, the same photo can appear in a bunch of different albums without actually being duplicated. You can remove a photo from an album at any time, confident that you're not actually removing it from *Photos*—only from this virtual grouping.

All the albums you've created appear here, on the **Albums** tab, right at the top. Scroll the My Albums thumbnails horizontally to see your most recent ones, or tap **See All** to view all your albums.

> **TIP:** To create a new album, tap + at top left, and tap **New Album**. Type a name for it; tap **Save**.
>
> Now you're looking at your All Photos screen: an endless field of thumbnails. Go nuts tapping them, adding a blue checkmark each time; then tap **Done**. They're now in your new album!
>
> To remove an album, start on the **Albums** tab; by My Albums, tap **See All**. Tap **Edit** and then the ⊖ button on the album's corner. This process *never deletes any photos*— only the imaginary album that contained them.

Photos itself creates many of the photo sets on the Albums tab. For example:

- **Recents** holds the latest pictures.

- **Favorites** are those you've honored by tapping the ♡ button.

- **Shared Albums** (page 207) get their own row here.

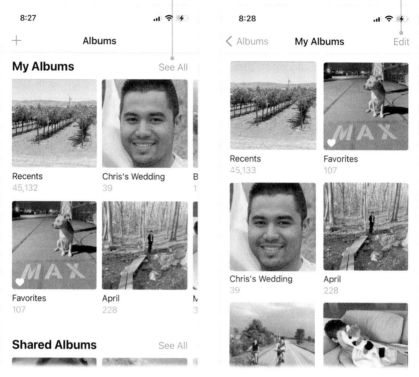

*Tap to see the complete list of albums—and the **Edit** button.*

The Albums tab

- **People.** Photos can actually tell who is *in* your photos. We're talking facial recognition good enough to recognize the same child at ages 4, 10, and 16.

  It does, however, need a little assistance from you to get started. When you tap **People**, you see thumbnails of the faces Photos has found so far. You can tap the ♡ to denote the most important people as Favorites; they appear at the top of the People screen, with extra-big thumbnails.

  Now, Photos does a reasonable job of gathering up images of what it thinks is the same person. But even the best AI in the world can't guess at their *names.*

  To fix that, tap a thumbnail to view all the pictures of that person. Tap **Add Name** to supply this person's name, so Photos will know who it is later. When you tap **Next**, Photos might show you some other shots it *believes* to be the same person but isn't quite sure about. It needs your superior human judgment. Turn off the blue checkmarks for anybody it got wrong, and then hit **Done**.

You'll be really grateful for this feature the next time somebody asks for a couple of photos of you (or anyone in your collection).

> **NOTE:** Every now and then, Photos announces, "There are additional photos to review." It's telling you it's found new photos that it *thinks* might be one of the people it already knows about. It brings up the same blue-checkmark screen that appears when you're naming someone (see "People recognition"). Each time you work through these shots, Photos gets better at recognizing your friends and family.

The faces Photos already knows

Supply a name—and view Photos' other guesses at this person.

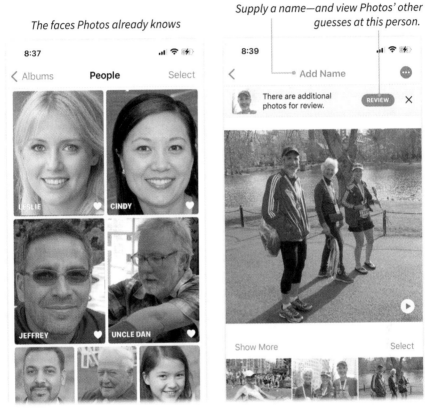

People recognition

- **Places** offers a map dotted with photo clusters taken in each location. Tap over and over again, zooming in each time, until you're looking at the actual photos taken in that spot.

- **Media Types.** These folders group your photographic exploits according to what *kind* of file they are. Most correspond to the modes in your Camera app: **Videos, Selfies, Live Photos, Portraits, Panoramas, Time-Lapse, Slo-Mo,** and so on. Kind of cool that you can easily find all your **Screenshots** in one place (page 435), all your **Screen Recordings** (page 437), all your **Bursts** (page 179), and so on.

- **Utilities** offers three miscellaneous folders that can be far more useful than their last-place positioning in the list might suggest.

Media Types

Imports shows every batch of photos you've ever brought into Photos, chronologically, scrolling back to the dawn of photographic time. **Recents** are all the photos that have entered your life electronically *lately*—from your Mac and from other people.

Hidden holds any photos you want to keep but would rather conceal from people peeking over your shoulder when you're scrolling through your shots. To hide them, select the photo and tap ⬆→**Hide**.

As for **Recently Deleted**: Photos has its own little trash that saves you from yourself. Any photo or video you delete in Photos doesn't actually self-destruct for 30 days. You have a month to realize your terrible mistake. At that point, you can open **Recently Deleted**, tap the doomed photo, and then tap **Recover**. Photos puts the photo back wherever it came from.

Search

The fourth tab at the bottom of the main Photos screens is, of course, the familiar Q button.

Now, you might, at first, wonder what a search button is doing in a *photo* app. How can you search for an *image*? How can the phone know what it's a picture *of?*

Well, it does. This is Apple's AI team showing off again. You can, in fact, search your photos for common subjects like "cat," "mountain," "pizza," and so on.

You can also search the *metadata*—the textual information Photos stores behind the scenes for each picture, like when it was taken, where it was taken, and who's in it.

To begin a search, tap Q. On the Search screen, Photos starts you off with some ready-to-tap searches: **Moments** (**One Year Ago**, **Summer**, **Trips**, and so on), **People** (using that face-recognition thing), **Places**, **Categories** (**Animals**, **Sports**…), and even **Groups** of people.

Suggested searches before you even type   *Results when you type*

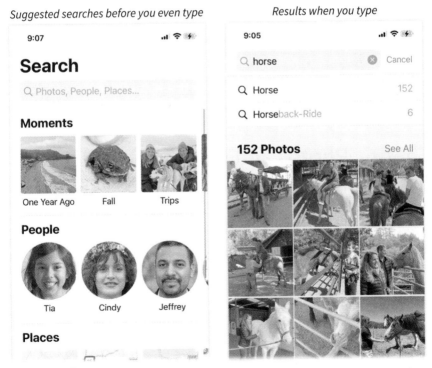

Searching Photos

To create a search of your own, you can type things like these:

- **Things.** *Field, cow, car, guitar, wall, restaurant, concert, fruit, teddy bear, salad, graffiti, money, dog, tennis, canoe…*

- **People.** *Alex, mom, me, harold toomey…*

- **Place or event names.** *Home, boston, weybridge road, toronto film festival, super bowl, carver high school, museum…*

- **Times.** *Last year, last week, yesterday, 2020, winter, evening, morning, november 2015…*

- **Captions or album names.** *Scrabble tournament, toby's birth, dad's bday…*

As you type, Photos builds a list of suggestions that match—both search terms (*dog, dog pound, hound dog…*) and photo thumbnails. If one of the search suggestions is what you have in mind (*dog*), tap it to see the matching photos.

At this point, you can refine the search by typing *another* search term—or tap one of Photos' suggested subcategories.

Photos' ability to understand what's *in* a picture isn't foolproof. When you search for *bear*, for example, the occasional cat photo might sneak in. But most of the time, it does an impressive job of pixel discernment.

> **TIP:** Unless you're paid by the hour, it's usually better to perform a photo search by voice. "Hey Siri: Show me all the photos from Vermont in 2019."

Sharing Photos and Videos

When you get right down to it, there are really only two reasons you take pictures. One, of course, is to capture a moment so you can recall it later. But the other is for sharing, to show *other* people what you've experienced.

iOS 14 offers about a thousand ways to give a photo or video a life beyond the Photos app. You can send it, share it, post it, copy it, and much more. All you have to do is follow a two-step process—with about a thousand options along the way.

Choose What to Send

If you're just sending one photo or video, tap its thumbnail. If you're sending a bunch, start on any screen full of thumbnails—for example, on the Photos

Selecting photos and videos

tab or in an album. Tap **Select**. Now you can tap the thumbnails of the outbound photos, or drag through several at a time. The ones with blue checkmarks are the ones you'll send.

> **TIP:** Don't forget that you can spread or pinch two fingers to adjust the sizes of the thumbnails, which makes it much easier to see what you're selecting.

Choose the Sharing Method

This is really the key to the whole operation: Tap the ⬆ button. You've just opened what Apple calls the share sheet, shown on the next page. It lists every conceivable option for processing the selected photo(s).

At top: a scrolling row of other photos. The idea is that you can refine your choice of outgoing photos—adding some, omitting some—by tapping.

> **NOTE:** If you're holding the phone horizontally, perform this step first and then tap **Next** to see the options described next.

Next row: icons that represent people you share things with most often—and *how* you share with them. You might see a Messages icon for Alex, a group Messages button for your golf buddies, a Mail button for Toby, an AirDrop icon for your own laptop, and so on. Tap one of these icons to send the photos, without having to re-specify the people and the sending methods. This row is just an incredible time- and tap-saver.

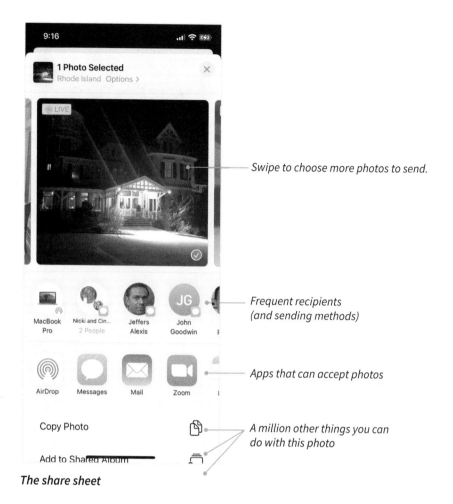

9:16

1 Photo Selected
Rhode Island Options >

LIVE

——— Swipe to choose more photos to send.

MacBook Pro   Nicki and Cin...   Jeffers Alexis   John Goodwin

——— Frequent recipients
(and sending methods)

AirDrop   Messages   Mail   Zoom

——— Apps that can accept photos

Copy Photo

Add to Shared Album

——— A million other things you can
do with this photo

The share sheet

Next row: Apps that can accept photos, like **Messages**, **Mail**, **Notes**, **Facebook**, **Books**, **Messenger**, and so on. If you tap **Messages** or **Mail**, the next screen asks you to specify the recipient(s).

> **TIP:** You can edit this row of app icons—get rid of the ones you never use, add new ones you'd like to use, and rearrange them. Tap **More** at the end of this row, and then tap **Edit**; see page 4 for details on using the resulting configuration screen.
>
> Unfortunately, YouTube and Vimeo aren't among the options here. To post videos to those services, use the YouTube, Vimeo, or iMovie apps.

Next: a long list of commands that do other things with your shared photos. Some of them are incredibly useful. Unfortunately, few people scroll down far enough to discover them, for the same reason that hardly anyone scrolls past the first page of Google search results.

- **Copy Photos** puts the selected pictures or videos on the invisible iPhone Clipboard, so you can paste them into another app. (To see the **Paste** command, open the other app and then double-tap.)

- **Add to Shared Album** adds these pictures or videos to an album you've shared (page 207).

- **Add to Album** puts them into one of your existing photo albums—or into a new one you create on the spot (page 197).

- **Duplicate** gives you a second copy of the photo, which you're free to mutilate without messing up the original.

HOW TO COMPRESS THE PIX OR VIDS YOU'RE SENDING

Right there, at the top of the share sheet, below "7 Photos Selected" (or whatever), is an **Options** button. It's so tiny you might miss it. It's useful only when you're sending photos and videos by Messages or with AirDrop.

The **Send As** options are Apple's way of acknowledging that pictures and videos are *big*, file size–wise. If you choose **Individual Photos**, the files you're sending may eat up a lot of cellular data and storage on your recipients' phones, which they may not appreciate.

If you choose **iCloud Link** instead, you'll send them only a web link. It opens a private web page that displays your pictures, where your grateful recipients have the option of downloading only the photos they actually want—or none of them. The link expires after 30 days.

If you choose **Automatic**, Photos chooses between those options all by itself. It sends an iCloud link if you're sending more than three photos (or a big video); the rest of the time, it sends the actual photos.

The Options screen also lets you turn off **Location** (the data that indicates where the photo was taken), which is

	Options	Done

Send As

Automatic ✓

Individual Photos

iCloud Link

When sending with Messages only, Automatic selects the best format based on file size or number of items being shared. Sending as an iCloud link will share a URL to view or download photos or videos.

Include

Location

All Photos Data

convenient if you're a scurrilous low-down no-good two-timer.

You can also turn on **All Photos Data**, which means you'll send not just the photo, but also descriptions and keywords you may have applied, and even your *editing history*! In other words, if the recipients have Apple devices, they can use *their* Photos app to undo or change any editing you've applied! ✦

- **Hide** makes these photos disappear from Photos, so there's no chance of your being embarrassed when scrolling through your pictures while someone's watching. The only place you can see them is in the Hidden album described on page 200.

- **Slideshow** is awesome. It begins an instant slideshow of your photos and videos, complete with a soundtrack. (The music even gets softer when one of your videos plays, so you can hear the original sound.) To change the music, speed, and crossfade effect, tap the screen and then tap Options.

- **AirPlay** displays this photo or video on your Apple TV (or another AirPlay receiver), so your admirers can get a better look.

- **Use as Wallpaper** slaps this photo onto the background of your home screen, your lock screen, or both (page 85). You're offered the Move and Scale screen, where you can shift or enlarge the image to fit the screen. (Available only if you've selected a single photo or video.)

- **Save as Video**, available only for a Live Photo (page 181), converts it into a stand-alone three-second video, which you can share with non-Apple people.

- **Copy iCloud Link** puts, on your Clipboard, a link that you can text or email to someone else. If they're Apple people, they get a button that lets them add the photos to their own Photos apps. If they're not, they get to see the photos on a web page, complete with a Download button so they, too, can have your pictures forever.

  This is a simple, quick way to share photos; just remember that anyone with the link—even if they're friends of friends, or *enemies* of friends—can have your pictures.

- **Create Watch Face** turns your photo into a clock face for an Apple Watch.

- **Save to Files** drops these photos as graphics icons into the Files app (page 327).

- **Assign to Contact** lets you use this photo, or the facial part of it, to represent somebody in your Contacts. Once that's done, you'll see this image every time that person calls, messages, or FaceTimes you. (Appears only when you've chosen a single photo.)

- **Print** prints the picture on any AirPrint wireless printer (page 432).

- **Edit Actions** lets you manage the list of commands you've just read—by hiding the ones you'll never use. See page 4 for details on using this configuration screen.

Other apps can add their own icons to this menu list, too. For example, you may see **Save to Dropbox**, if you use that service. And if you've made any Shortcuts (page 451) that handle photos or videos, their names show up here, too.

Taken together, these share sheet options offer an insane amount of freedom in bringing your photographic (or videographic) masterpieces to the masses.

Shared Albums

The most obvious way to send a photo to somebody else is to email it or send it as a text message, as described above.

Unfortunately, sending by Messages or Mail sends scaled-down images, not the full-size originals. Sending your photos is also a one-way process. It offers no *interaction* with the recipients; they can't give you a thumbs-up, leave comments, or contribute their own photos.

That's why Photos also offers Shared Albums. Using this method, your audience can respond to your shots the way they would to photos on Facebook or Instagram: They can click a **Like** button, leave comments, and even contribute their own pictures.

> **NOTE:** This feature works only if **Shared Albums** is turned on in **Settings→Photos**.

Here's the process:

1. **Choose the pictures and videos.**

   Use the **Select** button as described on page 202.

2. **Create the shared album.**

   Tap ⏏ and then **Add to Shared Album**. When the iCloud sheet appears, type in a Comment, if you like. Tap the Shared Album menu at the bottom, and then tap **New Shared Album**.

3. **Name your album.**

   This is the album name other people will see.

4. **Tap Next. Invite your audience.**

   Tap ⊕ to view a miniature copy of your address book. Choose the email address or phone number of each person you're going to invite to see your photos.

1. Tap to begin the process of creating and naming a new album.

| 4 Items Selected | | |
| McLaren Park Options > | | × |

| Cancel | **iCloud** | Post |

Comment (optional)

4 Items

| Shared Album | Devil's Den > |

| < iCloud | **Add to** |

+ New Shared Album >

SFO Staycation!
You and Nicki Pogue

Trout Valley Preserve
You and 2 more

| Cancel | **iCloud** | Next |

McLaren Park hike!

2. Once you invite your audience and add a description, the album is ready to go live.

| Cancel | **iCloud** | Next |

To: Jeffers Alexis Steven Apper | ⊕

| Cancel | **iCloud** | Post |

Memories of this afternoon!

4 Items

| Shared Album | McLaren Park hike! |

Creating a shared album

In general, only members of the Apple ecosystem—people who have their own iPhones, iPads, or Macs—get the full Shared Album treatment, and only their iCloud email addresses and phone numbers show up here.

> **TIP:** You *can* still share your photos with unfortunates who don't have Apple devices; read on.

5. **Tap Next and then Post.**

At this point, everybody you invited gets a notification in their Photos apps (iPhone, iPad, Mac) that you shared an album. And you see the name of the shared album on your **For You** tab, so you can have a look at what you shared.

There's one more change, too: On your Albums screen, a thumbnail for your new shared album appears under Shared Albums. That's important, because the only way to make changes to the album (or delete it) is by tapping this thumbnail.

Now you can add or remove photos and videos. Any photos you delete here (long-press; tap **Delete**) disappear from everybody else's machines. And any you add (tap +) show up on their screens within a few seconds.

If you tap the ⊙ at the top of the screen, you unlock a few more important Shared Album controls:

- **Rethink the invitees.** Here you can invite new people, or delete old people, from the invite list.

- **Prevent others from contributing.** If you and some of your audience were at the same party, concert, or trip, they can contribute some of their *own* photos to your shared album. Pretty cool: new angles, new subjects.

  Unless, of course, you fear either mischief or a sullying of your artistic vision. In that case, turn off **Subscribers Can Post**.

- **Post the album on the web.** If you turn on **Public Website**, then even people not blessed with Apple equipment can look at your pictures. "Public" here does not mean the teeming masses will have access to your photos; it means only that you can now add the email addresses of non-iCloud

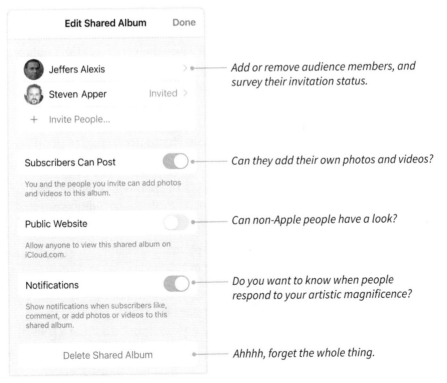

Shared Album options

people. Their invitations will include a web link that opens up a private web page featuring your magnificent photography. (They won't be able to add comments or contribute their own photos. That's what they get for being second-class ecosystem citizens.)

- **Delete the whole thing.** To stop sharing your album with the world, tap Delete Shared Album.

Editing Photos and Videos

Over the years, successive versions of Photos have crept closer and closer in power to professional touch-up programs like Photoshop. And these days you can apply most of those same changes—rotation, cropping, brightness, saturation, contrast, and so on—to *videos* as well as photos.

Here's the best part: *No editing you do here is ever permanent.* It's all non-destructive editing. That is, you can return to a photo minutes, days, or even years later, and restore it to its original, unedited state.

> **NOTE:** To do that, open a photo, tap **Edit**, and then tap **Revert**. After asking if you're sure, Photos restores the photo to its original condition.

To enter editing mode, open a photo or video and then tap **Edit** at top right. At this point, the Photos editing window feels like a separate app. It has its own layout and a totally different look. And there's a *lot* going on here.

For starters:

- **You can hold the phone in either orientation.** The controls jump to the side or the bottom of the screen.

- **At any point during your editing,** you can tap the photo to compare your edited masterpiece with the original image. Professional editors call that kind of comparison "A/B testing."

- **Also at any point, you can zoom in** on the photo by spreading two fingers. Professional editors call that "zooming in."

At the very bottom of the screen (or the right side, if you've turned the phone), Photos offers three icons: Adjust Color (⦿), Filters (⊛), and Crop/Straighten (⊡). (You may see an additional button if you're editing a Live Photo, a Portrait photo, or a video.)

These three buttons aren't anywhere near equally important. You'll probably use ⦿ daily, ⊡ occasionally, and ⊛ almost never.

Adjust Color

Filters

Crop/
Straighten

Edit mode

TIP: You can draw or type on your photos, too. In Edit mode, tap ● and then **Markup**. Now you have access to iOS's standard annotation tools (page 7).

Adjust Color (⦂●⦂)

Here Photos presents a scrolling row of round adjustment buttons: for exposure, contrast, saturation, and so on. As you drag this row, you feel a little click as you land on each one.

For each adjustment, you're supposed to drag the slider scale to specify how *much* of each effect you want. (You feel another little click every time you pass the zero point.) As you drag, you see the effect on the photo in real time, a yellow "amount" ring grows around the adjustment button itself, and digits inside the ring show how much of the effect you've applied (on a 100% scale).

> **TIP:** To hide the effect of an effect (that makes sense, right?), tap its button (**Exposure**, for example). Tap again to restore the zero point.

Here's what the controls do:

- **Auto-Enhance** (⸕) is a one-tap magic wand. It auto-adjusts *all the other adjustment* controls in one fell tap—color, brightness, contrast, saturation, and so on—using huge servings of artificial intelligence as its guide. (You can even look over the other adjustment dials to see how they've been tweaked.)

  The result *almost* always looks better. If you don't think so, tap ⸕ again to turn Auto-Enhance off.

- **Exposure** adjusts the overall brightness of the photo.

- **Brilliance** reveals hidden detail in the darkest and brightest spots. In other words, it manages Highlights and Shadows simultaneously.

- **Highlights** recovers lost details from very bright areas.

- **Shadows** pulls lost details out of very dark areas.

- **Contrast** deepens the most saturated colors.

- **Brightness** brightens only the dark and middle tones. Unlike Exposure, it doesn't brighten parts that are already bright.

- **Black Point** shifts the entire dark/light range of the photo upward or downward.

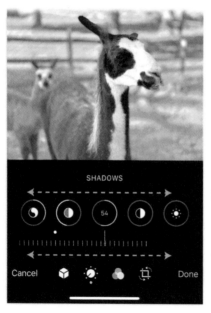

Drag to choose an adjustment...

...and then dial in how much.

Adjustment sliders

- **Saturation** adjusts the intensity of the colors. Turn Saturation down all the way to get black-and-white; nudge it slightly to lend a little pop to a bland photo; turn it up all the way to get a hallucinogenic circus.

- **Vibrance** is the same as Saturation, except that it leaves skin tones alone.

- **Warmth** makes a photo warmer or cooler overall by shifting all colors along the yellow-to-blue spectrum.

- **Tint** affects colors on the green-to-magenta spectrum.

- **Sharpness** tries to crispify a photo. It can't save a truly out-of-focus or blurry shot, but it does enhance the difference in brightness between adjacent color areas.

- **Definition** tries to make a photo clearer by adjusting the contrast of its midtones.

- **Noise Reduction** helps to remove "noise" (random, grainy colored speckles) that often show up in low-light photos.

- **Vignette** brightens or darkens the four corners of the photo. It can be a subtle (or blatant) way to guide the viewer's eye to the center of the photo.

- **Remove Red Eye** (⊘) shows up only on flash pictures of people that exhibit the *red-eye* problem: creepy, glowing red pupils of their eyes. It's the result when the flash reflects off the retinal tissue at the back of the eyeball.

  To use this control, tap it. Zoom in (spread two fingers) and then tap with your finger inside each red eye. A white ring encircles the pupil, and the app turns the red in each eye to black.

At any point, you can tap Cancel to abandon your editing altogether, or Done to save the edited photo or video and close the editing controls.

Filters (⊛)

These nine effects affect the photo's overall color impression. They can make it washed out, oversaturated, warmer, cooler, or black-and-white.

As with picture adjustments, each one offers an adjustable intensity scale. (The first one, Original, removes whatever filter you've applied.)

Crop/Straighten (⛶)

Using this tool, you can crop out the edges of the photo, rotate it, or flip it horizontally. In iOS 14, you can even perform these operations on a *video*, which is a blessing for anyone whose phone has ever incorrectly recorded one sideways. Here's how it all works:

- **Flip Horizontal** (◮) creates a mirror image of the photo or video.

- **Rotate** (◼) turns the picture or video 90 degrees. It always rotates counterclockwise, so you may have to tap a few times to get the rotation the way you want it.

- **Constrain Cropping** (◼). Cropping involves shaving off outer chunks of a photo (or video!), usually to make the composition better: to trim away dead space above the subject's head, for example, or to frame the subject more tightly, which is usually a good idea.

Just drag inward on any edge or corner of the white border. You can recenter the image by dragging it.

If you tap ▥, you get a scrolling list of preset aspect ratios (rectangular proportions), including **Original, Freeform, Square, 16:9, 7:5, 4:3,** and so on. Each confines your dragging to those canned proportions. That's useful when, for example, you intend to order prints of your photos at standard sizes, like 4 × 6, 5 × 7, or 8 × 10. (The iPhone's standard photo size is 4 × 3, which doesn't fit evenly on most of those sizes.)

- **Straighten (⬬).** Drag the slider to tilt the photo or video, to fix bad composition.

- **Vertical, Horizontal Perspective (▲, ◀).** These sliders make one edge of the photo or video seem to lean toward you or away, as though you're chang-

Cropping basics

ing your viewing angle.

Editing Live Photos

A Live Photo, as you know from page 181, is a still photo with a three-second video attached. When you open one for editing, a fourth adjustment button (◎) appears. When you tap it, you get a few special buttons just for Live Photos: a ◀») button that turns off the sound; a ◎ LIVE button that converts the

---

THE MIRACLE OF ICLOUD PHOTOS

The service called iCloud Photos may be one of Apple's greatest inventions. If you turn it on, you give permission for Photos to store your entire video and photo collection online, on Apple's servers.

And why is that such a big deal? Because it means every computer, phone, or tablet you own is always up to date with all your photos, photo albums, *and* whatever editing and organizing you've applied to them. You never have to wonder: "Was that picture on my Mac or on my iPhone?" They're all everywhere.

The *editing history* of every photo is even synced. If you cropped a photo on the phone, for example, you can open it up in Photos on the Mac and change or modify that cropping.

It also means that the most precious files you own—your photos and videos—are always backed up, wirelessly and in real time. Your house could fall into a sinkhole, engulfed in flames, and you wouldn't care. You'd know that your photos were still safely stored online. (OK, you might care somewhat.)

iCloud Photos also means you have access to your entire photo collection from any computer or phone in the world. You just log in at iCloud.com, click **Photos**, and there they are.

Maybe most useful of all, using iCloud Photos can grant you an iPhone whose

storage drive never gets full. If you turn on a feature called Optimize iPhone Storage, then whenever your phone does start getting full, Photos quietly removes the huge, space-eating, full-size originals of your photos from the phone—and replaces them with scaled-down copies that occupy far less disk space. When you're working in Photos, you probably won't even know the difference, because you can still work with them, edit them, organize them, and zoom into them until they fill your screen.

But if you ever want to export, print, or share those photos, Photos downloads the full-resolution originals, quickly and invisibly, for you to work with.

All these benefits may sound joyous and effortless, and they are, but beware: They cost money. Apple gives you 5 gigabytes of storage space with your free iCloud account, which is nice. But unless your photo collection is very small, 5 GB is not enough to hold it. You'll have to upgrade your iCloud storage. Apple charges $1 a month for 50 gigabytes, $3 a month for 200 GB, or $10 a month for 2 *terabytes*. (Unless you take so many photos that your family hates you on vacations, the $1 plan is probably plenty.)

If you decide to try iCloud Photos, choose **Settings→Photos**. Turn on **iCloud Photos**.

It's really darned cool. ✯

Live Photo into a dead one (a regular still image); and a trimming strip along the bottom.

> **TIP:** The *key photo* is the frame of the "video" that serves as the Live Photo's thumbnail in the Photos app. Drag your finger along the filmstrip, and then tap **Make Key Photo**, to choose a new one.

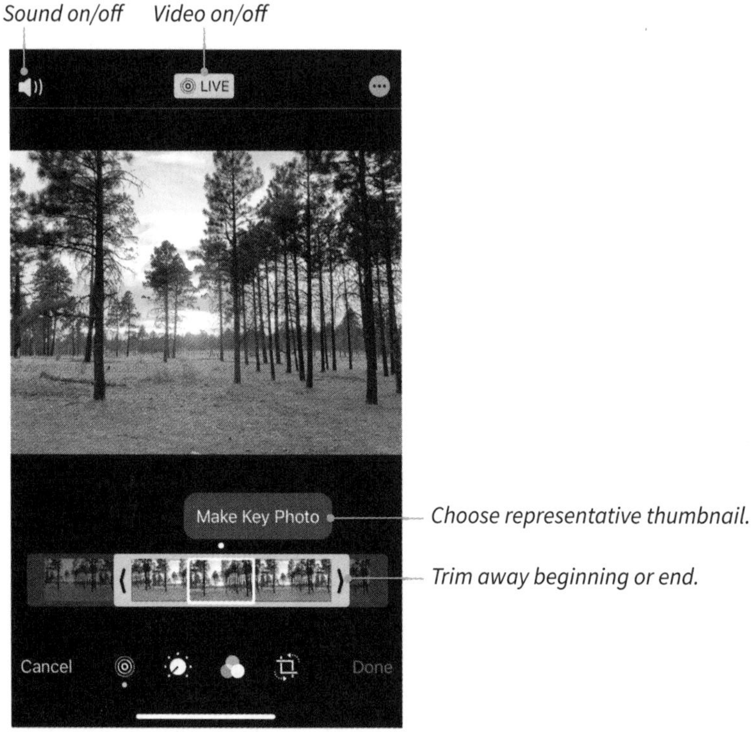
Sound on/off
Video on/off
LIVE
Make Key Photo
Choose representative thumbnail.
Trim away beginning or end.
Cancel
Done

Editing a Live Photo

Weirdly, some of the most useful Live Photo editing controls do not, in fact, appear on the Edit screen. (Hit **Done** or **Cancel** to get out of there, if necessary.)

Instead, to find them, open a Live Photo's thumbnail as usual—and *swipe upward*. Here you have four special video-playback effects:

- **Live** is the usual Live Photo mode: a three-second video that plays once and then stops.
- **Loop** makes the three-second video play over and over, with a crossfade between repetitions. This can be funny or annoying, depending on how comic the shot is.

- **Bounce** plays from beginning to end, and then backward to the beginning, like a yo-yo. Pure hilarity when applied to a pie hitting someone in the face.

Special Live Photo effects

- **Long Exposure** works best on shots where the background is stationary but the subjects are moving: people, cars, water. It impersonates a camera whose shutter was left open for a long time. It converts a river into a silky, milky channel of white, or converts nighttime taillights into cool orange trails.

> **TIP:** The Long Exposure effect doesn't do much for the Live Photo as a *video*. It's all about the still image that results when you turn off the ⊚ LIVE badge.

Editing Portrait-Mode Shots

You can make adjustments to your blurry-background Portrait-mode shots at the moment you take them, as described on page 172. But usually, you'll have more time to make the changes afterward.

Once you tap **Edit**, all the usual controls are available: color adjustments, filters, cropping, and so on. But two new buttons appear at the top-left corner.

When ⊙ is selected, your studio-lighting controls appear, exactly as described on page 172. Feel free to play around with your virtual studio lights.

Tap to dial in less blur...or more blur.

Portrait mode

But the ⨍ button at the top of the screen may be even more useful. It produces an f-stop slider. Drag to the right (or drag up) to choose more background blur—what photographers think of as a lower f-stop (wider aperture). Drag to the left (or down) to bring the background more into focus.

Dial up the blur that works for now; you can always change your mind later.

Editing Video

The Photos app can edit videos in all the same ways it can edit photos, which is pretty fantastic.

When you open a video and tap **Edit**, three of the buttons work exactly as described starting on page 210: :◐: for color and brightness; ⊛ for the nine color filters; and ⊕ to rotate, flip, or straighten videos. Even the two perspective buttons (▲ and ◄) work on videos.

> **TIP:** You crop in on a video for exactly the same reasons you'd crop a photo: to focus attention on the subject, or to eliminate dead space around it. Remember the discussion of the 4K video format on page 187? Here's a solid reason to shoot in 4K: You can crop away a lot of the outer frame and still have enough resolution left for high-definition playback. (If your original video was in high definition, on the other hand, and you cropped away a lot of margin, what's left might start to look pretty janky.)

But in Edit mode for a video, there's a fourth button: ▆◀. It presents a "filmstrip" version of the entire video. Here's how you use it:

- **Drag your finger across it** to "scrub" through the video, as though you have an infinitely powerful rewind/fast-forward control.

- **Tap the ▶ button** to play the video to check your work.

- **Drag the white (and) handles inward** to trim the video, making it shorter. For example, you might use these handles to chop out dead air at the beginning.

> **TIP:** Grabbing these handles is a little tricky. Try pausing with your finger down before you start dragging. Also, if you hold the phone horizontally, the filmstrip gets longer and therefore offers better precision.

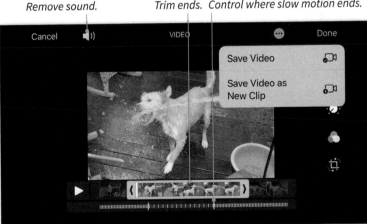

Remove sound.   *Trim ends.*   *Control where slow motion ends.*

Editing video

- **If it's a slow-motion video,** you can drag the inner handles to govern when the slo-mo part begins and ends.

Tap **Done** to immortalize your edits. Photos offers you two options: **Save Video** (meaning "Modify the original video") and **Save Video as New Clip** (meaning "Now you'll have two clips—the original and this edited one").

It doesn't really matter which you tap. Here again, everything you've done is non-destructive; you can return to this video later to readjust the trimming handles, colors, crops, and so on.

> **TIP:** Photos is not a full-blown video editor. It doesn't let you split a clip into pieces, or string several clips together, or add titles, music, and effects. But Apple's own iMovie can do all that. It's free, and it's waiting for you on the App Store at this very moment.

PART THREE

The iCommunicator

CHAPTER NINE

Phone Calls and FaceTime

Depending on your age, you might occasionally forget that the iPhone is, in fact, a *phone*. It's so good at taking pictures, and texting, and surfing the web, that you hardly ever see people actually holding it up to their heads and talking.

And yet, sure enough, it does make phone calls. (Or at least it does if you're reasonably close to a cell tower in town.)

Then, when hearing the other person isn't quite the next best thing to being there, you can hop into a FaceTime call—and see them, too!

Dialing from the Phone App

The iPhone comes with an app—the Phone app—dedicated to placing phone calls. But here's a little secret: It's a lot faster to make calls using Siri. Say, "Call Alex" or "Call Toby's cell" or "Dial 951-6223." (Actually, you don't even need to say "Dial." Just say, "951-6223." Siri knows what to do.)

But if you're feeling luxurious and want to explore the options in the Phone app, by all means open it up.

> **NOTE:** If you see a tiny circled number on its icon (❷), iOS is trying to tell you that you've missed that number of calls and voicemail messages.

The first four buttons at the bottom of the Phone app all represent ways to place calls: **Favorites, Recents, Keypad,** and **Contacts.**

Favorites

Favorites is your speed-dial list. You can put up to 50 frequently dialed people on this list, ready to call with a tap.

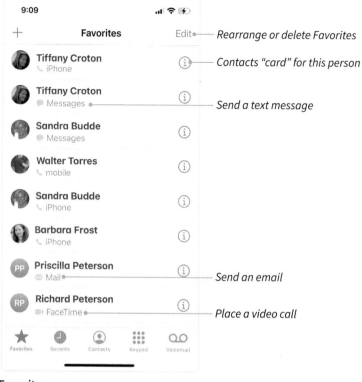

9:09

+   **Favorites**   Edit— *Rearrange or delete Favorites*

Tiffany Croton
iPhone   ⓘ — *Contacts "card" for this person*

Tiffany Croton
Messages — ⓘ — *Send a text message*

Sandra Budde
Messages   ⓘ

Walter Torres
mobile   ⓘ

Sandra Budde
iPhone   ⓘ

Barbara Frost
iPhone   ⓘ

Priscilla Peterson
Mail — ⓘ — *Send an email*

Richard Peterson
FaceTime — ⓘ — *Place a video call*

★   ⏱   ⊙   ⠿   ◎
Favorites   Recents   Contacts   Keypad   Voicemail

Favorites

What may throw you at first is that these aren't just phone numbers. You can designate *any* communications method as a favorite: somebody's email address, texting address, FaceTime address, or even the address for an internet calling app like WhatsApp, Skype, or Cisco Spark. Your Favorites list can wind up as a combination of one-tap buttons for dialing, messaging, or emailing.

To add somebody to this Favorites list, begin on their Contacts card. (Either tap **Contacts** at the bottom of the Phone app or tap the ⓘ next to somebody's name on the **Recents** or **Favorites** screen.)

> **NOTE:** And what if this person isn't in Contacts yet? You have to add them before you can make them a favorite.

Scroll down and tap **Add to Favorites**. If you have more than one scrap of contact information for this person—an email address and a couple of phone numbers, for example—the iPhone now asks you which one you want to enroll as a favorite. Tap accordingly. (A gray star on the Contacts screen denotes a contact method that's already listed in your Favorites.)

If you want to list two different communications methods for the same person on the Favorites screen, you'll have to repeat this process and create two different Favorites entries.

Managing your Favorites list is easy, once you discover the **Edit** button at top right:

- **Reordering favorites.** Drag the grip strip (≡) up or down to organize your Favorites list.

- **Deleting favorites.** Swipe leftward across the undesired name. Tap **Delete.** (You're not actually deleting a phone number—you're only removing it from the Favorites screen.)

Recents

On this screen, you get a list of lists: all the calls you've made, answered, or missed. You get to see the names or origin cities of the callers, if the phone can figure them out; a little handset icon indicates calls *you* placed. Calls you missed appear in red and also have a list of their own (the **Missed** tab).

Recents list

Tap a name or number to call back.

> **TIP:** Not everybody you speak to on the phone is in your Contacts. Yet, in one of its most brilliant attempts to be helpful, the iPhone takes a *guess* at some people's identity—by matching their phone number with numbers it finds in your email (for example, in the signature line of that person's email). You might see, for example, "Maybe: Chris Pendergrass" below a phone number in this list. It's usually right.

If you tap the ⓘ next to a call on this list, you open a fancy Contacts page that not only identifies all this person's contact information, but also reveals a list of calls to and from this person.

You can erase a single call from the Recents list by swiping to the left and tapping Delete. Or, if you're feeling a little spied upon, you can erase the *entire* Recents list by tapping Edit→Clear→Clear All Recents.

Keypad

Here's your basic 10-digit dialing pad. Of course you can use it to dial a number manually; tap the 🕿 to place the call. Probably more often, you'll use the keypad to navigate voicemail systems and respond to prompts like "For Italian, press 37."

> **TIP:** If you tap the 🕿 *before* you touch any number keys, you recall whatever number you dialed most recently, so you can dial it with another tap on the 🕿.

Once you've dialed, no matter which method you used, either hold the iPhone up to your head, put in the earbuds or AirPods, turn on the speakerphone, or put on your Bluetooth earpiece—and start talking!

Contacts

iOS includes a standalone app called Contacts, which is its master repository for all the contact information for everybody in your circle. But tapping Contacts at the bottom of the Phone app opens the same app.

At the very top of Contacts is a very special card: yours. Tap it and make sure all your contact information is up to date and correct, because all kinds of iOS features rely on the information here. This is how Maps knows where to take you when you say, "Give me directions home." This is how Calendar knows how far away your workplace is from some appointment. And this is what makes it easy to send your electronic business card to somebody: Tap My Card, scroll down, and tap Share Contact.

To find anybody else in Contacts, you can use the search box, you can scroll, or you can tap the little alphabetic index letters down the right side (or drag your finger down them).

Once you find the person you're looking for, tap the name. Here before you is all of person's contact information, as well as one-tap buttons to **message**, **call**, **FaceTime**, **mail**, or **pay** this person.

Drag or tap to jump through the alphabet.   *Tap a name to see all the contact info.*

Contacts

Here's a quick tour of Contacts:

- **Import an address book.** Mercifully, you don't have to enter every name and address by hand. The iPhone can import whatever existing contacts list you have on your Google, Yahoo, AOL, iPad, or Microsoft Exchange account. (Open **Settings→Mail→Accounts→Add Account**, choose the service name, and then enter your login information.)

- **Add an entry manually.** In Contacts, tap +. The New Contact screen appears, where you can enter contact information about somebody new: name, phone number, email address, physical address, birthday, social-media accounts, notes, and on and on.

THE UTILITY AND CONFUSION OF GROUPS

At the top-left corner of Contacts is a button called **Groups**. It reveals *subsets* of people, like **Work Buddies**, **Family**, **Bowling League**, and so on.

The real payoff for setting up groups like this is managing email. Every time you want to write to the entire family, for example, you can just choose **Family** instead of manually adding each person's email address.

Unfortunately, iOS doesn't offer any way to set up these groups (at least not without downloading an app for the purpose). You have to create them on your Mac, PC, or Exchange server; at that point, they show up in Contacts, under **Groups**.

What you *can* create on the iPhone: individual internet accounts like Google, Yahoo, AOL, and Microsoft Exchange. Each of these has its own address book, and each of them shows up on the Groups screen as well.

And that explains a common bit of confusion: Every now and then, you won't be able to find somebody you're certain is in your Contacts list. It's probably because the account or group containing that person's name is *turned off* at the moment.

Done

Groups

ICLOUD

- ✓ All iCloud
- ✔ Announcements
- ✓ Team B
- ✓ The Kids
- ✔ Work

GMAIL

- ✔ All Gmail

YAHOO!

- ✓ All Yahoo!
- ✔ Contacts

Show All Contacts

To see for yourself, tap **Groups** at the top-left corner. On the list of accounts, make sure the appropriate group not only has a checkmark but is solid blue. Or, if you want to make sure you're seeing everyone, tap **Show All Contacts**. ✶

The iPhone capitalizes names automatically as you type, and it formats phone numbers with hyphens and parentheses. You can even enter phone numbers like 1-800-GOT-JUNK; when it comes time to dial, the phone knows what to do.

TIP: When you're entering phone numbers, you can use the # symbol to represent a pause. That's handy when you're building in access numbers, extension numbers, or voicemail passwords into a phone-number box.

When you tap add phone, you get a box for entering a phone number. But you also get a label for that number. Tap it to choose a different label for this number from Contacts' impressive list, which includes not just home, work, and mobile but also options like home fax and pager. You can even tap Add Custom Label to make up entirely new ones (like Lame 2005 flip phone).

Each time you add a new phone number, street address, email address, or web address, Contacts creates another new, *empty* field below it. That way, you'll never run out of room to record additional numbers or addresses.

TIP: At the very bottom of the New Contact screen, you can tap **add field** to add a new miscellaneous info box. Your options include **Nickname**, **Maiden name**, **Pronunciation first name**, **Job title**, and so on. Some of them could really save your bacon on a job interview or a date.

At the very top of the screen, you can tap Add Photo to add a picture of this person. If they're nearby, you can tap 📷 and take a picture on the spot; if you already have a picture somewhere, you can tap 🖼️ and choose it from your Photos collection. You can also use any emoji you like to represent this person. (If you have an awful boss, for example, you're welcome to choose the 👹 or even the 💩.)

If getting an actual photo isn't practical, you can recreate the person artistically using the Memoji tools here (page 264). At the very least, tap the ✏️ to dress up the generic two-initials-in-a-circle representation.

Whatever picture you choose for this person will appear on your screen whenever they call, text, or FaceTime. That picture also appears next to their name on the Favorites screen, and, if you like, in Messages or Mail.

- **Editing a Contacts card.** To edit somebody's card, tap their name in the Contacts list and then Edit. You're right back on a clone of the New Contact card, ready to add or update the information.

- **Deleting a Contacts card.** Somebody dies, somebody moves away, somebody breaks up with you. It happens.

  Open the person's card, hit Edit, scroll all the way to the bottom, and tap Delete Contact→Delete Contact.

- **Sharing a Contact.** Somewhere in the world, people still exchange contact information by handing little pieces of cardboard to each other.

  But if the person you've just met has a smartphone (iPhone or Android), it's infinitely faster and more accurate to open your card in Contacts and tap Share Contact. You can send this contact "card" by AirDrop, Messages, Mail, and so on; see page 203 for a description of the standard share sheet.

  Most of the time, you'll probably want to share *your* card, but you can share *anybody's* contact information this way.

Visual Voicemail

The final tab at the bottom of the Phone app is Voicemail. This is the famous *visual* voicemail, a 2007 Apple invention that's now standard on smartphones. It displays voicemail messages in a scrolling Inbox, just like emails or text messages. You can listen to them in any order, share them with other people, and even read a written transcription of what each message says.

> **NOTE:** If you've never visited the screen before, the iPhone asks you to make up a numeric passcode for accessing your voicemail. Frankly, you'll never need it. (It's for *dialing in* for messages from another phone.) Here's where you can record your greeting, too—your "I'm unavailable—leave me a message" message. (At any time, you can tap **Greeting** at the top left of the Voicemail screen to re-record your voicemail message.)

New voicemail. Tap to listen.　　　　　　　　　*Play through speaker*

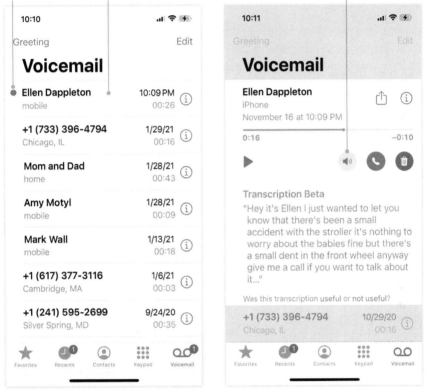

Visual voicemail

A blue dot denotes a message you haven't heard yet. Each message shows the date it came in, how long it is, and what number and city it came from. If it came from someone in your Contacts, you see the person's name instead of the number.

When you tap a message here, it reveals an array of useful buttons: ⬆ (send this message to someone else), ⓘ (see the Contacts card for this person), ◀)) (turn on the speakerphone when playing the message), ● (call the person back), ⊖ (delete the message)—and, maybe most important, ▶, which begins playing the message aloud.

> **NOTE:** Well, aloud enough for you to hear by holding the phone to your head. For obvious privacy reasons, the iPhone doesn't use the speakerphone unless you tap the ◀)) button to turn it on.

You can drag along the scrubber bar to skip ahead or back. And, of course, you can look at the iPhone's rough transcription of the message contents.

You may find some entertaining gibberish here and there—it is, after all, a computer trying to understand the people calling you on bad phone lines—but it's usually enough for you to get the gist of the message.

Deleting Messages

Here's an easy way to delete a message you no longer need: Swipe left across it and then tap **Delete**.

But if you're really on a roll, and you want to purge a bunch of messages at once, tap **Edit** (top right). Tap the circles of all the messages you want to target, and then tap **Delete**.

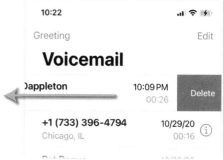

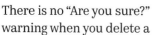

Swipe to delete

There is no "Are you sure?" warning when you delete a voicemail message, but don't get nervous. "Deleting" a message only puts it into a **Deleted Messages** *folder* at the very bottom of the voicemail list. There it sits for about a month, giving you plenty of time to reconsider. (To recover the message, tap **Deleted Messages**, tap the message, and then tap ◉.)

Only if you let 30 days go by does the iPhone finally vaporize the message for good.

Answering Calls

When someone calls your iPhone, you'll know it. You hear the ring, you feel the vibration, and you see the notification.

What that incoming-call notification looks like, however, can vary:

- **If the iPhone is asleep or locked:** The screen awakens, showing the name of the caller (if they're in Contacts) or the number (if not). Swipe across the words **slide to answer** to answer the call.

Phone is locked. You get a full-screen photo, because the contact card is recent.

Phone is locked. You get only a tiny photo, because the contact card is older.

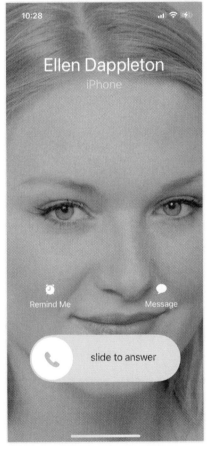

Phone is awake. Tap to expand full-screen.

Three looks for an incoming call

NOTE: If the caller is in your Contacts, their photo appears, too. But for some people, it's just a tiny photo in a little circle, and for others, it's a full-screen photo big enough to see from across the room. What's the difference?

Believe it or not, the difference is how long ago you added (or edited) the person's Contacts photo. People whose cards you've created recently get full-screen photos; older Contacts get the little ones. If you'd like their photos to appear full-screen, too, all you have to do is *edit* their photo in Contacts. On their card, tap **Edit**; under the photo, tap **Edit**; tap the photo and tap **Edit**; move it or enlarge it a tiny bit—and then tap **Choose**, **Done**, **Done**, and **Done**.

From now on, that photo fills the screen when they call.

- **If you're using the iPhone:** A notification appears at the top of your screen.

Yes, the *top* of your screen—iOS 14 no longer hijacks the entire screen when someone calls.

NOTE: Not everyone is a fan of the new incoming-call banner. The old full-screen incoming call screen was easier to spot from across the room—or when the phone was on the bathroom counter and you were still in the shower.

Fortunately, the new design is optional. If you don't like it, open **Settings→Phone→ Incoming Calls**, and switch from **Banner** to **Full Screen**.

Tap the green **Accept** button.

ADDING A CONTACT ON THE FLY

You're standing there at the end of the flight, the corporate off-site meeting, or the date. It went well. You've made a new friend. You need to record their phone number.

Here's by far the fastest way: Open the Phone app, tap **Keypad**, and type in the number. But you're not going to dial it; you're going to tap **Add Number→ Create New Contact**. The New Contact card appears, ready for you to enter the first and last name. But the phone number you dialed is already safely in place. ✦

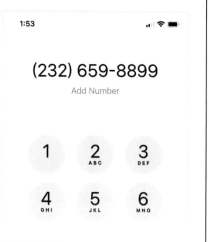

- **If you're wearing earbuds:** The music or podcast fades and pauses. Answer by squeezing the clicker on the earbud cord, squeezing the button on your AirPods Pro, or double-tapping a regular AirPod. Or answer the call on the phone itself, as described above.

After the call, repeat the same step to hang up. Your music or podcast resumes.

> **TIP:** If your hands are greasy, busy, or nonfunctional, the iPhone has a solution: After a few seconds of ringing, it can answer incoming calls automatically! You turn this on in **Settings→Accessibility→Touch→Call Audio Routing→Auto-Answer Calls**. You get to specify how quickly the phone picks up.

Once you've answered, you can do other things on your phone without interrupting the call. That's a blessing, since you may need to check your calendar, refer to an email, or (if the call is really dull) play a game.

Rejecting Calls

On the other hand, sometimes you *don't* want to answer the call. Maybe you're busy. Maybe you're sleepy. Maybe you can see who it is.

- **Silence the ring.** Often, you're willing to take the call, but you need a moment to leave the meeting, the train seat, or the bathroom. In those situations, press any button on the phone's edges—sleep switch, volume keys—to make the ringing stop. (Flicking upward on the notification banner does the same thing.)

You haven't declined the call; the caller still thinks it's ringing. But at least there's no noise to annoy everyone around you.

You can then answer the call when you're in a more appropriate place.

> **TIP:** *Any button* on the edges of the phone silences the ring, instantly. That's a valuable fact when your phone starts ringing at a bad time—in a movie, in a meeting, during a funeral service. Reach into your pocket or purse and just *squeeze the phone*. You don't have to look at it. One of your fingers will hit one of the buttons. Nobody needs to know that *you* were the idiot who forgot to turn on Do Not Disturb.

- **Dump it to voicemail.** If you don't answer the phone within four rings, it goes to voicemail.

You can also send it to voicemail sooner. If the phone is locked, double-click the side button; if it's awake, tap **Decline**. If you're wearing earbuds, squeeze the clicker for two seconds. The call instantly goes to voicemail.

- **Respond with a text.** You can also auto-reply to an incoming call with a prepared text message that says something like, "Sorry, I can't talk right now." You might consider that a more considerate and communicative response than just dumping the caller to voicemail.

On the lock screen, tap **Message** and choose among the three messages you want to send. (If you're using the phone when the call comes in, you have to tap the banner to view the full-screen Incoming Call options.) You can edit the three canned messages in **Settings→Phone→Respond with Text.**

The incoming call goes straight to voicemail, and the caller's cellphone gets your text.

> **NOTE:** Of course, Do Not Disturb While Driving is an even more automated way to achieve the same result whenever you're driving. See page 51.

You have two ways to decline a call more gracefully.

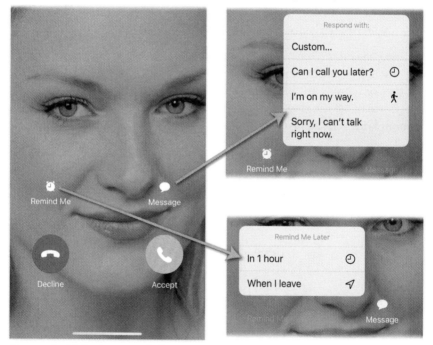

Respond with text—or remind me

- **Remind Me Later.** If the caller isn't on a cellphone, then the respond-with-text option isn't very useful; landlines and business phones generally can't get text messages.

If you tap **Remind Me** instead, the phone will remind you to call this person back later. Your choices are **In 1 Hour**, **When I leave**, **When I get home**, and **When I get to work**. (The **home** and **work** options are available only if the iPhone *knows* your home and work addresses; see page 343.)

Fun with Phone Calls

Once you're on a call, the call screen offers some useful call-manipulation options.

> **NOTE:** If you don't see these options, because you don't see the full-height call screen, tap either the incoming-call banner, if it's still there, or the 📞 on the left "ear" of the screen. (If you're using a home-button phone, tap anywhere on the green strip at the top of the screen.) That's how you get the call screen back.

- **mute** cuts your mic so the caller can't hear you. (You can still hear them, though.) That's useful when you need to have a quick in-person conversation with someone, run the blender, or flush the toilet. Tap again to unmute yourself.

- **keypad** reveals the standard touch-tone keypad, for use when you need to "Press 1 for more options." Tap **Hide** to return to the call screen.

- **audio** or **speaker** lists whatever ways you might have to take the call right now. **iPhone** means "Hold the phone up to my head"; **Speaker** turns on the speakerphone. You might also see the names of your earbuds, AirPods, car Bluetooth system, Bluetooth speaker, or even Mac laptop, any of which

---

FIGHTING BACK AGAINST ROBOCALLERS AND SPAMMERS

The average American is bombarded by robocalls and telemarketers 17 times a month. (Other Americans only *wish* they'd get that few.) Fortunately, the iPhone has a built-in defense—called Silence Unknown Callers.

It's kind of great. Turn it on in **Settings→ Phone→Silence Unknown Callers→ Silence Unknown Callers**.

From now on, anyone the iPhone considers unknown—people who are not in your Contacts, *and* you haven't communicated with them recently by email, text, or phone—goes straight to voicemail.

But the App Store is full of apps that do a smarter job of auto-blocking calls from robocallers and other phone spammers: Robo Shield, Truecaller, Nomorobo, RoboKiller, and so on. They work by comparing the incoming caller's number with a list of known phone spammers' numbers.

Why not fight evil with technology? ✦

let you continue this call hands-free. You can freely switch among these sources during a call.

- **add call** lets you add a second person to the call, creating a conference call.

When you tap it, you put the first person on hold and open your Contacts list. Find and tap the person you want to add; place the call; say hello.

During all of this, the first caller is still on hold, as you can see at the top of the screen. Now you should tap either **swap** (to put the second person on hold while you talk to the first person) or **merge calls** (to start the conference call).

> **TIP:** At this point, there's nothing to stop you from tapping **add call** again, and again, until you've got up to five people on the same call.
>
> If you need to dump someone from the call—we've all met *that* kind of person—tap the ⓘ and then that person's **End** button.

Calling one person

Creating a conference call

Options on the call

Additional Incoming Calls

Suppose you're on a call—and *another* comes in. (Hey, it could happen!)

On the screen, you see these buttons:

- **End & Accept.** Hangs up on the first call; answers the new one.

- **Hold & Accept.** Puts the first call on hold (presumably after you say, "Oh, sorry, hang on—I've got another call") and answers the new call. Now you can hop between the two calls, or hit **merge calls** to combine them.

- **Send to Voicemail** dumps the new call to voicemail; the caller doesn't know you did that manually.

Call Forwarding

The iPhone can auto-forward incoming calls to your iPhone so they ring on a *different* phone. Maybe, for example, you'd like to be able to answer calls on your home-phone handsets. Or maybe your iPhone is in for repair, so you're using a temporary phone but you don't want to miss any calls.

Here's how you turn on call forwarding:

- **AT&T.** In Settings→Phone→Call Forwarding, turn call forwarding on, and then tap in the new phone number.

- **Verizon, Sprint/T-Mobile.** Open the Phone app; tap Keypad. Dial *72, plus the number you're forwarding calls to. Then tap �. (To turn off call forwarding, dial *73, and then tap �.)

FaceTime Video

FaceTime is Apple's video-chat app. It's available on iPhones, iPads, and Macs, which makes it easy (and free) to make high-quality video and audio calls to anyone else in the Apple universe. It's perfect for letting parents or grandparents see the kids, for asking your spouse at home which kind of olive oil you're supposed to get at the store, and for long-distance relationships.

Making the Call

Apple has made sure that you can start a FaceTime call whenever the whim strikes you:

- **Use Siri (page 137).** Just say, "FaceTime Dad" or "FaceTime Alex." That's the fastest way.

- **Often, a FaceTime call grows out of a text-messaging conversation.** You know: "Will you bring my sweater when you come?" "I'm in the closet. Which sweater?" "The gray one." "There are three of them!" "OK, show me on FaceTime."

  From Messages, tap the icon of the person you're texting (top of the screen), and then tap ⊙.

- **You can also start a FaceTime call from somebody's card** in Contacts (tap ▮◀).

- **There's even a dedicated FaceTime app,** which is basically Contacts but dedicated to starting FaceTime calls.

No matter how you place the FaceTime call, what happens next is the same: The other person's iPhone, Mac, or iPad rings, and a message indicates that you're trying to reach them for a FaceTime session. If they tap Accept, the call begins. You see them on your screen; they see you on theirs. (Each of you also appears on your own screen, in a smaller rectangle, so you can make sure you don't look ridiculous. You can drag that inset to another corner, which is handy if it's covering up something you want to see.) This is state-of-the-art videoconferencing.

During the Call

In iOS 14, by the way, the FaceTime video no longer "pauses" when you switch to a different app while on a call. Instead, a picture-in-picture miniature remains on the screen as you switch out of FaceTime, and your conversation partner can still see and hear you. You can drag that window around the screen, or pinch and spread to resize it (see page 391).

Or, if that little inset is in your way, flick it off the side of the screen; it becomes a ❭ tab peeking out, so you can retrieve it when necessary. Tap it to return to full-screen FaceTime.

> **TIP:** In the early testing versions of iOS 13, Apple introduced a new option into FaceTime: An algorithm that warps your eyes so that you appear to be looking at the camera—instead of the middle of the screen, which is where you're usually looking. The idea was to make you look less shifty and distracted.
>
> Testers found it to be super-creepy; it sometimes produced a visible bend in people's glasses or brows, so Apple killed it. But it's back again in iOS 14, as an option in Settings→FaceTime→Eye Contact.
>
> Don't get excited, though; Apple dialed back the effect so much that it's barely noticeable.

The FaceTime picture is either portrait (upright) or landscape (widescreen). Tap anywhere on the screen to bring up a panel of four essential buttons:

- **Mute yourself.** When you need to make an unseemly noise (or somebody in the background is making one for you), tap the screen and then 🎤.

- **Flip the camera.** If you need to show the person something in front of you, don't turn the phone around. Instead, tap 📷. Now they can see the view from your back camera, and you can see, on your screen, exactly what they're seeing.

- **Special effects.** Tap ⭐ to play with a world of crazy effects you can add to your own screen image while FaceTiming—filters that make you look like a comic-book character or pencil sketch (😵), custom text you can add to the screen (Aa), shapes that can float around your head (〰️),

Basic controls (tap the screen)　　　　　*Advanced controls*
　　　　　　　　　　　　　　　　　　　　(swipe up on basic controls)

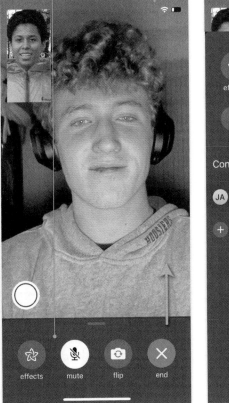

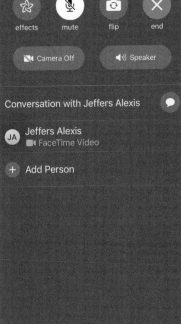

FaceTime controls

and Memoji or emoji stickers you can slap on your own image. Page 265 shows an example.

- **End the call** by tapping ⊗.

This little button panel is, in fact, taller than it seems. Drag upward on it to reveal a few more useful options:

- **Pause the video.** You can "mute the video" manually, which is handy when you need to tend to something the other person doesn't really need to see. To do that, tap the screen to make the controls appear, drag upward to reveal the hidden buttons, and tap **Camera Off**. The other party can still hear you, though.

  Tap **Camera On** to resume video.

- **Speaker.** You might wonder why there's a **Speaker** button. Surely nobody does FaceTime calls with the speakerphone off! If you're holding the phone up to your ear, *you* can't see anything, and the other person sees... your ear.

  In fact, the purpose of this button is to open a menu of other audio sources, like wireless earbuds, so that you can switch *between* them and the speaker.

- **Switch to texting.** Tap ● to jump into a Messages conversation with the person you're FaceTiming.

- **Add someone.** Tap + to add more people to the call (read on).

> **TIP:** If you see a white ◎ button, then you and your friend both have **FaceTime Live Photos** turned on in **Settings→FaceTime**. What it means is that if you tap the ◎, you take a Live Photo, and you've just preserved three seconds of this magical FaceTime conversation for posterity.
>
> Don't think you can take this shot sneakily, though; a message appears on both screens, so everybody knows what you've done. You can, however, take a *regular* screenshot (page 435) without the other person finding out.

Group FaceTime

FaceTime isn't exactly Zoom. Nobody is going to use it to make presentations to 100 people at a time.

Still, up to 32 people can join a FaceTime call, as long as they're using iOS or macOS versions from 2018 or later.

One natural way to begin such a session is from a group chat in Messages. Tap the group name at the top of the screen and then ◎; you've just sent each member of the group an invitation. Anyone who taps Join becomes part of the call.

Or start in the FaceTime app. Tap +, and add the names or numbers of the participants. Use the ⊕ to search Contacts for each additional callee. Finally, tap ◎ Video. Their phones, Macs, or tablets start ringing. Once they tap Accept, the call begins.

> **TIP:** You can add someone else to an existing FaceTime call, whether it's a group or not. Tap the screen to open the buttons panel, and then swipe up to see the full set of options (see the figure "FaceTime controls" on page 241). Tap **Add Person** and start typing the person's name until you can tap it; then tap **Add Person to FaceTime**.

When someone speaks, their tile grows.

You

Group FaceTime

Each person's face appears in its own micro-window on your screen, which enlarges itself when they're speaking. (If there are more than four participants, the fifth and additional people appear in a scrolling row at the bottom of the screen.)

If you tap someone's box, it gets a little bigger and identifies their name or iCloud ID. If you *double*-tap someone's box, it fills the window so you can really have a good look.

If you prefer a little whimsy—or wish to remain obscured—tap ✿; tap the little monkey (the Animoji icon); and choose an animal Animoji or Memoji to replace your head in the video call. See page 264 for more on these cute effects (or cutesy, depending on your age and cynicism level).

FaceTime Audio

Even today, 10 years after the dawn of the video-call era, most people still make phone calls. Audio, no video. Not for technological reasons, but for human ones. Video calls burden you with looking good and seeming put together; audio is often less of a hassle.

FaceTime audio calls work just like FaceTime video calls. To start in Messages, tap your correspondent's icon, ⬤, and then FaceTime Audio. In the FaceTime app, tap the +, type in the name of the person—or people—you want to call, and then choose ⬤ Audio.

You'll discover that FaceTime audio calls sound amazing. The audio has a clarity and presence that sounds more like a CD than a lousy cellphone call.

> **NOTE:** When someone calls *you*, a notification appears, and you hear a ringtone. (It's whatever sound you choose in **Settings→FaceTime→Notifications→Sounds**.) You can even answer the call on the lock screen if your phone is asleep, without having to enter your password. You "pick up" by tapping **Accept**.
>
> If you click **Decline**, the caller sees a "not available" message for you. And if you never want people to bug you with FaceTime calls, open **Settings→FaceTime** and turn FaceTime off.

CHAPTER TEN

Text Messages

T ext messages have blown past emails and phone calls as today's
preferred means of communicating. No wonder, then, that Apple
has invested mightily in its Messages app. The best part: It lets
you text any kind of cellphone, of course. But when you're texting fellow
Appleheads, it offers endless special effects—Memoji! falling confetti!
voice recordings!—to keep life from getting dull.

In iOS 14, Apple has delivered even more bells and whistles—including
the ability to pin your favorite conversations to the top of your
Messages screen, as well as some sanity-saving improvements to
managing group texts.

Welcome to Messages

The Messages app can send and receive two kinds of text messages:

- **Regular text messages,** the ones that every cellphone on earth can send.
  These show up in green bubbles in Messages. They're subject to all the
  usual limitations of text messages: For example, each one is limited to 160
  characters, with no choice of fonts or styles. But, hey—*any cellphone on
  earth.*

- **iMessages** are Apple's custom versions of text messages. In the Messages
  app, these messages show up in blue. They're limited to exchanges
  between Apple devices; you can't send an iMessage to an Android phone.

  On the other hand, that's about the only limitation. iMessages are like
  superhero text messages. They have no practical length limit—they can
  be pages and pages long. You can animate their delivery with special
  effects—confetti falling, balloons rising, and so on.

Standard SMS texts: green and labeled "Text Message"

Apple-device iMessages: blue and labeled "iMessage"

Text messages vs. iMessages

The iCloud service synchronizes your iMessages across your Apple machines, too. You can start texting on your phone while you're on the way home and then sit down at your Mac and pick up in midsentence. Your Messages chats look identical on every Apple gadget.

And because they're sent over the internet, rather than through the cellular network, iMessages don't count as text messages on your cellular plan.

More on the miracle of iMessages later in this chapter.

> **NOTE:** iMessages are a feature of a free iCloud account. If you haven't already signed up for one of those, Messages invites you to do so the first time you open it.

How to Chat

On the main Messages screen, a list of past conversations awaits, each represented by the name and icon of the person you were chatting with and a snippet of the most recent message you exchanged. When you tap an individual conversation, you open your entire history of texts with that person, scrolling all the way back to your first exchange.

> **NOTE:** If it raises your hackles to know that Messages remembers your entire history with everyone, you can opt out. Tap **Settings→Messages** and scroll down to the Message History section; here, you can specify how long you want to Keep Messages (**30 Days, 1 Year, Forever**). You can also delete an entire conversation by swiping left across a conversation's name and tapping the 🗑.

Most of the time, you'll begin a chat session by tapping one of your *existing* conversations—and then just adding to it.

Conversations and messages

To message somebody for the very first time, on the other hand, tap ☑️ above the conversations list. In the **To** box, type the person's name or phone number. As you type, a list of potential matches appears. The names in blue are Apple people who can have an iMessages conversation with you; the ones in green are not.

> **NOTE:** Either an email address or phone number can be somebody's iCloud address. Most people have recorded both, so you can search for them by either bit of data.

If it's easier, tap ⊕ to open a miniature version of your Contacts app, so you can find somebody's address using its tools.

Once you've chosen your correspondent, tap in the box at the bottom of the window and begin typing. (The box starts out saying either **Text Message** or

iMessage, to let you know whether this is an Apple person or not.) After each utterance you type, tap ⬆ to send it.

In time, your Messages conversation will come to look like a bubbular screenplay. The last 50 exchanges appear here, but you can scroll up, up, and up to see every text you've ever exchanged with this person.

> **TIP:** That business—scroll down, wait for messages to load, scroll down again—gets exhausting when you're trying to return to some weeks-old message. Fortunately, there's a glorious shortcut: Tap the left "ear" of a Face ID phone, or the top bar of the screen (where the clock appears) on older phones, over and over again. Each time you tap, another batch of older messages appears.

PINNING MESSAGE CONVERSATIONS

In iOS 14, Messages includes a tiny change that makes a big difference. You can now *pin* certain conversation partners at the top of the conversations list—that is, install their icons there for good. In previous iOS versions, conversations could slide way down in the list as more recent exchanges displaced them. Now you never have to hunt for the people you love. (Well, at least not in Messages.)

To pin a conversation, whether it's an individual or group chat, just swipe right all the way across it.

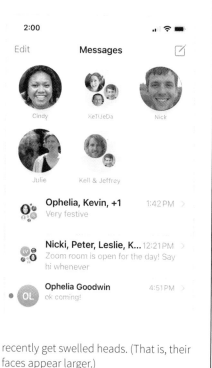

That person or group now enters a special VIP district for pinned people, above the conversations list. If you've signed into an iPad or Mac (with macOS Big Sur) with the same iCloud account, your pinned people appear there, too, courtesy of iCloud syncing.

Pinned conversation icons are especially informative. They display blue dots when they contain new messages; they display the animated ellipsis (•••) when someone's typing; Tapbacks (page 256) show up as they're applied; and so on. In group chats, people who've spoken up more recently get swelled heads. (That is, their faces appear larger.)

If one of these people or groups falls out of favor with you, just long-press their face in the Pinned area and choose **Unpin.** ✦

And now, a few tips to make your messaging experience more pleasant:

- **Hide the keyboard.** You can swipe down anywhere on the messages to hide the keyboard, if it's in your way.

- **Tap an address or number.** If, in the course of texting, someone sends you a street address, you can tap it to open it in Maps. And if someone sends a phone number, tap it to open a Call button.

- **Search your messages.** Searching your chat conversations in Messages has never been easier. Tug downward on your conversations list and tap inside the search box; instantly, the app shows you the latest links and photos people have sent to you, for easy skimming.

Before you type   *After you type*

Searching Messages

From here, you can tap in the search box and type your query. Messages displays all matching texts, web links, photos, documents, and so on.

- **Check the time stamp.** You can see, in gray type, the date and time that each conversation began—but that time stamp appears only at the beginning of each texting volley. To view the time stamp of the individual messages, drag leftward anywhere on the screen.

- **Delete one message.** You can delete a single message within the conversation, though the technique is a little clunky. In the conversation, long-press the message you want to delete. From the shortcut menu, tap **More**. A checkmark appears on the offending text message.

  At this point, you can select any other messages you want to nuke simultaneously; when you're ready, tap 🗑 to delete it or them. (Note that it disappears only from your copy of Messages. You can't delete one of your inflammatory texts from the other guy's phone. You've been warned.)

> **TIP:** On the list of conversations, you can also delete an entire conversation—your entire chat history with this person—by swiping left across it and then tapping 🗑.
>
> Or, if you're really feeling purgey, tap **Edit→Select Messages**; tap all the conversations you want to vaporize; and then tap **Delete**.

- **Forward some texts.** That same menu (long-press a message, tap **More**), features a ⇗ button—just what you need when you want to share the selected text—or a bunch of them—with a third party.

- **Mark all as read.** When you reenter internet coverage after some time away (plane, boat, sleep), it's sometimes teeming with blue dots, representing messages you missed in a bunch of different conversations. If you'd rather not deal with it, start on the conversation list. Tap **Edit→Select Messages→Read All**. The new messages are all still there, but the blue dots are gone.

- **Back out to the conversation list.** If you're in a conversation, you can return to the list of conversations by tapping ‹ at top left—or by swiping inward from the left side of the screen.

Incoming Messages

If somebody texts *you*, a notification appears, and you hear the classic iPhone *dingggg*! (You can change the text sound in **Settings→Sounds & Haptics→Text Tone**—but why?) If the phone is awake, you can see who sent the message and even read the beginning of it, right there on the notification banner.

If it's locked, you may have to unlock the phone before you can see that information.

Here's how you can handle that notification:

- **Ignore it.** Flick it up and away. Or, if it's on your lock screen, just wait; the phone will go to sleep again. You can always read the notification later.

- **Answer it.** Long-press the banner. It opens to reveal the keyboard; you can answer the text without even unlocking the phone or opening the Messages app. (See "Notifications" on page 43.)

- **Open it.** Tap the notification to open the incoming text in Messages. (You may have to unlock the phone, if it wasn't already unlocked.)

> **TIP:** The Messages app icon, at the home screen, bears a "badge" (**2**) that tells you how many new text messages are waiting.

- **Silence it.** If you swipe left on the notification, the **Manage** button appears. This is a quick way to make future text-message notifications silent—or to turn off text notifications altogether.

iMessage Specials

Whenever you're corresponding with another iCloud account holder, messaging becomes infinitely more flexible and powerful. Let us count the ways:

- **Informative feedback.** iMessages are inherently more informative than regular text messages. For example, if your correspondent is typing a message but hasn't yet sent it, an animated ellipsis (···) appears next to their speech bubble on your screen. It lets you know they're working on a reply and not just ignoring you.

> **NOTE:** The ··· dots remain on the screen for 60 seconds and then disappear. If they go away, it doesn't necessarily mean your friend changed their mind about replying. It could just mean they're writing a really long answer.

Similarly, Messages tells you when the internet has successfully delivered your message to the other person's phone or computer by displaying the word *Delivered* beneath it. That doesn't mean the other person has actually seen it, only that it has arrived on their machine. Either they or the machine might be asleep at the moment.

If you see the word *Read* beneath something you've sent, though, that means they *have* seen it, which can be an important piece of information.

What about you? Do you want the other person to know when you've seen *their* texts? You can turn **Send Read Receipts** on or off in **Settings→ Messages.** But you can also turn it on or off individually for each correspondent. You might want your love interests to know when you've seen their texts, for example, but you might not think your boss needs to know. The per-person override switch appears when you tap your correspondent's name at the top of the conversation's screen and then tap ⓘ.

- **Inline replies.** Ordinarily, texts appear in Messages in the order they're sent. That may seem logical, but it can lead to hilarious or disastrous confusion if, because of typing lag, somebody's reply to topic A winds up appearing below topic B.

*1. Long-press a message; tap **Reply**.*

2. Reply to the text, even if it's pretty old.

3. Your side chat shows up in both its original spot and at the end of the chat.

*4. Tap **4 Replies** to open the side chat on its own screen.*

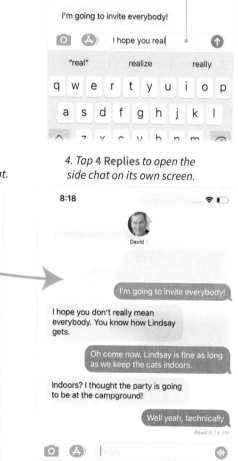

Inline replies

But in iOS 14, you can attach replies to one *particular* text message, even if it's weeks old. The replies become a subchat all its own. To create your own subchat, long-press (or double-tap) the message and tap **Reply**.

> **NOTE:** If your conversation partners aren't using iOS 14, they don't see your reply as indented; it just appears at the bottom of the conversation as usual.

- **Live web and video links.** When somebody sends a web address in an iMessage, it shows up with a picture thumbnail of the actual web page. If it's a link to a YouTube video, you can play it right in Messages by tapping the thumbnail.

Tap to play right here.

Tap to open in YouTube.

Live web links

- **Adjust the text size.** In Settings→Display & Brightness, scroll down to Text Size. The slider here lets you make the text bigger or smaller.

- **Where are you?** If your correspondent has turned on **Share My Location** on their phone, you can see a map of their present location by tapping ⓘ (see page 330).

- **Capture incoming data.** What happens in Messages doesn't have to stay in Messages. If someone sends you a photo or video, you can tap the small preview to view it full-screen—and at that point, you can tap ⬆ to save it into your Photos app.

Similarly, if someone sends you a phone number or other contact information, you can add it to your address book. Tap an underlined phone number and then hit **Create**, or tap an address to see it on a map—and then tap **Create New Contact** or **Add to Existing Contact**.

You can also capture individual text messages. Long-press one and then choose Copy; it's now ready to paste into another app.

You can also send a bunch of texts to somebody else. Long-press one message; tap More; choose the messages; and then hit ⤳. All the selected messages go along for the ride in a single combined message to a new person.

The Info Screen

When you tap your correspondent's name/icon at the top of the chat screen, three buttons appear. You already know about audio and FaceTime.

The info button, though, opens up a special screen that probably should have been called Miscellaneous. It's everything you might want to know or do related to the person you're texting.

- **Contact.** The buttons at top—call, video, mail—give you more ways to contact your texting buddy. Or tap info to open their Contacts card, packed with even more ways to contact them.

- **The map.** After a moment, you get to see where your correspondent is, on a map. (This works only if the other person has decided to share or send their location, described next.)

- **Send My Current Location** sends a map of your location to the other person. That's useful when you need to be picked up or you're trying to find each other in a busy area.

- **Share My Location** is similar, but better if you're moving around. It sends your location to your chat buddy—and then updates that map for a period of time you specify (One Hour, Until End of Day, or Indefinitely). It's very useful if you're moving around town and want to make it easy for friends to catch up with you.

  (When you turn this on, the button then changes to say Stop Sharing My Location.)

- **Hide alerts** stifles whatever notifications ordinarily pop up when this person texts you. It's a useful option when you're trying to focus, or when someone's just texting way too much.

- **Send Read Receipts** turns on read receipts (page 252) for this person.

Done

Jeffers Alexis

call   video   mail   info   —— *Ways to contact*

—— *Current location*

Send My Current Location

Stop Sharing My Location   57m left

You are currently sharing your location from "David's iPad". Share from this device...

Hide Alerts   —— *Notification control*

Send Read Receipts

Photos   See All

     —— *Photos, links, documents*

The info panel

- **Photos, Links, Locations, Attachments.** Messages rounds up all the photos and other stuff you've ever exchanged with this person. It shows you a few of the most recent in each category, but you can tap See All to view the rest.

To open one of these attachments, tap it. You can also long-press for choices like Save (a photo to your Photos app), Copy, Share, or Delete.

> **TIP:** If somebody messages you a *lot*, as in harassing you, you can block them. Tap their icon at the top of the screen, and then ⓘ→info→**Block this Caller**. (You can review your blocked callers, and even give them a reprieve, in **Settings**→ **Messages**→**Blocked Callers**.)

Six Ways To Be More Expressive

Mere mortal cellphones get to send typed text back and forth. But for you, the sky is the limit to your expressiveness:

- **Tapbacks.** In everyday texting, people spend an extraordinary amount of time sending standard reaction responses like "Whaaaaaa??" "haha," "!!!!!" and "LOL." But thanks to Tapbacks, you can respond to somebody's comment with one of six standardized reaction icons: a heart, thumbs-up and thumbs-down icons, a question mark, exclamation points, or a "ha ha."

  To see your reaction choices, double-tap (or long-press) the message. When you tap the desired icon, it appears instantly on your screen and your buddy's.

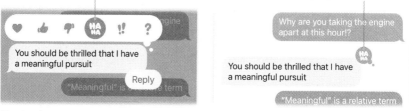

Double-tap a message; choose your tapback.   *Both of you see it now.*

Tapbacks

> **TIP:** Stamping a Tapback icon on a text isn't permanent. If new information comes to light, you can change your reaction. Long-press the message again and choose a different icon—or tap your original one again to remove it.

- **Animated fanciness.** Send messages with full-screen animations: confetti falling, balloons rising, lasers shooting, fireworks blazing, and so on. To see all the options, type your message and then long-press the ⬆ button.

  The options on the Bubble tab animate only your text message; tap one for a little preview of how it will look. You'll discover that Slam, Loud, and Gentle animate the typography of your text. Invisible Ink conceals your message with animated glitter until your recipient drags a finger across it. It's a great way to reveal dramatic news or send very personal messages; you're granting your recipient the opportunity to choose a moment of privacy to reveal what you said.

  The choices on the Screen tab, though, fill your recipient's entire Messages background with animation—and sound. Swipe across the screen to preview each one: Love, Balloons, Confetti, Lasers, Celebration,

Animate one message, or the whole screen. *What the recipient sees*

Sending effects

Echo, Spotlight, and Fireworks. (In fact, if you send a text containing the words "Congrats," "Happy birthday," or "Happy New Year," Messages adds one of these animation effects *automatically*. No charge.)

Once you've chosen your animation style, tap ⬆ to send it on its way.

> **NOTE:** You can choose one of these effects even if you're not sending to a member of the Apple cult. But they won't see the animation; they'll just receive the possibly confusing phrase "sent with Slam effect," "sent with Confetti," or whatever.

- **Jumbo emoji.** No matter what kind of message you're sending, don't forget that you can use the ☺ on your keyboard to insert emoji symbols (page 119). Emoji really shine in text conversations.

But in iMessages, if your message consists of one, two, or three emoji and no other text, you get to send them at three times their usual size.

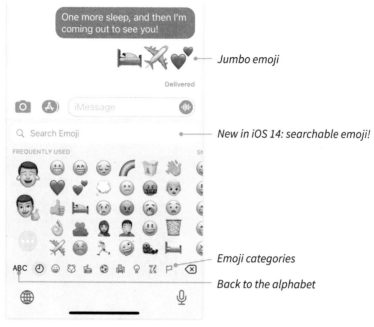

One more sleep, and then I'm coming out to see you!

Delivered

Jumbo emoji

iMessage

Q Search Emoji — *New in iOS 14: searchable emoji!*

FREQUENTLY USED

ABC

Emoji categories

Back to the alphabet

Emoji

- **Audio texts.** Sometimes what you have to say is too long, too emotional, or too nuanced to send as text. Or maybe you can't type at the moment because you need your eyes and one hand free—for example, when you're going up stairs. In those situations, you might welcome the fact that you can make a voice recording and send it right in the middle of your chat.

Oh my gosh you have to hear this

The entire city is celebrating

00:08

Delivered

00:10

Raise to listen

00:03

Audio texts

Hold down on the ⫿⫿⫿ beside the text box and speak to record; lift your finger to stop recording. Now you can tap ⊙ to listen to your message, 🔼 to send it, or ✖ to cancel it.

(In iOS 14, you can use Siri to send audio texts, too. Say, "Send a voice message to Alex"…and then speak it when she prompts you.)

> **NOTE:** Audio recordings take up a decent amount of storage space. Therefore, Messages ordinarily deletes them after two minutes. If you want to hang onto one, though, tap **Keep** before it disappears.

- **Make a sketch.** A picture is worth a thousand—well, you know. Turn your phone 90 degrees, tap the ꝏ key, and presto: Your phone is now a $1,000 whiteboard. Anything you draw, you can send.

Turn the phone horizontal; tap the Doodle key.

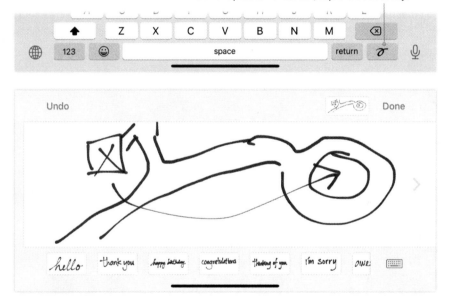

Insta-whiteboard

- **Jump onto a call.** At any point, you can tell Siri, "Call him" or "FaceTime her." That's a useful escape hatch when the conversation is getting too heated, too complicated, or too slow for typing back and forth.

Or, if it's not a good time to use Siri, do it the long way: Tap the person's name/icon at the top of the conversation screen. Now tap 📞 to place a call—either a **Voice Call** (that is, a standard phone call)—or a **FaceTime**

Audio call (page 244). You can also hop onto a FaceTime video call by hitting the ◎ in that same panel.

Actually, those are only the *built-in* ways to be more expressive over texts. As it turns out, they're only the beginning of the start of the tip of the iceberg. Wait till you get a load of the Messages App Store (page 263).

Sending Photos and Videos

If you tried to compare how many text messages people send and how many photos they send via text, plain text would win, but not by much. Photos and videos are incredibly useful tools when you're texting. You can show someone what you're seeing. You can send a picture of the restaurant menu to someone who's running late. In a store, you can ask your spouse at home which paper-towel brand you're supposed to buy.

Because photography is the Texting Tool of a Thousand Uses, Apple has seen fit to install two buttons right next to the texting box: One for taking new pictures, and one for choosing them from your Photos app.

Take a Photo

Tap ◙ to open the Camera app, complete with every mode, button, and tool described in Chapter 7 except Time-Lapse. (When you're texting, nobody's got time for that.)

You even get a button that *isn't* in the Camera app: ✿ (Effects). On the Effects screen, you can adorn a video or photo with "stickers" (still or animated), photo filters, stamped text, speech balloons, arrows and squiggles, Animoji, and so on.

> **TIP:** Once you've stamped one of these elements onto a photo, you can enlarge it by spreading with two fingers, move it by dragging with one, and rotate it by spinning two fingers on the screen.

You can add these superimposed goodies either before or after snapping the picture:

- **Before.** Mark up the viewfinder using these tools, and then tap the ✖. (**Done** takes you back into Messages without accomplishing anything.) Now take your photo or video.

- **After.** Take a new photo or video, exactly as described in Chapter 7. After you snap it, you wind up at an editing screen that offers the **Effects** button,

Dress up your photos

the standard **Edit** button (page 210), and the **Markup** tools (page 7). You can thoroughly fix up—or ruin—your photo before sending it.

Tap ⬆ to send the photo or video immediately, or **Done** to place it into the text box so that you can add a comment before sending it. (Messages also stashes a copy of your photo in Photos, so you can use it again if necessary.)

Sending Existing Photos

To send a photo or video that's already in Photos, tap the ⬤ button beside the texting box, and then tap the first button in the apps drawer.

It's first for a reason: It's the app most people use most of the time. It opens a copy of your Photos app, so that you can choose a picture or video—or several—to send to your texting pal.

Each thumbnail you tap jumps into the message box, ready to send (with or without adding a text comment). Or you can tap the ⊗ in the thumbnail's corner if you change your mind about sending it.

Tap **All Photos** to open a fuller-blown photo selector, complete with your albums and a search box.

And if you scroll down a little, you'll find **Sharing Suggestions**: sets of photos from various recent events that contain photos of people your Photos app can identify. The idea is to make it simple for you to send photos of people who were with you at some event *to* those people. See page 207.

The Apps Drawer

When attempts at creativity like emoji and audio recordings just feel so 2017, Apple's got your back. It has created an entire App Store just for add-ons to Messages. Software companies can create micro-apps that let you exchange all kinds of stuff over iMessages: payments, restaurant reservations, movie schedules, and so on.

To see your starter options, tap ⓐ next to the typing box. Here's how this "apps drawer" works:

- **To see the apps' names,** long-press one of them. (Drag horizontally to scroll through them.)

- **To use an app,** tap it to open its panel in a compact form. Swipe up to make it fill your screen.

- **To delete, hide, or rearrange apps,** scroll all the way to the right end of the apps drawer. Tap More to see your entire list of installed mini-apps.

  This management screen works just like the one described on page 4. Note, in particular, that you can swipe left across an app's name to reveal

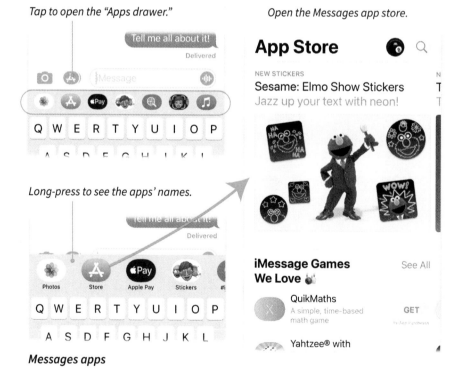

Tap to open the "Apps drawer."

Open the Messages app store.

Long-press to see the apps' names.

Messages apps

the Delete button (or, for Apple apps, Remove from Favorites), or turn off its switch to hide it without actually uninstalling it.

Apple starts you off with a selection of apps, which you may find either thrilling or forehead-slapping, depending on your age group and your tolerance of texting fanciness.

Here's what you'll find in the apps drawer on a new phone.

Photos

Here's the access to all your photos and videos, as described in the previous section.

The App Store

The second app in the drawer opens the Messages App Store (at right in "Messages apps"). Here's where you can add even more options to your apps drawer.

This store's design and operation is just like the regular App Store (page 377); here again, some apps are free, and some cost a couple of bucks. Lots of them are "stickers"—pictures or animations you can drag upward onto texts in your conversation, thus commenting on them in a visual and unmistakable way.

Other apps give you in-chat access to apps like Kayak (to book flights), Yelp or OpenTable (to research or book restaurants), Airbnb (to reserve lodging), Cash App or Circle Pay or Venmo (to send money to friends), Doodle (to find a mutually free time to meet), and so on. There are, of course, games by the boxcarfull.

Apple Pay

Here's Apple's favorite way for you to send money directly to other fellow Apple people, right in Messages. ("Hey, you still owe me $10 for pizza." "Oh, right—here you go!") See page 404.

Stickers

These are cartoon images of you—well, of whatever Memoji you've created (read on). They're your Memoji with various facial expressions and reactions. Tap one to slap it into the text box, ready to send to express your delight, laughter, tears, boredom, or fury at whatever your texting partner just said.

#Images

It's become popular these days—and hilarious—to respond to something somebody has said with an animated GIF: a short, looping, silent video clip, usually a moment from a movie or TV show where the actor is exhibiting exactly the right reaction.

This app (🔍) is a searchable database of thousands of reaction GIFs. Type to search for the kind of GIF you'd like to send: *slow clap, popcorn eating, disgusted,* or whatever.

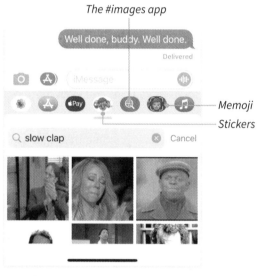

Reaction GIFs

Memoji

A Memoji is Apple's name for an emoji that looks like you (well, a cartoon version of you). Thanks to Memoji, you don't have to reply to someone's text message using a generic smiling (or frowning, or crying) face; you can use *your* face.

A Memoji can also be the basis for Animoji, the feature of Face ID phones that outfits you with an uncanny digital, animated mask in video chats (described in the next section).

Tap Memoji in the apps drawer. (If you can't see these icons' labels, long-press one. You want to avoid the Stickers button for now, which looks very similar.)

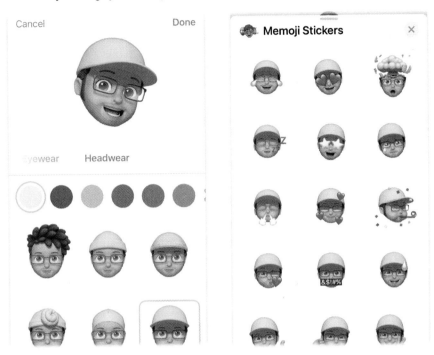

Once you design your Memoji...   *...you get an army of expressive stickers.*

Memoji stickers

The Memoji panel starts you off with various cutesy animal characters, but—unless your parents had a particularly unfortunate genetic mix—they don't look like you.

Swipe right until you reach the **New Memoji** button (+). Tap it.

Meet the Memoji editor: a row of buttons that list various aspects of your appearance. Swipe horizontally through the screens—**Skin**, **Hairstyle**, **Brows**, **Eyes**, **Head**, **Nose**, **Mouth**, **Ears**, **Facial Hair**, **Eyewear**, and **Headwear**—and tap the variation you think looks most like you.

> **NOTE:** That's what most people do, but there's no law that says your Memoji has to look like you.

By the end of the setup, you've created a little bubble-nosed, baby-headed version of yourself (or your idealized self). Tap **Done**.

You're welcome to build as many Memoji as you want. And to edit, duplicate, or delete one, tap it, tap ⬤, and then tap **Edit**, **Duplicate**, or **Delete**.

From now on, when you think the only perfect response to somebody's text is your Memoji head, tap 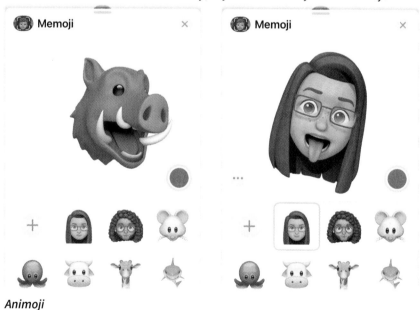, choose **Stickers**, and then swipe to the variation you want. You now have two options:

- **Send as a picture.** Tap the version of your avatar that has the perfect expression. It pops into the Messages text box. You can type a comment or just tap the ⬆ button to ship it off.

- **Stamp as a sticker.** You can stamp your Memojiface onto a text bubble in the chat, as a reaction to what they said. Long-press until the Memoji sticks to your finger; without lifting, drag upward onto the text bubble.

Animoji (Face ID Phones)

Apple is especially proud of this feature, which is available only to Face ID phones. Animoji are cartoon animal faces—or Memoji faces—that are super-imposed on your actual face in a video chat or Messages, and *animated*. The TrueDepth camera studies your facial expression in real time—detects the movement of 50 different muscles, including your mouth, eyes, eyebrows, tongue, head angle, and so on—and makes your cartoon Memoji perfectly follow you. It's truly freaky.

Apple offers a bunch of canned critter faces: bunny, piggy, panda, unicorn, alien, a robot, and, inevitably, animated poop. Any Memoji you've created appear here, too.

You can use one of the critter Animoji, or you can animate your own Memoji.

Animoji

Before you ever start "broadcasting," you can try out your Animoji selection. Talk, move, express, holding your phone up like a mirror. Swipe upward for a larger view.

So what can you *do* with this feature?

- **Send as a picture.** At the moment when your avatar has a great expression, tap it. It pops into the Messages text box, ready to send.

- **Stamp as a sticker.** As with emoji, long-press your Animoji until it sticks to your finger; drag it directly onto the text bubble.

- **Send as a video.** Here's the real magic. You can record a video, with sound, up to 30 seconds long.

  Tap ◉ and make your little cartoon head perform (remember to talk!). When you send it, the video goes out as a plain old video; your buddies don't need iPhones to play them.

> **TIP:** After recording a video—but before sending it—you can see how a different Animoji character might do the same performance, without having to re-record. Just tap another critter.

- **Use it as your new head in FaceTime chats.** You can wear one of your Animoji as a digital mask during your FaceTime video calls. Maybe you think it'll entertain your conversation partner; maybe you just look like hell today.

Music

Here you can see the names of songs you've recently played on your phone. Tap one to send a text saying, "I'm listening to [name of song]." That's all this does.

Digital Touch

This weird little app, modeled on the one on the Apple Watch, lets you create another kind of custom animations: hand-drawn doodles that appear, line by line, on the recipient's phone, iPad, or Mac, just as you drew them. You can draw freehand or draw onto a photo or a video. Here's a tour:

- **Color picker.** Tap to choose from among seven "paint" colors, or long-press to open a more complete color dial.

- **Draw.** Drag on the black canvas to make a doodle. There's no eraser or Undo; do your best. Nobody's expecting Picasso here.

When your doodle is done, tap ⬆ to transmit it. If the other person uses iMessages, they see your drawing "played back," line by line as you drew it.

- **Deface a photo or video.** The ⬤ icon opens a mini-Camera app. You can tap either the white shutter button for a still photo, or the red one to shoot a video. Tap 📷 to use the front camera.

You can draw on the photo after you've taken it. For a video, you can draw before you shoot it or (more usefully) *during* the recording. Use the same "choose a color, then draw with your finger" technique described already.

> **TIP:** You can also add Digital Touch animations (described next) to a photo or video. Here again, you have to do your Digital Touching *before or during* shooting, not after.

STORING MESSAGES IN ICLOUD

You might not think that text messages would eat up much storage space—and you're right.

But remember that what makes Apple's version, iMessages, so special is that they're *not* just bits of text. They can be audio recordings, videos, photos, files, web links, Memoji—all kinds of multimedia goodness that, while superb for self-expression, eat up space like crazy. Messages never deletes anything you send or receive (unless you've changed its factory settings), so over the years, this stuff really piles up.

You're not stuck in that situation, however. You can, if you wish, offload all that Messages history to iCloud storage space rented from Apple. (See page 409 for the convenient price list.)

Doing so has three immediate advantages. First, you instantly save many gigabytes of space—not just on your iPhone, but also on your iPad and Macs. That's because, until you turn this feature on, every one of your Apple gadgets maintains a *duplicate* copy of every photo, video, and audio clip you've ever sent or received in Messages.

Second, there's no more waiting for your other devices to "catch up" with all the messages you exchanged using your iPhone during, for example, a plane ride.

Third, when you get a new Apple machine, Messages presents to you your entire history of text messages and attachments. (Without Messages in iCloud, you'd see only your current and future messages. You couldn't scroll back into the past.)

If that sounds good—and if you don't mind the amount of iCloud storage space this feature will consume— open **Settings→[your name]→ iCloud→Messages**. Assuming it's plugged into power and on Wi-Fi, the iPhone begins backing up your Messages history, which can take quite a while.

You have to turn this on separately on each device; on the Mac, for example, you do that in the Messages app (**Messages→Preferences→iMessage→ Enable Messages in iCloud**). ✦

When you send the resulting photo or video, iMessages recipients will see your drawings "played back" line by line as you drew them. (Non-iMessages people just get a finished, frozen sketch-on-photo/video.)

Once you've created a doodle, photo, or video, tap ● to send it, or × to cancel it.

- **Send Digital Touch animations.** There's no button for this feature—no user-interface evidence at all; consider it a secret bonus for insiders like you.

  Turns out you can create a cool-looking animation to send while texting just by tapping or pressing your fingers on the black canvas. You can create a ring of fire (tap as many times as you want), a fireball (long-press), a lip-kiss (tap with two fingers), a beating red heart (long-press with two fingers), or a breaking heart (long-press with two fingers, and then drag downward).

> **NOTE:** Messages sends a Digital Touch animation instantly after you perform your finger tap or drag. You don't have the opportunity to look it over before it goes. In matters of the heart, proceed with caution.

Choose a color. Draw something.

Tap, two-finger tap, or long-press to make these animated stamps.

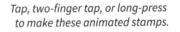

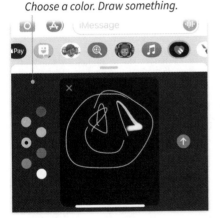 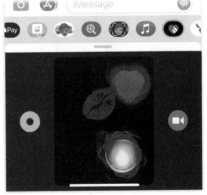

Digital Touch

Group Chats

Messages is perfectly capable of accommodating conversations among more than two people. When you tap the ✑, just enter several people's names or phone numbers in the To field. (Or, if you've set up a group as described on page 228, you can type the group's name.) Group chats are highly handy for

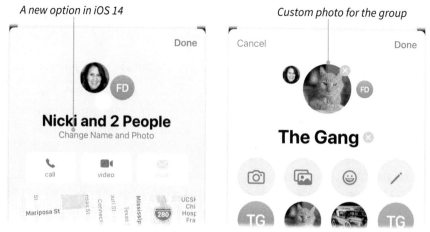

A new option in iOS 14

Custom photo for the group

Group text setup

orchestrating get-togethers, collaborating on work or homework, or making any kind of plans.

> **TIP:** In busy group chats, you may feel deluged by notifications. Sure, you can turn on Do Not Disturb for the chat—but now you might miss an important conversation involving you!
>
> In iOS 14, there's a solution. Whenever you type somebody's name preceded by an @ (as in @Marley), that typed name shimmers—and that person gets a notification, even if the group chat is muted. (They can mute those "you were mentioned" notifications, too, though, with the new **Settings→Messages→Notify Me** switch.)

You can give your group a name and a photo, which everybody in it sees at the top of the chat. Tap ⓘ and then tap **Change Name and Photo**. (This works only if every member of your group uses an Apple device.)

CHAPTER ELEVEN

Mail

The invention of the smartphone led to an explosion of communications channels—FaceTime, Messages, Facebook Messenger, WhatsApp, WeChat, and so on. But no app has attained the universal reach, the old-school literacy, or the solid paper trail of good old email.

The iPhone's email app is called Mail. And it's complete and well-integrated with the rest of iOS.

Some people access their email through apps for their specific services, such as Gmail, Yahoo Mail, Microsoft Outlook, or even AOL, or use their websites for doing email. But Mail is worth considering as a front end for all those services, too. It consolidates all your email accounts into one place.

Mail Setup

The first time you open the Mail app, it asks what kind of email service you have: iCloud, Microsoft Exchange, Google, Yahoo, AOL, or Outlook.com. If indeed you use one of these free services, then setting up your email is as simple as supplying your email address and password.

> **NOTE:** If it ever comes time to add another account, you can return to this Add Account screen by choosing **Settings→Mail→Accounts→Add Account**.

If, on the other hand, you get your email from some oddball service not listed here, tap Other. You'll have to supply the login specifics manually, which may include your name, password, and mail-server details.

Account options

Once everything is set up, Mail goes online and starts retrieving mail. In most cases, you'll even get all your *old* mail—read and sent—so what you see here perfectly matches what you would see on your email service's app or website (for example, Gmail.com).

Mail Basics

Mail automatically checks for new messages on a schedule you set in Settings→Mail→Accounts→Fetch New Data. It starts out set to Push, which fetches mail automatically as it appears. If you need to conserve battery life or data, you can set it to check manually or on an infrequent schedule.

Or, inside the Mail app, you can force a check. Swipe down on any screen and hold until you see the ⁂ icon.

When new mail arrives, a notification appears. You can turn those notifications off on a per-account basis in Settings→Notifications→Mail—but don't shut them down until you read page 42. Turns out you can dispatch many email messages right from their notification banner—Trash, Mark as Read, Open, and Manage commands are all one swipe away.

At the home screen, a circled number "badge" (❷) appears on the Mail app's icon, letting you know how many new messages are waiting. (You can turn it off, account by account, in Settings→Notifications→Mail→[account name]→Badges.

The Mailboxes List

The opening Mail screen lists all the folders that contain your email. You get a separate heading for each email account you use (iCloud, Gmail, and so on). You can see all the folders within an account by tapping its name.

> **TIP:** If you have more than one email address, you get a folder called **All Inboxes**. It contains all the incoming messages from all your accounts in one place.

All your accounts on one screen   *Expand to see the mailboxes within.*

2:18		all 🔋
		Edit
Mailboxes		
📫 All Inboxes	16431	>
⭐ VIP	ⓘ	>
📭 iCloud	7	>
📭 Gmail	3662	>
📭 Yahoo!	12762	>
🏳 Flagged		>
ICLOUD	7	>
GMAIL	3662	>
YAHOO!	12762	>
	Updated 3 minutes ago	✏️

2:15		all 🔋
		Edit
Mailboxes		
📫 All Inboxes	13212	>
⭐ VIP	ⓘ	>
📭 iCloud	7	>
📭 Gmail	13205	>
📭 Yahoo!		>
🏳 Flagged		>
ICLOUD		⌄
📭 Inbox	7	>
🗋 Drafts		>
✈ Sent		>
🗙 Junk		>
🗑 Trash		>
	Updated Just Now	✏️

Accounts list

Within each account heading, you'll generally find folders like these:

- **Inbox** is mail you've received. If you have more than one email account, tap **All Inboxes** to see all the mail, consolidated from all the accounts.

- **VIP** holds mail from the most important people in your life—those you've designated as VIPs (page 289).

- **Flagged.** As you work through your email messages, you can apply colorful little flags to them, meaning whatever you want them to mean. This folder rounds up all the messages, from all your accounts, bearing any flag.

- **Outbox** lists messages you've written but haven't yet sent.

- **Drafts** holds messages you've started writing but aren't ready to send.

- **Sent** contains copies of all the messages you've sent.

- **Trash** holds your deleted messages. This is only a waiting room for deletion; messages you've put here remain here until you select them and delete them permanently.

- **Junk** appears when you use Mail's spam filter (page 290).

And, of course, you see any mail folders you've created yourself: To Do, Old Flames, Delegate, and so on.

Mail offers a huge list of optional, but useful, folders, like Unread, Attachments, and Today (today's mail). To see them, tap Edit at top right. Turn on the folders you'd like to see, turn off the ones you never use, and drag the ≡ handles up or down to change their order. Tap Done.

> **TIP:** The Mail app is a series of nested lists. It presents a list of your accounts; tap one to see the list of folders; tap one to see the list of messages; tap one to open a message.
>
> Technically, you're supposed to tap the ‹ in the upper-left corner to backtrack, or use the backtrack menu described on page 12—but it's much more fun to swipe rightward across the screen.

Composing Messages

To write a new email message, tap ☑. Now fill in the various pieces of the New Message window:

- **To.** Enter the recipient's email address. As you type, Mail automatically offers a list of potential email addresses, to save you time and typos. It draws its guesses from the list of people in your Contacts app, plus recent Mail correspondents. Once you see the one you want, tap it.

If you're sending to more than one person, type a new name or address after each previous name autofills.

Type to find an addressee, or tap to open Contacts.

A universe of formatting and color awaits. Self-control is welcome.

2:25	2:27
Cancel	Cancel **Next week** ↑
New Message	Subject: **Next week**
To: dan ⊕	I am **so excited** about the possibility of meeting you next week. Your recent work exploring the psychological effects of typography has truly shaken up the email world.
DM **Dan** McCabb dan@bigproductions.com	
🧑 **Dan** Woo dwoo@optonline.net ⊘	--*Alex* -------------------
DG **Dan** Gorne dgorne@gmail.com	Alex Bodardy CEO, Founder, President
DH **Dan** Howley	**Format** ✕

Keyboard:
q w e r t y u i o p
a s d f g h j k l
⇧ z x c v b n m ⌫
123 😊 space @ . return

Format toolbar:
B *I* U S̶
Chalkduster › A A ⚫
☰ ☰ ☰ ☰ ☰
|◂ |||▸ ☰◂ ▸☰

Composing a message

- **Cc.** On the Cc (carbon copy) line, enter the addresses of people who should get a copy of this message, even though they're not the primary recipients.

- **Bcc.** Any addressee you enter here, on the Bcc (blind carbon copy) line, gets a *secret* copy of the message. The other recipients don't see any evidence of it.

- **Subject.** Here's the title of your message.

- **From.** Obviously, you know who this message is from—it's you. But you might have more than one email account; tap here to choose which one you want for sending this message.

Now that you've set up the administrative details, you can compose the actual message. You can swipe, tap, or dictate the text.

By no means are you limited to plain black text on a white background; this is Apple, after all. A wonderland of format, color, and visual options await when you tap ‹ above the keyboard:

- Aa opens the Format panel (see "Composing a message" on the previous page), which teems with controls for type formatting (bold, italic, underline, and strikethrough styles), fonts (typeface, size, color), paragraph styles (numbered or bulleted lists, left/right/center justification), and indentation/quoting levels.

- ▣ lets you choose a photo from your Photos collection.

- ⊙ opens the Camera app so you can take a photo or video to attach.

- ⬀ is how you send a file attachment. It lets you grab any file on your iCloud Drive or in the Files app (page 327).

- ⊞ opens the iPhone's document scanner (page 359) so you can scan a document and attach it.

- ⊘ opens the iPhone's standard annotation (markup) tools (page 7) on a blank canvas. You can go nuts drawing or sketching. The ⊕ button even lets you add text; insert your signature; magnify a part of the drawing; and add arrows, squares, ovals, and speech bubbles.

Tap to open the formatting rabbit hole.

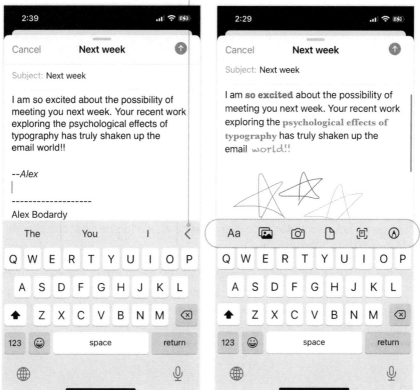

Photos, scans, drawings

As you type, the iPhone spellchecker marks questionable words with a dotted underline.

If you run out of time or inspiration, tap Cancel and then Save Draft. You've just put your unfinished message into the Drafts folder. You can complete and send it whenever you have the time and interest.

> **TIP:** If all you want to do is check another message for reference while you're writing this one, swipe down from the top of the screen. Your half-finished message becomes a tab peeking up from the bottom of the screen, freeing you to fish around in the Mail app, looking up what you need to reference. Bring back the message-in-progress by tapping its tab.

But if you are ready to send the message, tap ⬆. Mail sends the message and puts a copy in your Sent folder.

Signatures

Unless you step in, the iPhone stamps "Sent from my iPhone" at the bottom of every message.

That's a signature. Usually it's the person's name and contact information, a pre-apology for smartphone-keyboard typos, a graphic, or what the sender considers to be a hilarious quote.

To make up your own, choose Settings→Mail→Signature. You can make up one signature for All Accounts or a different one for each account (tap Per Account). Compose your signature in the text area. You can use bold, italic, or underline formatting—select the text, tap the ▶ to bring the **B***I*U button into view—and even emoji.

Reading Email

When you tap an Inbox on the start screen, Mail presents a tidy list of all your waiting mail; blue dots indicate messages you haven't yet opened. For each message, you see the sender's name, the subject line, and—in gray—the first few lines of its contents. That preview is meant to help you scan for any messages that need your attention immediately.

Flick upward to scroll the list; tap a message to open it.

If the type is too small to read, spread two fingers to zoom in, exactly as you would on a photo.

And if you'd like to see more of the usual header details—**To, From,** and so on, as well as the complete list of addressees—tap the abbreviated header.

To move on to the next (or previous) message in the Inbox, tap ⌄ or ⌃ in the upper-right corner. Or jump back to the inbox (or whatever mailbox you're in) by swiping across the screen to the right.

Conversation View

Unless you've changed the settings, your **Inbox** uses what Apple calls *conversation view* and what the rest of the world calls *threading*. The idea here is that all the messages that are part of a single back-and-forth conversation appear as a single item in the main message list. Mail is smart enough to cluster messages in the same conversation, even if somebody changed the subject line along the way.

You can spot a conversation right away, because it displays a ⊙ at the right edge of the list instead of the plain ⟩ indicator.

When you tap a conversation, it expands to reveal *all* the messages. To save you a lot of wading through clutter, Mail hides all the duplication and replies to replies.

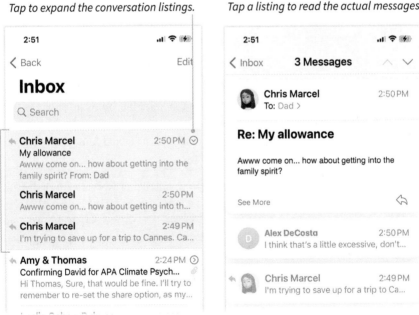

Tap to expand the conversation listings.   *Tap a listing to read the actual messages.*

Conversation view

In one regard, conversation view can be confusing. When it's turned on, Mail attaches all older emails on this topic to the most *recent* email. So if a particular conversation has been going on for six weeks, earlier messages no longer appear in their proper chronological place in the message list. Instead, they're grouped with the one that came in today.

If that drives you crazy, you can turn off **Settings→Mail→Organize by Thread**. And sure enough: Now every message appears in its proper place, no longer grouped by conversation.

Filters

When you tap ☰ at the bottom of the message list, Mail hides all the messages you've opened and seen. You're looking at only the unread messages. You've just turned on a *filter*; the other messages still exist, but you've hidden them.

While the filter is on, you can tap **Unread** to reveal other ways to filter the list. You can view only the flagged messages, only messages written directly to you (hiding the ones where you were only copied), only messages where you *were* copied, only messages with attachments, only messages sent today, or only messages from people on your VIP list. Or you can view messages that meet any combination of these criteria.

Tap ☰ to turn off the filter.

TIP: If an incoming message comes from somebody who's important enough to add to your Contacts list, tap their name (in the **From** line). A new Contacts card appears, auto-filled with the phone number and address grabbed from the person's email signature (if any). Tap **Create New Contact** to save the new card.

The advantages of putting somebody into Contacts: From now on, Mail can autocomplete their name when you're addressing an outgoing message and will no longer flag anything from this person as junk mail.

Opening File Attachments

When someone sends you files attached to an email, a little 📎 appears next to the message's name in the message list and also next to the sender's name when you open the email.

A document icon also appears at the bottom of the message. Tap it to download and open it. Now you can zoom in, pan around, or rotate the phone 90 degrees, exactly as though the document were a web page or a photo.

On the other hand, if you long-press the attachment's icon, a shortcut menu appears. It offers many more options for processing this attachment.

The commands available depend on what kind of document it is, but often you'll get **Quick Look** (reveals what's inside that attachment—the full PDF file, photo, Word document, spreadsheet, or whatever it is), **Save to Files** (adds the document to your Files app, as described on page 327), **Share**, and **Copy**.

If it's a graphic, you also get **Save Image** (puts the graphic into your Photos app). If it's a PDF or a graphic, **Markup and Reply** lets you annotate it. If it's a Word document, you get buttons for apps that can open Word docs, like **Mail**, **Dropbox**, **Evernote**, and so on.

That's a lot of power in a rather hidden spot.

Data Detectors

As you work on your email in the coming months, keep an eye out for a gray underline that appears beneath an address, phone number, date, time, web address, or flight number. It indicates that Mail has recognized what kind of information it's looking at.

If you *tap* the underlined text, you open the appropriate app. Maps opens to show you an address; a **Call** button offers to dial a number.

But if you long-press an underlined info-bit, you get some truly useful pop-up displays. For contact information, you get buttons that let you call, message, or FaceTime this person; you can also **Add to Contacts**.

If Mail finds a discussion of an appointment, it can **Create Event** or **Create Reminder**—or, if you're not quite ready to commit, it offers to **Show in Calendar**, revealing the time slot of the appropriate day.

> **NOTE:** Because so few people realize they can long-press, Mail also tries to catch your eye with a banner that says "Siri found 1 event"; tapping it, too, gives you the option to drop the thing onto your calendar.

For a flight number, you get a map of the route and a **Preview Flight** button that reveals every scrap of information about the flight: its times, its terminals, and even its baggage-claim details.

Mail has found a proposed appointment.   *Long-press for more options.*

Data detectors

Surviving Email

Some people don't do anything at all with their incoming messages. Their Inbox counter grows to tens of thousands or even millions; they just ignore most of it. Lucky, lucky people.

The rest of us have to *do* something with our incoming messages. For us, there's the ↩ at the bottom of the screen. It offers far more than just replying. Read on.

Quick Swipes

Hey, you've got a touch screen! In Mail, Apple has exploited that fact by inventing various swipes that let you charge through your Inbox, dispatching messages as you go, without even opening them. For example:

- **Delete: Full left-swipe.** Drag your finger all the way left across an Inbox message to delete it.

- **Flag: Partial left-swipe options.** If you swipe leftward only partway across a message, you reveal a set of three buttons. One of them is **Flag** (page 286).

- **Reply panel: Partial left-swipe.** If you swipe leftward partway and then tap More, you open the standard Reply panel. It lists just about everything you can do: **Reply, Reply All, Forward, Trash, Flag, Move, Mute, Notify Me,** and so on.

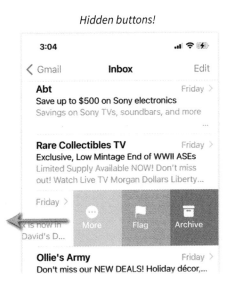

Hidden buttons!

Swipe gestures

- **Mark as Unread: Full right-swipe.** Drag rightward all the way across a message to make it look like a new one (unread), so you won't forget about it. (If it's already unread, a rightward swipe marks it as read.)

> **TIP:** In **Settings→Mail→Swipe Options**, you can fiddle with the positions of these left-swipe/right-swipe buttons.

- **Preview: Long-press.** Long-pressing a message in a list presents you with a sneak preview of what's in it *and* a list of commands for processing it (**Reply, Forward, Move Message, Trash Message,** and so on). Tap anywhere on the background to return to the message list.

> **TIP:** The commands panel doesn't fully fit on the screen until you drag upward on it.

Long-press...   *...and then slide up to see the whole list.*

Long-press options

Replying

To answer a message, hit ◁ and choose **Reply.**

Reply All (◀) is a little different: It sends your response to everybody originally addressed in the message, not just to the person who composed it. Reply All has gotten a lot of people into a lot of trouble over the years; be careful with it.

In either case, you get a new outgoing email message, already addressed, with the subject line filled in ("Re: [Whatever the original subject was]"). The message you're responding to appears at the bottom of the window as quoted text (in a special color and set off by a vertical bar).

You can, if you wish, use the Return key to create blank lines in that quoted message and splice your own responses into those gaps. Your correspondent can see which parts are the original message and which are your response, thanks to the vertical line and colorized text of the original.

> **TIP:** Here's a fantastic trick: You can drag through a portion of a message—the part you want to talk about in your response—*before* you hit ↩. When you do that, Mail pastes only *that portion* of the original message into the reply. Whoever reads your answer will know exactly what part of the original message you're talking about.

Now it's just like composing a fresh email message. Attach files, change the subject line, add or delete recipients, whatever. When you're finished writing your response, tap ⬆.

Tap to open the Reply panel…

…which offers a lot more than just the Reply command.

struck up a little deal with my sister.
She's supposed to take care of the trash
for $.25 a week.

See More   ↩

| DP | Alex DeCosta | 2:45 PM |
| | I think that's a little excessive. I notic… |

| | Jeffers Alexis | 2:45 PM |
| | I'm trying to save up for a trip to Ca… |

🗑   📁   ↩   ✏️

struck up a little deal with my sister.
She's supposed to take care of the trash
for $.25 a week.

| DP | **Alex DeCosta** | ✕ |
| | $1.05 a week is my final offer. Take… | |

↩ Reply   ↩↩ Reply All   ➡ Forward   🗑 (2)

Flag 2 Messages   🏳

Mark as Unread (2)   ✉

Move 2 Messages…   🗂

Archive 2 Messages   🗃

Replying

Forwarding

Forwarding a message, of course, means sending one you've received along to a third person. To do that, tap ⤺ and select **Forward** (➡).

A new outgoing message appears, ready for you to edit and address before sending. Usually you'll want to add a little comment at the top of it to explain why you're forwarding it. You know: "Dear Alex: See? I was right all along."

Filing

The folders in the Mailboxes list include such old favorites as **Inbox**, **Sent**, and **Drafts**. But you can create folders of your own, too, for organizing mail. You can create one called *Flight Confirmations*, one called *Stuff from Dad*, and another called *Deal with These Later*. You can even create folders *inside* folders.

To create a new folder from the Mailboxes list, tap **Edit** and then **New Mailbox**. Mail asks you to name the new mailbox. You can specify which existing folder, in which email account, this one goes under.

When you tap **Save**, a new icon appears in the Mailboxes list, ready for use. You can file any message in a message list by long-pressing it and choosing **Move Message**. Or, if the message is open, tap ⬓ and choose the destination folder.

Flagging

Mail offers seven colorful flags you can apply to messages to mean anything you want. You can use them to make certain messages stand out in the lists. But you can also use them to round up deeply scattered messages from your various folders and accounts into one place (the **Flagged** folder in your folders list).

> **TIP:** If you're not much of a flagger, you can hide the Flagged folder so it doesn't eat up space. Tap **Edit**, and then tap **Flagged** to turn it off. Tap **Done**.

If a message is open, flag it by tapping ⤺, **Flag**, and then the color you want. (You can see the **Flag** button in "Replying" on the previous page.) If the message is in a list, long-press it and then tap Mark→Flag. (In a list, you can flag a bunch of messages at once by tapping **Edit**, selecting them, and *then* tapping Mark→Flag.)

Use the same commands to unflag a message.

Searching

Mail isn't just an email-processing app; it's also, in a way, a database. Therefore, it's ideally suited to *finding* email in your ocean of thousands of sent and received messages.

Tug downward on any message list to reveal the search box.

When you tap there, the keyboard appears, and so do a few ready-to-roll searches like **Flagged Messages** and **Messages with Attachments**.

Begin typing. Mail hides all but the matching results, letter by letter. No need to specify which fields to search (From, To, Subject, Body); you're searching everywhere. You're also searching all of your mail folders at once—unless you tap **Current Mailbox**.

Mail offers some canned searches.

Mail reveals messages that contain your query in any parts of the messages.

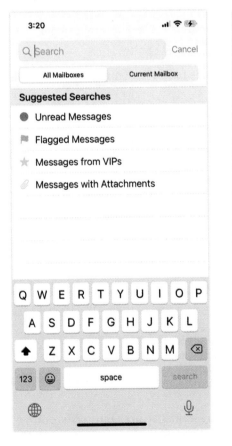

Searching Mail

Search results can take their sweet time to arrive, because some of the messages are no longer on your phone (but they're still online).

Just below the search box, Mail shows search refinements it thinks you might find helpful. For example, if you've typed *birthday*, Mail offers search shortcuts like Subject contains: birthday, Mailboxes: birthday, Attachment name contains: birthday, and so on.

When you tap one of those refinements, Mail places a little bubble in the search box—a *search token*. You can create complex searches by creating several tokens in a row. Maybe the first one limits the search to messages with a certain word in the subject line; a second one might limit the search to a certain date. Each token you add starts a new Mail search.

Following a Conversation

Sometimes, one of your back-and-forths is important enough that you'd like to be notified every time someone adds new thoughts to it.

If the message is open, tap ⤺ and then Notify Me. If you're composing or replying to a message, tap in the subject line and tap the ⌂.

From now on, any new reply to this message produces a notification banner on your screen.

Deleting

To delete a message in a list, swipe left across it, all the way to the left edge of the screen. Or, if the message is open, hit ⤺ and then Trash (or Archive).

Mail's Trash is like the wastebasket in your office: Things you put in there aren't really gone until you *empty* the trash. You have, in other words, a safety net, a little protection against mistakes. If you change your mind, tap Trash in your Mailboxes list; rescue a message by long-pressing it and then tapping Move Message. Put it back into your Inbox or any other folder.

To empty the Trash, open it up and hit Edit. Select the messages you never want to see again (or Select All if you don't want any of them) and hit Delete.

Some kinds of email accounts can auto-empty the Trash folder on a schedule. Open Settings→Mail→Accounts→[account name]→[the email address]→Advanced→Remove. If yours is one of them, you can choose Never or After one day (or one week, or one month).

(For iCloud accounts, that ludicrously long sequence is slightly different: Go to Settings→Mail→Accounts→iCloud→iCloud. Scroll to the very bottom, and then tap Mail→Advanced→Remove.)

(Gmail auto-empties its Trash every 30 days; you can't change that setting in Mail on the iPhone.)

> **NOTE:** Some email services, including Gmail, offer an **Archive** button instead of a **Delete** button. It's roughly the same thing, except that your deleted messages land in an **Archived** folder and don't auto-delete.

Unsubscribing

Whenever a legitimate business sends you helpful marketing mail (they would never call it spam!), there's usually an Unsubscribe button in tiny type at the very end of the message. Tapping it gets you off that mailing list.

When Mail detects the presence of that Unsubscribe link, it makes life easier for you by presenting a much more visible Unsubscribe banner at the top of the screen.

(None of this works on the scummier sort of spam that most people get—only mail from brand-name commercial businesses.)

VIPs

In the worlds of high society, business, and politics, a VIP is a very important person. On the iPhone, it's somebody whose *email* is very important.

That might be your boss, your spouse, your kids, or your state lottery notification department.

The VIP mailbox in your folder list contains all the mail you've been sent—even deleted messages—from your VIPs.

Tap the ⓘ to see the list of VIP names so you can add more. This is also how you demote someone from the VIP list; swipe left across the name and then tap Delete. (There's also an Add to VIP button on every person's Contacts card, for your convenience.)

> **TIP:** Thanks to iCloud syncing, your VIP setup appears automatically on all your other iPhones, Macs, and iPads.

If your VIPs are very, *very* important, you can direct the iPhone to alert you with a notification when mail from them arrives. Tap the ⓘ next to the VIP mailbox and choose VIP Alerts.

Dealing with Spam Mail

Spam is the worst. It's junk mail—marketing emails you didn't ask for.

Mail has a spam filter that's supposed to weed it out, but you have to train it. Long-press any message you consider junk and choose Move Message→Junk. (If you've already opened the message, tap ↩ and then Move to Junk.)

In each case, Mail not only moves it to the Junk folder but also learns to identify that kind of message as spam the next time.

On the other hand, if Mail flags a legitimate email as spam, you do the opposite. Open the Junk folder, long-press the miscategorized message, and choose Move Message→Inbox.

By the way: Did you ever wonder how the spammers got your email address in the first place?

Frankly, you gave it to them. Spammers use automated software robots that scour every public internet message and web page, recording email addresses they find. If you've ever provided your email address on any web page—signing up for a newsletter, buying something online, joining a discussion board—you've subjected it to spam.

The solution is to maintain two email accounts. Use one for all that public stuff. Use another for private correspondence—and never, ever type it onto a web page.

CHAPTER TWELVE

Safari

You may be too young to remember life before the web. You couldn't look anything up—you had to *call* places to get directions. You couldn't read articles on a screen—you had to subscribe to newspapers and magazines made of paper. You couldn't order stuff online—you had to visit stores in person. Life was a living hell.

Safari is the iPhone's web browser. Its rivals, like Chrome, Edge, and Firefox, have plenty of fans. (And in iOS 14, you can choose a different browser to be the *default* on your phone—the one that opens when you tap a link in an email, on Facebook, or on Twitter. Open Settings→Chrome, for example, and tap **Default Browser App→Chrome**.)

But Safari is fast (even faster in iOS 14), easy to use, and ingeniously integrated with the rest of the phone in ways that can save you time and hassle.

> **NOTE:** On the phone, screen space is tight. So shortly after you start reading a page (or immediately after you scroll down into one), all the controls *disappear*: the address bar at the top and the buttons at the bottom. Safari is trying to give you more screen space for reading. To bring them back, scroll up a little (drag down with your finger).
>
> On the other hand, if the toolbar appearing and disappearing makes you nuts, tap the AA in the address bar and turn on **Hide Toolbar**. Now those elements are hidden at all times, even when you scroll up or open a new link.

Address/search bar   Reload page

1:53   .ıl 📶 🔋

AA   🔒 en.m.wikipedia.org   ↻

Welcome to Wikipedia,

the free encyclopedia that anyone can edit.
6,194,160 articles in English

The arts   Biography   Geography   History
Mathematics   Science   Society   Technology
All portals

From today's featured article

Super Mario World is a
1990 platform game
developed and published
by Nintendo for the Super
Nintendo Entertainment
System (SNES). The story
follows Mario's quest to
save Princess Toadstool
from the series antagonist Bowser, and his

Shigeru Miyamoto

<   >   ⬆   📖   🗔

———— Back to previous page

———— Share page, mark page

———— Bookmarks, history

———— View all open tabs

Safari layout

Five Ways to Start Out

The first time you open Safari, it has no idea what web page you want to view.
You have to tell it, using one of these five methods.

Use the Start Page

The wide box at the top of the screen is the address bar, and it's all-
important. For starters, it's where you type the web address (the URL) of the
page you want, like www.nytimes.com or www.whitehouse.gov.

But once you've tapped into the address bar, before you can even type a
single letter, you'll discover the first way to begin a surfing session: Tap a
button on the Start page. That's the page full of bookmark icons that appears
when you've tapped into the search bar (or opened a new Safari window or
tab) but haven't typed anything yet.

The Start page gives you quick access to all kinds of websites Apple has
concluded you care about.

Start page   *Typing-saving suggestions*

2:04	..ll 🔋 ■
Search or enter website name	Cancel

Favorites   Show More

Finances Folder	CBS Folder	Sheets Folder	YouTube DP
Nest	Trains	Schools Folder	Maps

Frequently Visited

The New York Ti...   YouTube

Siri Suggestions   Edit

Four Seasons Total
Landscaping on Twitter
twitter.com
🔵 From Messages · Nicki Pogue

2:05	..ll 🔋 ■
app	⊗ Cancel

Top Hit

Apple
apple.com

Google Search

Q app

Q apple   ⓡ

Q apple store   ⓡ

Q apple watch   ⓡ

Bookmarks and History

Detect Webex Application
appleinc.webex.com

iPhone 12 Pro and iPhone 12 Pro Max – A...
apple.com

On This Page (1 match)

Find "app"

Starting a search

> **TIP:** If you long-press any of these icons, you get a panel offering commands like **Copy**, **Open in New Tab**, **Edit**, and **Delete**. If you long-press a folder of Favorites (page 297), you get an **Open in New Tabs** command. It opens all the sites in this folder at once.

Here's what's here:

- **Favorites** is a subset of your bookmarked sites.

> **TIP:** Actually, you can request any folder of bookmarks to display its contents here; it doesn't have to be the folder *called* Favorites. A list of all your Bookmarks folders appears in **Settings→Safari→Favorites**; tap the one you want to be your new Favorites folder on the Start page.

- **Frequently Visited.** The latest slice of your History.

- **Siri Suggestions.** These are links to sites that iOS found in your Messages chats, recent Mail messages, and stories you've read in the News app. You know how you scroll through your messages or email, and people have sent you links, and you say to yourself, "OK, I'll click that later"? The Siri Suggestions section makes that easier.

Each section shows a few icons in its category, plus a **Show More** button to see more.

Type an Address

The second way to begin a web session: Type the web address (URL) of a site you want. The actual process goes like this:

1. **Tap inside the box.**

Whatever was already there is highlighted, ready to replace.

> **TIP:** No matter how far down a page you've scrolled, you can jump directly to the address bar. Remember this tip: *Tap the top*—the very top edge of the screen. (On a Face ID phone, tap one of the "ears." On a home button phone, tap the status bar.)

2. **Type or paste the address.**

You can leave off the http:// and www; Safari will add them for you. If you're trying to get to http:/ www.amazon.com, for example, you can just type amazon.com.

Long-press the period.

Don't bother deleting whatever address is already there, either. Once the address bar is ready for typing, *just type.* You'll type right over whatever was there.

URL suffixes

Safari offers a secret menu of web-address suffixes like *.com*, *.net, .org, .us,* and *.edu.* To see them, long-press the period key on the keyboard. If you want *.com,* just release your finger—no need to tap anything else.

Otherwise, this address bar works just like the one in any other web browser. Tap inside it to make the keyboard appear.

3. **Tap Go (on the keyboard).**

That's your Enter key. Or tap **Cancel** to hide the keyboard *without* "pressing Enter."

In a matter of moments, the web page appears.

At this point, the Back button (<) lights up. It returns you to whatever page you were just on before this one. So does swiping inward from the left edge of the screen, which is useful when the Safari toolbar buttons are hidden.

And at *this* point, you can use the Forward button (>) or swipe inward from the *right* edge to revisit the page you were on when you tapped <.

You can also *long*-press the < or > buttons to see a pop-up menu of *all* the pages you've visited during this browsing session.

NOTE: You can interrupt the downloading of a page by tapping ✕ in the toolbar.

Once the page has loaded, that same button turns into a Reload button (↻). Tap it to re-download the page. That's useful when a page isn't looking or working quite right, or if you want to update it—if it's a news or sports page that gets updated constantly.

Do a Search

Safari's address bar doubles as the search bar. It's built-in Google.

TIP: Or built-in Bing, built-in Yahoo, or built-in DuckDuckGo. Choose your favorite search in **Settings→Safari→Search Engine**.

Just tap in the address bar and start typing what you're looking for (like *kidney stones* or *how old is the moon*).

As you type, a page of suggestions appears; the idea is to save you time and typing. It incorporates websites you've visited recently, websites you've bookmarked, tabs you have open, Google autocomplete suggestions, and appearances of this search term *on the current web page.*

You can ignore those suggestions, though. If you finish your query term and tap **Go**, Safari opens the Google results page.

Choose a Bookmark

Just as you can stick a bookmark into the pages of a book to find your place later, so can you bookmark pages of the web that you intend to revisit. That way you don't have to remember (or type) their addresses.

In a browser's case, though, you can create as many bookmarks as you want—dozens or hundreds. You can make *folders* full of bookmarks.

To see whatever bookmarks you've got so far, tap ⌗; on the screen that opens, tap the ⌗ tab. Tap a bookmark to revisit that site.

Here's your complete bookmarking guide:

- **Save a bookmark.** With the web page in front of you, long-press the ⌗

Long-press to add a bookmark.

possess similar tarsal spurs, so it is thought that, rather than having developed this

chara ⋯⋯⋯⋯⋯⋯⋯⋯⋯ simply
inher **Add Bookmarks for** ⌗ :edents.
4 Tabs
Rathe ⋯⋯⋯⋯ ⋯⋯e
platy **Add Bookmark** ⌗ what
was (
chara **Add to Reading List** ∞ a model

<   >   ⬆   ⌗   ▯

Name the bookmark; file it away.

Cancel   **Add Bookmark**   Save

W   Platypus venom - Wikipedia ⊗
        https://en.m.wikipedia.org/wiki/Platypus_ve...

LOCATION

☆ Favorites

Making a bookmark

icon; from the shortcut menu, choose **Add Bookmark**.

TIP: If you have several tabs open, you can also tap **Add Bookmarks for 7 Tabs** (or whatever). You've just created a single bookmark that opens a whole bunch of tabs at once, which can save you time if you like to read the same sites every morning.

On the Add Bookmark screen, you can type a shorter or clearer name for the page. If you tap the **Location** option, you can choose a bookmarks folder that will contain your new bookmark. The folder options always include **Favorites**: a special, exalted folder containing bookmarks you *really* love. Favorites, as you may recall, show up on the Start page.

TIP: Actually, here's a more direct way to "favorite" a web page: Tap ⬆ and then **Add to Favorites**.

You can, of course, choose any other folder to hold this bookmark—you might have different folders for different interests or projects—or tap **New Folder** to make a fresh one.

- **Organize your favorites.** Once you've opened the list of bookmarks, you can delete one (or a folder full) by swiping fully leftward across it.

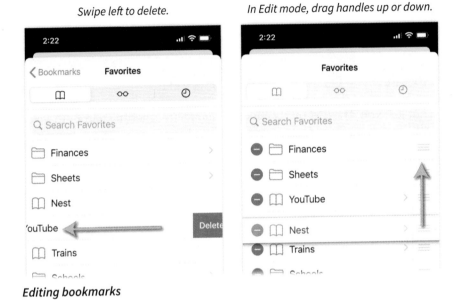

Editing bookmarks

For more ambitious organizing, tap **Edit**. Now you can delete something by tapping the ⊖ and then **Delete** to confirm. Rearrange the list by dragging the grip strip (≡) up or down in the list. (You can't move or delete the Favorites folder.) Edit a favorite's name, folder location, and web address by tapping it. Tap **Done** when you're finished.

Open the History Menu

Safari keeps track of every web page you've visited in the past week or so. It lists them in a History list, organized by time period, so you can revisit them later without having to remember how you got there. Because these entries are, in a way, a flavor of bookmarks, you see them by tapping the ▭ button—but now tap ⏱.

The first question many people have about the History list, of course, is: "How do I erase it?" You may not want anyone to see what you've been up to, either on your phone or on your Mac. (Remember that your bookmarks and history list auto-sync between your Apple machines; see page 394.)

To delete one entry, swipe fully left across its name. To erase bigger swaths of your History menu, tap **Clear**; then you can choose **The last hour**, **Today**, **Today and yesterday**, or **All time**.

In addition to the History menu entries, the Clear History command also vaporizes all the cookies (web page preference files) you've accumulated, your downloading history, and any cache files (web page pieces stored on your disk for quicker access the next time you visit).

Now your secrets are safe with Safari.

Five Ways to Magnify a Page

It's not just you. The text on the modern web really is getting harder to read. Partly that's because today's screens have higher resolution than ever—more little pixels packed into less space—and partly it's because your eyes are getting older.

Make the type smaller or larger.   *Much better!*

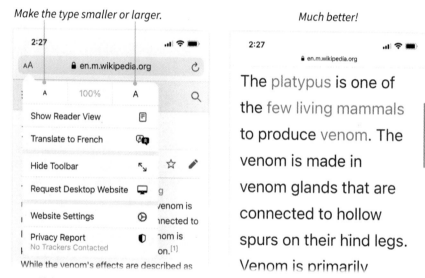

Magnifying the type

Fortunately, Safari offers all kinds of ways to magnify web pages:

- **Magnify the text for all websites.** In Settings→Safari→Page Zoom, tap a percentage enlargement that feels right for all your web surfing.

- **Magnify the text on this web page.** Tap AA in the address bar. The A and **A** buttons adjust the text size for this site. (To jump back to original size ("100%"), tap the percentage number between the two buttons.)

- **Magnify everything.** Spread two fingers on the glass, as though you're stretching out the Safari page on a sheet of rubber. You can pinch to shrink the page down again. This is a fantastic trick that also works in Mail.

- **Turn the iPhone.** If you rotate the phone 90 degrees so it's horizontal, Safari magnifies the web page to fill the wider view. In many cases, that's just enough bigger that you can read it.

- **Double-tap.** Very few websites still appear on your phone the way they would on a desktop computer; most now auto-detect that you're using a phone and present a specially formatted, scrolling view that spares you the zooming. But if you do come across a page that's formatted for computers (or you use the **Request Desktop Website** option described on page 310), you can double-tap any block of text or any image to enlarge *only it* to fill the screen.

Once you're zoomed in, drag or flick your finger to move around the page.

AutoFill

What we can do on the web these days is a miracle: Read the news, keep up with our friends, order groceries.

What's less of a miracle is how many of those activities require typing in your name, address, phone number, email address, and credit card number. Over and over, until your fingers are numb, calloused stumps.

Tap to auto-enter your contact info.

AutoFill at work

Apple's answer to that tedium and repetition is AutoFill. It's a feature that automatically completes those forms for you.

Contact Information

To set up Safari to fill in your name, address, and phone number, open Settings→Safari→AutoFill. Turn on Use Contact Info. Tap My Info, and then, from Contacts, find your own "card" to tell Safari *which* contact information belongs to you.

Next time a web page asks you to input your address, smile nonchalantly and tap the AutoFill button that appears over the keyboard. Safari auto-enters all your contact info.

Or at least it's supposed to. It doesn't work on some oddball sites, and if it leaves a box empty here and there, tap ∧ and ∨ to hop from one to the next, rather than tapping and scrolling manually.

Credit Cards

Safari can fill in your credit card information when you're shopping online, too—another time-saving blessing.

To add a new card...

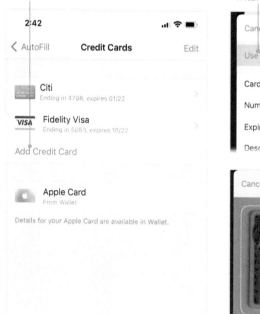

...take a picture of it.

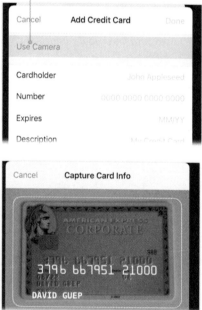

AutoFill for credit cards

To store your credit card details, choose Settings→Safari→AutoFill; turn on Credit cards. Tap Saved Credit Cards. Once you prove that you're authorized to be in here (Face ID, fingerprint, or passcode), your list of stored cards appears. If you're just starting out, it's probably empty.

Tap Add Credit Card. Now you have four pieces of data to fill in about your credit card: Cardholder (your name), Number, Expires (MM/YY), and Description (Citi Visa or whatever).

But wait a minute—why are you fussing around with little digits? Your phone has a *camera* on the back!

This is supercool: Tap Use Camera. Aim the camera at your credit card. The iPhone auto-recognizes your name, the card number, and the expiration date, and even proposes a description of the card.

> **NOTE:** Mercifully, iCloud autosyncs your card information to your Macs and iPads, too, so you won't have to fumble with your wallet no matter what device you're using.

One thing Safari does *not* ask for, and does not memorize, is the three- or four-digit security code, sometimes called the CVV or CVV2 code. You have to type that in every time you use this card on a web page. Apple says that extra step is a security thing, but you might consider it an annoyance thing.

The next time you're on a website that's requesting your credit card number, tap AutoFill Credit Card. (If you've stored more than one, Safari asks which card you want to use.) Presto: Safari blithely fills in the credit card information for you, saving you time and hassle.

Passwords

As you know, today's websites insist that you make up a long, unguessable password—capital and lowercase letters, numbers and symbols, and a few obscure punctuation marks, please. And you're not supposed to reuse a password. And you're supposed to change them all every month.

Here's a much better idea: Let Safari memorize the passwords for you—and even make them up for you.

It happens at the moment you first sign up for an account on some site. You've just filled in your contact information and made up a password. You've tapped OK, Next, Save, Continue, or whatever the next button is. At this moment, Safari offers to save this new login information so you'll never have to type it manually.

THE MASTER PASSWORD LIST

Behind the scenes, Safari quietly compiles a master list of all the passwords it's memorized—not just of websites, but of network servers too. You can look them over, long-press them (for copying or sharing with someone else), and even change them, in **Settings→Passwords**, or ask Siri to "Show my passwords." (You have to supply your face, fingerprint, or passcode before you can see them. You know—for security purposes.) The list looks something like the one shown below at left.

But Safari does more than just memorize your passwords. It can actually guide you toward better password hygiene. For example, if you tap a password, Safari may offer a SECURITY RECOMMENDATION like "Easily guessed" (shown below at right). (In **Settings→Passwords**, tap **Security Recommendations** to view all your password weaknesses in one place.)

Or maybe you've used the same password for more than one site ("Reused password"). That's theoretically risky,

because if the bad guys acquire one of your passwords in a data breach, suddenly they have access to more than one of your accounts. Since Safari can remember and fill in your passwords for you, what's the harm in making them all unique?

In iOS 14, Apple takes this Safari password-guidance business one step further into Brilliance Land. When a big data breach does occur, Apple's servers know about it. They become aware of which passwords have fallen into nefarious hands.

Safari runs a comparison of your saved passwords against the list of stolen ones. If it finds a match—if you are one of the unlucky victims whose passwords have been stolen—Safari lets you know while there's still time to change the compromised passwords.

Apple wishes to stress that all this computation is performed right on your phone. At no time does Apple ever have any of your passwords. ✧

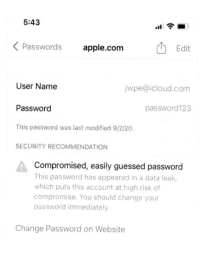

Log in to contribute.

Frank2023

coDfar-8zomqe-syww	Strong Password
coDfar-8zomqe-syww	Strong Password

Enter your email address

∧ ∨   Done

iPhone created a strong password for this website.

This password will be saved to your iCloud Keychain and will AutoFill on all of your devices. You can look up your saved passwords in Settings or by asking Siri.

Use Strong Password

Choose My Own Password

Enter your username

Enter your password

Log in

password for this website
Frank2023y

Q	W	E	R	T	Y	U	I	O	P

…and let Safari do the work.

Frank2023y

••••••••••••••••••

Log in

Forgot your password?

Auto-suggested passwords

Actually, Safari goes one better. When you tap into a Password field as you create your account, it offers to make up a password *for* you. It's a ridiculously long, complicated, and unique one, impossible for a human to memorize (or to guess). But since Safari is going to do the memorizing for you, what's the downside of accepting its unguessable suggestion? Tap **Use Strong Password**.

You don't have to worry about what will happen when you try to access the site on your iPad or Mac, either. Your other Apple devices *also* know this password and autofill it, thanks to the iCloud syncing feature in Settings→ [your name]→iCloud→Keychain.

You don't have to use Safari's suggestion, of course; you can tap **Choose My Own Password** and make up one of your own. (Safari will scold you, though, if you make up a really bad password—like "password.")

Next time you're logging in to this site, the QuickType bar above the keyboard offers that password, so you can enter it with one tap. (If you have multiple accounts for the same site, you get multiple QuickType buttons.)

Authentication Codes

If you hate nuisance and red tape, you're going to love this one.

These days, whenever you try to sign into an account where security really matters—your bank, your insurance company, your corporate network—sup-

The site texts you a verification code...

...but Safari can type it in for you.

Text code auto-enter

plying your name and password isn't enough. Most of these sites also use two-factor authentication (page 395): When you try to log in, the website texts a numeric code to your phone, which you're supposed to type when you sign in. You stare at the notification banner, trying to memorize those six digits, chanting them over and over, before the banner slips away.

In Safari, there's no need. The instant that Messages receives the security code, the code appears in the QuickType bar above the keyboard. Tap it to slap it directly into the site (or the app) that's requesting it. So awesome.

Tabbed Browsing

Every modern browser lets you keep multiple pages open simultaneously in a single window. On a full-size computer, it's called tabbed browsing, because each open page has what looks like a file-folder tab.

Safari, over the years, has become the mecca of tabbed browsing. To enter the page-juggling world, tap ⧉. If you're holding your iPhone upright, you see something like 3D floating pages; if it's horizontal, you get thumbnail buttons of the pages. You can tap one to open it full-screen, or you can get all administrative like this:

Swipe fully left to close a tab

Tabs from your other Apple machines

Long-press to close all tabs

Tabbed browsing

- **To create a new tab,** tap +. You get what appears to be a new blank browser window, containing your Start page. Tap an icon, enter an address, or use a bookmark, and off you go.

- **Rearrange the tabs** by long-pressing one and then dragging it up or down.

- **Close a tab** by tapping the ✕ in the corner, or by swiping a page to the left. It vanishes into the ether. (Or close all your tabs by long-pressing the Done button and then tapping Close All 573 Tabs, or whatever the number is.)

> **TIP:** You can set up Safari to close your tabs automatically after a day, a week, or a month in Settings→Safari→Close Tabs.

- **Search the tabs** by tugging downward to reveal the hidden search box. Safari can find matches among your open sites' titles and addresses.

Coolly enough, you can also see whatever tabs you had open on your other Apple gadgets, even if they're turned off. They show up here, at the bottom of the page-juggling screen, sorted by device. The idea is that you can start reading some article on your Mac and then pick right up on your phone when you have to leave the house. (If the feature doesn't seem to be working, maybe it's turned off. On the Mac, it's in System Preferences→ Apple ID→iCloud→Safari. On the phone or iPad, it's in Settings→[your name]→ iCloud→Safari.)

Nine Safari Specials

Apple is fully aware that people spend a lot of time in their web browsers, and that there are many web browsers to choose from. So over the years, its engineers have gone nuts fine-tuning Safari and adding grace notes to its feature list. Here are some of the best.

Long-Press a Link

By now, you've probably figured out that you tap a link on a web page to open *another* web page. But if you *long*-press a link, you get a preview of the page that the link will open, along with a partly concealed menu of commands. Drag upward to view all of your options: Open, Open in New Tab, Download Linked File, Add to Reading List, Copy (that is, "Copy the site address"), and Share.

If none of these options seems useful at the moment, tap anywhere on the background to back out.

The Reading List

When you spot a web article worth reading but you don't have time to finish it now, save it to your Reading List. The phone can download and memorize the entire thing so you can call it up when you have more time—even when you don't have an internet connection, like on a plane or a deserted island.

To add a page to the Reading List, tap ⬆ and then **Add to Reading List**.

Later, when you have time to do some reading, tap 📖 and then the center tab (∞). You can use the **Show All/Show Unread** button at the bottom of the screen to view all of your saved stories, or only the ones you haven't yet read.

Blinking, Ad-Heavy, Distracting view   *Reader view*

Reader view

iCloud synchronizes your Reading List, too, to your other Apple machines. So the next time you're being jostled on the noisy subway, clutching your phone, staring at some article that you'd rather read at home on your Mac in peace, by a crackling fire, with a glass of wine, keep the Reading List in mind.

Reader View

Websites have gotten pretty junky lately. They're filled with ads, banners, blinking things, self-playing videos, and other obnoxious detritus that distracts from what you're trying to read.

What you need is Reader view. When you tap ᴀA and then Show Reader View in the address bar, Safari hides everything junky from the web page you're trying to read. Only the text, headlines, and photos remain, against a simple, white background and clean, clear type. (You can use the ᴀA button in the address bar to change the background color and the font.)

> **NOTE:** You can't open Reader view until the page has fully loaded. Also, it may not always be available; for example, it doesn't appear on the landing pages of many sites, like the home page of a newspaper—only when you've opened an actual article. And it may not appear on sites that have already been formatted for cellphones.

To exit Reader, tap ᴀA→Hide Reader View. Reader view is the best.

The Downloads Manager

The web is full of interesting things to download—not just software, but photos, PDF documents, and other goodies. In Safari, the Downloads (⬇) button is a full-blown download manager. It appears when you're downloading a file, and offers a tiny progress bar to let you know how long this is going to take. You can tap the button to see the file's name, a list of previous downloads, a Cancel button (×), and even a ⵕ so you can search your phone to see where it wound up.

Saving Internet Graphics

It's easy to save a picture you find on a website. Long-press it; from the shortcut menu, choose Add to Photos, which sends it straight into your Photos app.

Or tap Copy, which grabs a graphic, ready for pasting into another app.

> **TIP:** You can also convert any web page—no matter how many screens tall—into a PDF document. Here's the trick: Take a screenshot, as described on page 435. In the resulting preview, tap **Full Page**. If you drag the tiny scroll bar, you'll see that you now have a very tall, scrolling, multipage document. You can mark it up (Ⓐ) or crop it (⊡)—and then, when you hit **Done**, you can tap **Save PDF to Files**. You've converted the web page into a PDF document in the Files app (page 327)!

Sharing a Page

It's easy enough to share a web page with someone. The ⬆ button opens the standard share sheet (page 203), full of ways to send its link: Mail, Messages, AirDrop, and so on.

Finding Text on Web Pages

You can search for words or phrases on a page, which is very handy if some article is long, cluttered, or badly designed. Just tap in the address bar, type your search word or phrase, and scroll down to the On This Page heading. Tap Find "salmon" (or whatever you searched for).

Safari highlights in yellow every occurrence of that phrase on the page and shows you how many matches it's found. Click the ∧ or ∨ buttons to jump from one to the next.

Put a Bookmark on Your Home Screen

One of the commands hiding in the ⬆ button isn't available anywhere else: Add to Home Screen. It deposits an icon for the current web page on your home screen, just as though it's an app. (Apple calls this shortcut a web clip.) You can rename it before tapping Add.

If you check a certain site every day, then over the years, web clips will save you literally *minutes* of time.

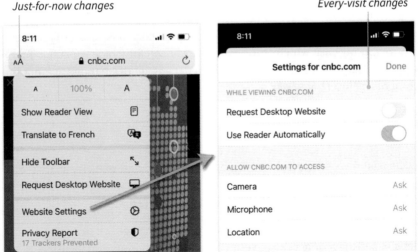

Just-for-now changes

Every-visit changes

Website settings

Settings for This Website

Safari can remember your settings for each website individually. To see your choices, tap ᴀA in the address bar and then tap Website Settings.

Your options include these:

- **Request Desktop Website.** Most websites these days recognize that you're using a smartphone, and they deliver specially designed layouts to fit your screen. They're usually smaller and more streamlined than the desktop sites, so you're saved loading time and cellular data—but they may not have all the features of the desktop versions.

  At any time, you can make Safari load the actual desktop version of such a site by tapping ᴀA→Request Desktop Website.

  But if there's some page you *always* want to see in its desktop version, here's your chance.

- **Use Reader Automatically** makes this page show up in the clean, uncluttered form of Reader view (page 308) every time you open it.

- **Camera, Microphone, Location.** Your camera, microphone, and location can all be useful on the web; for example, you can't use Zoom without giving it access to your camera and microphone. But you certainly wouldn't want criminals to invade your privacy through these channels. That's why, for each of them, for this website, you can Allow the feature every time, always Deny its use, or have Safari Ask you for permission each time you visit.

There's one more useful component to this website-settings feature: In Settings→Safari→Settings for Websites, Safari keeps a master list of all these features, along with which settings you've chosen for which websites. Taken together, this suite of options gives you a decent set of weapons against the worst of the internet sewer.

Privacy and Security

In the past few years, Apple has embraced privacy and security as its corporate mission on your behalf. And sure enough: Few web browsers offer as many features to protect you and your data as Safari.

You've already read about its password guidance (page 301) and its per-website settings for access to your camera, microphone, and location. But there's more.

Private Browsing

Every web browser these days offers private browsing. It's a mode that purports to maintain *no trace* of your web activity—no History entries, search histories, saved passwords, autofill entries, cookies, download history, cache files, and so on. You can surf the web in total anonymity, without leaving any tracks at all.

You can probably figure out why this feature is, ahem, so important to some people. Let's just say it: People use private browsing so they can shop for their spouse's birthday present without risk of ruining the surprise.

In any case, all you have to do is tap ⬇→**Private** before you open the questionable site. (Any other windows can remain open, whether they're in private mode or not.) Safari reminds you that you're in private mode by turning the address bar dark gray.

When you're finished with your clandestine research, just tap ⬇, close the tab, and tap **Private** again to go back to regular browsing.

Cookie Control—and Privacy Control

A *cookie* is something like a preference file that a web page deposits on your phone so it will remember your information the next time you visit. Most cookies are helpful, because they spare you from having to reenter your name, your preferences, and other information every time you're there.

But if you're worried that they're somehow spying on you, you can always turn on **Settings→Safari→Block all cookies.**

Otherwise, Safari accepts cookies from sites you actually visit but rejects cookies put on your phone by *other* sites—cookies from a fishy ad, for example.

Pop-Up Blockers

Most browsers these days, including the iPhone's Safari, come set to block the most obnoxious web-ads invention: pop-up and pop-under ads, which are ugly little windows that appear in front of the website you're trying to read—or behind it, so you don't see them until you close your window. Often, they're *phishing* attempts—scammers tricking you into surrendering your credit card info or passwords, like fake "Virus found" messages and fake "You've won a prize!" notifications.

You don't see pop-ups much anymore. If you *are* seeing them, make sure that Safari's blocker is turned on in **Settings→Safari→Block Pop-ups.**

While you're on that Settings page—and thinking about phishing—don't miss the **Fraudulent Website Warning** switch. Apple maintains a constantly updated list of known phishing sites, which masquerade as the login pages for banks and other important sites in order to capture your passwords. When this switch is on, you'll get a warning before you even get to see these pages.

> **TIP:** You can also install ad blockers from other companies, like Adblock Plus. Search the App Store.

The Privacy Report

In iOS 14, Safari makes a radical effort to protect you from *trackers*: software scripts and cookies whose sole purpose is to harvest information about you

Privacy Report	Done

LAST 30 DAYS

Trackers prevented from profiling you
186

Websites that contacted trackers
63%

Most contacted tracker
doubleclick.net was prevented from profiling you across 30 websites

Websites	Trackers

CURRENT WEBSITE
cnbc.com
22 >

WEBSITES BY TRACKERS CONTACTED
google.com
82 >

neighborfoodblog.com
80 >

businessinsider.com
74 >

nypost.com
71 >

Privacy report

and your activities. Most of this data is collected on behalf of advertisers for the purpose of targeting you with ads they think you'll be more likely to click. If you've ever done a search for, let's say, gym equipment and then discovered a bunch of ads for gym equipment in your Facebook feed later that day, well, that's trackers in action.

PART FOUR

The iOS Software Suite

CHAPTER THIRTEEN

The Built-In Apps

When you bought your new iPhone, Apple gifted you a
cornucopia of starter apps. On your home screens, you'll find
more than 40 apps that will let you get all kinds of things
done: staying in touch with people, editing photos, playing music and
movies, and generally organizing your life.

The major communication apps (Mail, Safari, Messages, Phone,
Contacts, and FaceTime), Apple's multimedia-store apps (App Store,
Books, Music, News, Podcasts, and TV), photography topics (Camera and
Photos), and the Shortcuts app get big write-ups elsewhere in this book.
This chapter offers mini-manuals for all the rest.

> **NOTE:** Many of the apps greet you, upon their first opening in iOS 14, with a
> welcoming "splash screen" that advertises some of their new features. To avoid boring
> you silly, this chapter doesn't mention them. Just tap **Continue** to open the app.

Calculator

For straightforward calculations, you don't need this app. You can just ask
Siri things like, "What's 15% of 24?" or "What's 1,723 times 8?" You can also do
math in the search box—to see it, drag downward on a home screen.

Still, the Calculator app has a few tricks up its sleeve. For one thing, it's two
calculators in one. When you hold your phone upright (in portrait orientation), you see a basic four-function calculator. But if you turn the phone
sideways, it becomes a scientific calculator, which you can use to compute
logarithms, square roots, trigonometric calculations, and other advanced
math.

Copy and paste also work with Calculator, so you don't have to remember the number to input it into another app. Long-press the answer and choose **Copy**, and then tap wherever you need it again and choose **Paste**. (You can do the reverse, too, when you need to paste a number from another app into Calculator.)

> **TIP:** You don't have to reopen Calculator to get access to the most recent result you calculated. Instead, on the home screen, long-press the Calculator app icon: There's your number, along with an option to copy it.
>
> If you have the Calculator (▦) installed in your Control Center (page 52), you can perform the same trick there.

The Calculator app

Calculator doesn't have an undo button, but if you make a mistake, you can swipe (left or right) across the number to delete the last digit—or press ⓒ to clear the whole thing.

Calendar

Calendar fulfills the usual function of a paper calendar—but exploits its electronic nature by entering recurring events automatically, popping up reminders when the time comes, and allowing you to share certain sets of appointments with other people electronically.

The Four Views

In the Calendar app, you can switch among Day, Week, Month, and Year views, which scroll endlessly into the future or the past. As you might guess, the larger the time period you see on your screen, the less detail you can see for each day. By the time you're in Year view, all you see are individual date numerals.

Switching among views is as simple as zooming in and out, one tap at a time. Starting in Year view, tap a specific month block; now you're in Month view, where you see the month up close, with gray dots beneath the dates where you have scheduled appointments. Tap the specific date to enter Day view, where your hour-by-hour calendar awaits. To zoom out again, tap the name of the month at top left, and then the year (or use the backtrack menu described on page 12).

Week view isn't part of that natural zooming sequence. To see it, turn your phone sideways (landscape orientation); now swipe left and right to see your weeks at a glance.

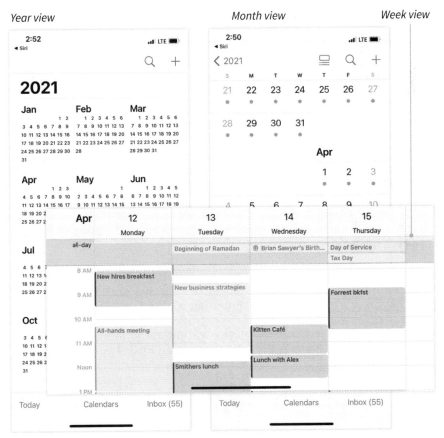

Calendar views

Subscribing to an Account

It's quite likely that, as a human being living in the 21st century, you already have an online calendar, courtesy of Google, Yahoo, Outlook, or Apple's own iCloud. The Calendar app on your phone is perfectly suited to showing those calendars to you.

To view which online calendars you have set up, or to add a new calendar, go to **Settings→Calendar→Accounts**. There you see a list of your accounts and which apps you've activated for each (**Mail**, **Contacts**, **Calendar**, **Notes**, and so on). To add an existing account's calendar appointments to your Calendar app, tap the account and turn on **Calendars**.

To add a new account, tap **Add Account**. The iPhone shows a list of services whose calendars it can import: **iCloud**, **Microsoft Exchange**, **Google**, **Yahoo**, **AOL**, or **Outlook**. (There's also an **Other** account option, for use with oddball services not listed here.)

Tap the service name, enter the email address and password you use to log in, ensure that **Calendars** is turned on, and tap **Done**.

At this point, the list of calendar categories on the main Accounts screen shows the name of the service you've just added. And all the appointments on that online calendar now appear in the Calendar app on your phone.

And you didn't have to manually enter a single appointment.

Recording an Appointment

When you do want to enter an event on the calendar manually, tap + in the upper-right corner of Year, Month, or Day view. (Or, in Day or Week view, long-press the specific hour block you want to schedule.)

You'll see the New Event screen, where you'll enter your event name and, depending on your level of fastidiousness, you can record the finer points of your new appointment.

Here are your options (shown at right on the next page):

- **Specify the calendar (category).** You can turn entire categories of appointments on or off—showing or hiding them on the calendar—with a tap. That's a big help in reducing calendar clutter. See page 321 for details.

  In any case, this menu lists all the categories you've created, so you can choose one for your new appointment.

- **Add location.** As you begin typing an address, Calendar offers to complete it for you. If you choose one of its suggestions, Calendar shows you a map of the place and the current weather.

- **All-day** means your event has no particular time of day associated with it. It's a holiday or birthday, for example. The event will appear as a banner on the calendar.

- **Starts, ends.** If this isn't an all-day event, you can use the **starts** and **ends** controls to dial up the exact time span.

- **Repeat.** If you tap here, you can turn this appointment into one that reappears on the calendar automatically—**Every Day, Every Week, Every 2 Weeks, Every Month,** or **Every Year.** Or you can tap **Custom** to input some wacky nonstandard repetition schedule.

  You can also tell Calendar when you want it to stop repeating this appointment—tap **End Repeat** and then choose **On Date** to specify a date. (You can also choose **Never,** meaning this appointment will show up on your calendar until the end of time. That would be a good option for, say, your anniversary.)

- **Travel time.** If Calendar recognizes the address you entered, and you've also entered your starting location—whose field shows up after you turn this feature on—it does something rather ingenious: It shows the amount of time it will take to get there. You see the time involved for all the available travel options. (Note to procrastinators: It's *travel time,* not *time travel.*)

  If you haven't entered an address for this appointment, or if you want to override Apple's "based on location" suggestions, you can choose one of the travel times in the menu, which range from **5 minutes** to **2 hours.**

- **Alert.** Your phone can get your attention—with a notification and a chirp— when you have an appointment coming due. Use this menu to specify how much warning you want, from **1 week before** to **At time of event.**

For all-day events like birthdays, the choices are more limited, but very useful: **On day of event (9 AM), 1 day before (9 AM), 2 days before (9 AM),** and **1 week before.**

Once you've set an alert, you get the option to set a **Second Alert**—a little insurance policy to make sure you don't miss that meeting with your boss.

> **NOTE:** In **Settings→Sounds & Haptics,** you can turn off the sound-playing aspect of these alerts. If you've flipped back the silencer switch on the side, calendar-alert sounds don't play, either.

- **Add invitees.** In the world of business meetings, this feature is bread and butter. In this box, type the email addresses of everybody who's invited to the meeting. (Or, if they're already in your Contacts, just type their names.) When you tap **Done,** everybody gets an invitation to the meeting.

  If they hit **Accept,** two things happen: The appointment gets entered onto their own calendars, and green checkmarks appear next to their names on this panel, so you can see who all will be there.

- **Add Notes, Add URL, Add Attachment.** This is your chance to store any random bits of information (or even a file from your phone) that will help you remember what this appointment is about.

When you're finished, tap **Add.**

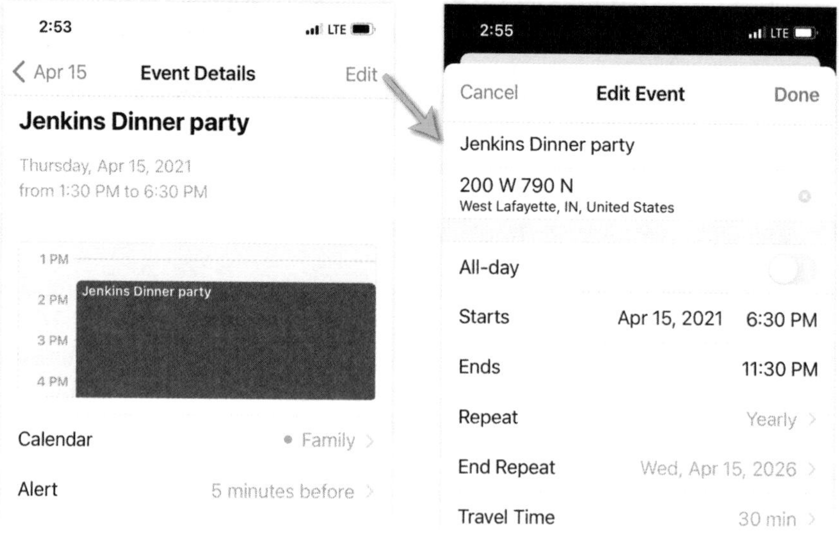

Editing appointments

Editing Appointments

To change an appointment's details, tap the event and then tap **Edit** (see "Editing appointments" on the facing page). When you're finished making your changes, tap **Done**.

To reschedule an appointment, it's much more fun to just *drag* the appointment to another date or time. You can do that in Day or Week view. Long-press the event until it darkens and a round handle appears at the bottom left. Now you can either drag that block of time to a new slot or use the tiny handle to make the block of time taller (longer) or shorter.

> **NOTE:** If you're editing a repeating event, Calendar asks if you intend to change only *this* occurrence or this and all *future* occurrences.

And to delete an event completely, tap it to enter the Edit Event screen, and then tap **Delete Event** at the bottom.

Calendar Categories

Apple chose perhaps the most confusing possible term for the color-coded categories you can use to organize your events: *calendars*. Ugh.

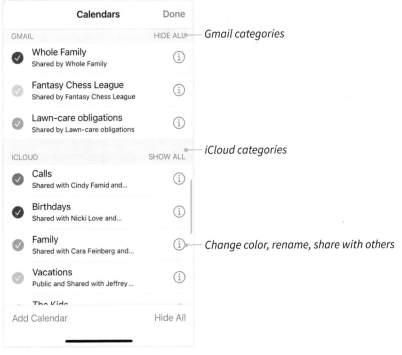

Calendar categories

You don't have to use this feature at all; it's fine to record all your appointments in a single category. On the other hand, you might want to use it to separate your calendar appointments into categories like Work, Family, Trips, and so on. The advantage is that you can hide or show all the appointments in a category with two taps: Tap **Calendars** at the bottom of your screen (in Year, Month, or Day view), and then the category's name.

To add a new calendar (category), tap **Calendars→Add Calendar**. Make up a name; tap **Done**. (If you ever want to rename a calendar—or change its color—tap ⓘ next to its name in the Calendars list.)

> **TIP:** Whatever changes you make to your categories on the phone get synced back to your Mac or PC, iPad, or other Apple gadgets. Unless, of course, you've turned off **Settings→[your name]→iCloud→Calendars**.

Publishing iCloud Calendars Online

Here's one of the payoffs of using the calendar categories feature: You can post a certain calendar online so other people can see it or subscribe to it, which adds its appointments to their own calendars. It's perfect for posting the schedules for a club, sports team, or group of colleagues.

To share a calendar, proceed like this:

1. **Tap Calendars. Tap ⓘ next to a calendar's name.**

   The Edit Calendar screen appears.

2. **In the Shared With section, tap Add Person and type the email addresses of anybody who should be able to see this category.**

> **TIP:** If you scroll down and turn on **Public Calendar**, you'll create a special web version of this calendar, viewable (but not editable) by anyone who has the web address that appears when you tap **Share Link**. You can send the URL to anybody you like, by mail or text message.

3. **For each participant, specify whether or not they're allowed to make changes.**

   Tap a name and turn on **Allow Editing** if you want to give them permission to change or delete things just as you can. Use these powers wisely.

4. **Tap Done.**

   Behind the scenes, your iPhone sends email invitations to the people you have blessed with access. If someone accepts the invitation, your shared calendar shows up on their phone; on yours, a green checkmark appears in the sharing panel next to their name.

> **NOTE:** From time to time, you might be the recipient of an invitation sent by somebody *else*. In that case, you see **Inbox (1)**, letting you know you have one invitation. Tap it, and then tap **OK** in the invitation to add that calendar category to your own calendar.

Clock

Actually, a better name for this app would be *Clocks*. It incorporates four different time tools: **World Clock**, **Alarm**, **Stopwatch**, and **Timer**. (The old iOS 13 bedtime-management feature has been moved; see page 422.)

> **NOTE:** Check it out. On the home screen, the Clock app's icon shows the current time. In fact, the second hand is actually moving. All the time. Including now. Talk about attention to detail!

World Clock

Never again will you call somebody in a different city and get the time zone wrong, waking them up in the middle of the night. This screen shows you the current time in as many different cities as you care to set up.

To add a new city, tap +. Scroll through the alphabetical list or use the search box; tap the city when you find it.

To delete a city or rearrange the ones you've got, tap **Edit**. Tap the ⊖ to remove a city (and then confirm), and drag the grip-strip handles (≡) up or down to change the order. Then tap **Done**.

Alarm

The iPhone makes a fantastic alarm clock. It always goes off on time, even if you've flipped on the silencer switch.

Truth is, you can live a long and happy life without ever messing with this tab. It's much faster to set your alarm by saying to Siri, "Wake me at 7:30 a.m." And to change the time by saying, "Change my 7:30 alarm to 8 a.m." And to cancel upcoming alarms by saying, "Cancel all my alarms."

If you truly prefer the long way, the **Alarm** tab of the Clock app shows a list of every alarm time you've ever set, even if none of them are scheduled to ring at the moment.

You can set up a new alarm by tapping + here. You can specify the time, which days of the week it repeats, what name you want to appear when it goes off, what sound it plays (you can even choose a song from your music collection), and whether you want a **Snooze** button to be available for nine extra minutes of heavenly rest.

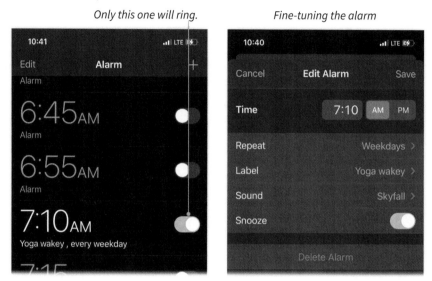

Setting an alarm

When you finally tap **Save**, your new alarm joins the existing ones in the list. Remember that it won't actually ring unless you see the green switch turned on. Apple figures that you might want to leave all your previous alarms in the list, easily reactivated, so you don't have to repeat the setup when you need them again.

Swipe left across an alarm's name to delete it, or tap **Edit** to change its time, name, sound, and so on.

> **TIP:** If your list of ancient alarms begins to get ridiculously long, you can tell Siri, "Delete all my alarms."

If you're ever wondering whether your alarm is set, look for the little 🕐 icon in the Control Center (Face ID phones) or in the status bar (home-button phones). Or just ask Siri: "Is my alarm set?"

When the alarm goes off, the only way to stop it is to tap **Stop** on the screen—a pretty small target, for sure. Apple wants to make sure you're really awake. On the other hand, if you think you can afford a little more sleeping, tap **Snooze** or press any button on the side of the phone.

Stopwatch

Tap **Start** to begin measuring how long something lasts, like a 50-yard dash, a scientific experiment, or the time it takes *that guy* to weave his new BMW into the conversation.

As the stopwatch rolls, you can tap **Lap** (as often as you like) to record how much time has elapsed since the *last* time you tapped **Lap**. Coaches might use this feature to clock each lap a runner makes around the track, for example.

> **TIP:** Swipe left to reveal a beautiful traditional *analog* stopwatch instead of the digital one.

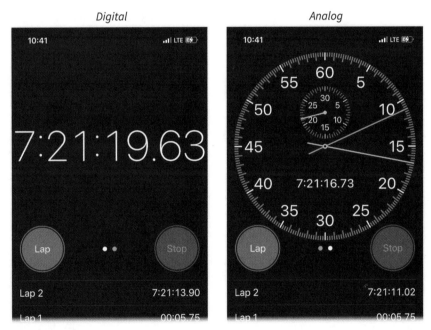

The stopwatch

You can switch into other apps, or even put the phone to sleep; the stopwatch continues to measure time. It's perfectly possible, in this way, to measure things that take hours, days, or weeks.

At any point, you can tap **Stop** to freeze the stopwatch. At that point, you can either resume by tapping **Start** or reset to zero by tapping **Reset**.

Timer

Whereas the **Stopwatch** counts up from zero, the **Timer** counts *down* to zero. It's perfect for measuring cooking times, the periods of sports games, and people's turns in chess.

Here again, using Siri is much faster: Say, "Set a timer for 20 minutes," for example. Or "Pause the timer" and "Resume the timer."

There's also a timer icon available on the Control Center (⏱). Long-press it to choose a countdown, from one minute to two hours.

But if you prefer to use the controls here, you can swipe the three "dials" to specify the amount of time you want to elapse before you're notified. When you tap **Start**, the timer counts down to zero and then plays an alarm.

> **TIP:** Tap **When Timer Ends** to choose the sound you want to play when the countdown reaches zero. If you choose **Stop Playing**, on the other hand, then at the end of the allotted time, the phone *stops* playing whatever music, podcast, or video it's been playing. In other words, you've got yourself a handy sleep timer.

Compass

Believe it or not, your iPhone has a built-in compass. That's good to remember the next time you're lost on a hike—without a map or a real compass, but somehow with a charged iPhone.

Just hold the phone parallel to the ground, and turn it until the red North arrow is aligned with the top of the screen. Now you know which way is which. The bottom of the screen shows your elevation, coordinates, and city.

If you tap the compass dial, you lock in the current heading. At this point, a bright-red band shows you how far off-course you are. (Tap again to unlock the heading.)

> **NOTE:** Technically speaking, there's a difference between *true* north (the "top" point of the Earth's rotational axis) and *magnetic* north (the actual magnetic north pole of the planet). Depending on where you are, the difference between the two can be as much as 23 degrees. In **Settings→Compass**, you can choose which version of north you want the Compass to show.

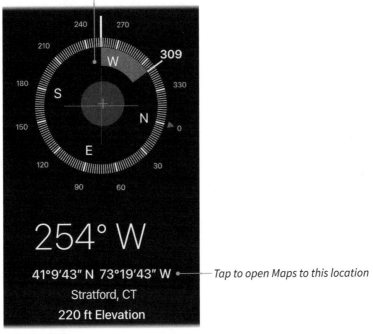

Tap to lock in a heading

Tap to open Maps to this location

The Compass app

Files

For many years, the iPhone didn't have an actual desktop, as a computer does. It didn't have files and folders that you could drag around and organize. When you got an email attachment, you couldn't save it onto your phone; there wasn't anywhere to save it.

Today, iOS comes with the Files app. Its purpose is to display (and let you organize) the files from three different sources: the phone itself, your iCloud Drive (page 408), and rival cloud-storage "disks" like Dropbox, Google Drive, OneDrive (Microsoft), Creative Cloud (Adobe), and Box.

To teach Files about your accounts on those services, tap **Browse**→⊙→**Edit**. Turn on the switches for whatever services you've already set up on your phone. That is, to add Dropbox here, you have to have the Dropbox app already installed and turned on. Once you've turned on the relevant switches, tap **Done**; your Locations list now shows those storage services.

At this point, here's what you can do in Files:

- **Search all storage sources simultaneously,** using the search box. It works on both the **Recents** tab (files you've used recently) and the **Browse** tab (everything).

- **Look at the contents of a "disk"** by tapping the service's name, like **iCloud Drive** or **Google Drive**. Of course, you're basically looking at a table of contents of what's in "the cloud" (online). But you can tap any of these icons to download the files directly to your phone.

- **Open a file** by tapping it. Your iPhone can open most common kinds of files: pictures, music, videos, PDF files, Microsoft Office documents, and so on.

If indeed you've opened a PDF document or a graphic, you can mark it up using the annotation tools described on page 7.

Files consolidates all your virtual disks.   *The commands menu*

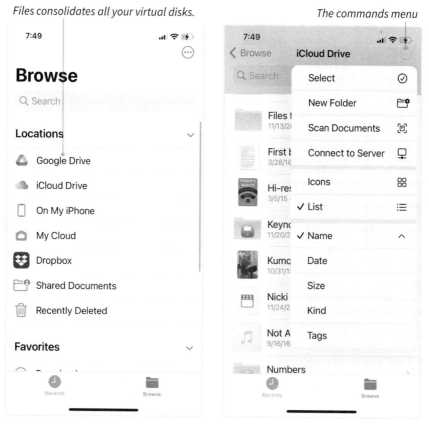

The Files app

If you tap some file that the iPhone doesn't know what to do with—a document from some Windows-based harp-notation program, for example—you won't be able to view it on the phone. It's still useful to have it in Files, though, because you can forward it to a machine that *can* open it.

- **Delete, move, or manipulate a file or folder** by long-pressing it. You get buttons like **Copy, Duplicate, Delete, Info, Move, Share, Quick Look** (view the document's contents without actually opening it), **Tags, Favorite,** and **Rename.**

  Move has a trick to it. Yes, it lets you move the selected files or folders into a different file or folder within (for example) iCloud Drive. But it also lets you move selected files and folders *between* virtual disk services. You can move a Google Drive file to your Microsoft OneDrive, grab a file on your Dropbox and copy it directly to your phone, and so on.

  The **Share** command in this menu is also powerful. It lets you send, to anybody, a download link for any file from any of your virtual-drive services.

- **Change the sorting order** by tapping ⊙ and then choosing **Name, Date, Size,** and so on.

- **Switch from icon view to list view** using the ⊙ menu.

- **Create a new folder** using ⊙→**New Folder.**

Tagging Files

If you've ever used Windows or a Mac, you may be aware that you can *tag* various files with color-coded labels that you make up, like *Urgent* or *McGillicuddy Proposal*. Later, you can round up all the files with a particular tag, from all over your computer, in whatever folders they reside, with one click.

You can tag files on the iPhone, too. The Files app doesn't care which virtual-disk service actually hosts them—Dropbox, iCloud Drive, OneDrive, or whatever. It's an equal-opportunity tagger.

To apply a tag, select a file or folder, or several (⊙→**Select**). Tap ⬆→**Add Tags**. Tap the name of an existing tag, or tap **Add New Tag** to choose a name and color for a new one.

> **TIP:** If you're tagging only a single icon, you can save a few steps by long-pressing it and tapping **Tags**.

Once you've tagged a file or folder, a colored dot appears next to its name—or several dots, if you've applied more than one tag.

Now comes the payoff: You can round up all the files, from all your services, that have a certain tag. On the Browse screen, scroll down to Tags, and tap the tag whose files you want to see.

Favorites

Tags too elaborate for you? Then mark a favorite file or folder as a *favorite* by long-pressing it and tapping **Favorite**.

From now on, it appears on the main **Browse** screen under Favorites for easy access. Here again, your favorites can come from all different syncing services.

Find My

It may appear as though the programmer who wrote this app fell asleep halfway through typing its name. Find My? Find My *what*?

As it turns out, what this app can find, on a map, is your iPhone, Mac, or iPad (a very helpful thing if you can't remember if you left them at the office)—and your friends. It can also show you your own current location on a map, in case you're wondering exactly where you are.

People

On the **People** tab, you can locate your buddies as you try to find one another in the city, or keep track of little kids or older parents. It's not exactly a spy satellite; you can't use it without the trackee's permission and awareness.

Find My expensive Apple gear

Find My friends

Devices

iPhone This iPhone		With You
Dad's iPad Stratford, CT · Now		0 mi
Dad's AirPods Pro Stratford, CT · November 26, 2020 at 6:38 PM		0 mi

People

Nicki Alexis St. Teresa of Avila Church · 1 minute ago		2,600 mi
Jeffers Alexis Houston, TX · Now		0 mi
Cindy Love Can see your location		

The Find My app

In fact, you have to let someone track *you* before you can track *them*. Say, for example, that you want to be able to see where your sister Sue is. Tap **Share My Location**; enter or choose Sue's name, and tap **Send**. Specify how long you want her to be able to find you: **One Hour**, until the **End of Day**, or **Indefinitely**.

Now Sue, wherever she is, gets a notification: "[Your name] shares location with you. Would you like to share yours?" She, too, can choose to share her location for **One Hour**, until the **End of Day**, or **Indefinitely** (or **Don't Share**).

If she fails to reciprocate, you can always ask for tracking privileges again later. On the **People** tab, tap her name and then tap **Ask To Follow Location**. She'll get another request, and another chance to let you in.

> **NOTE:** If you're using Family Sharing (page 401) and have **Location** turned on, your family members' names already show up here.

For the time period specified, either of you can tap the other's name in this app to see their position on a map. The app offers options to **Contact** this person, get **Directions** to their location, stop sharing your location, and so on.

Devices

Can't find your iPad? Happens to everyone. Tap **Devices** to see where your Apple phones, laptops, and tablets are on a map.

If a lost thing seems to be right where you are—meaning it's lost in the house somewhere—you can tap its name and then **Play Sound**. For the next two minutes, your device pings loudly so you can figure out which room it's in.

If you left your device somewhere in your travels—in a cab or a restaurant, maybe—scroll down and, under Mark As Lost, tap **Activate** and then **Continue**. Here's your chance to enter a phone number in case a good Samaritan finds the thing and wants to return it to you. Tap **Next** and add a message ("$35 reward if you return this MacBook! Thank you!"). Tap **Enable**.

Now, if someone picks up your device and tries to wake it up, your message and phone number are all they see.

> **NOTE:** At the bottom of the panel for your missing item, there's **Erase This Device**. That erases the lost device by remote control. In theory, you could use it if there are national secrets on your iPad (or just secrets that could get you into huge trouble).
>
> Realistically, of course, no spies can get into your iPad without your fingerprint, faceprint, or password. But if it helps you sleep at night, erase away.

If Find My can't find your gadget—because its battery has died—it will at least reveal its last known location. If it ever pops back online, you'll get an email. (Well, you will if you've turned on **Notify When Found**.)

In **Settings**→**[your name]**→**Find My**, by the way, three important options await:

- **Find My iPhone** leads to three further switches.

  First, there's yet another **Find My iPhone** switch. It's the master on/off switch for the whole phone-finding feature.

  Then there's **Find My network**. That's not a sentence ("Find my network!"). It's a network *called* Find My. And it consists of the hundreds of millions of people walking around at this very moment with iPhones, iPads, and MacBooks. They're unwittingly helping you find your lost phone, even if it's offline—no Wi-Fi, no cellular.

  Apple has built an encrypted, fully anonymous, Bluetooth-based beacon into every phone, tablet, and laptop. If a passing Apple gadget, carried by a total stranger, happens to pass by your lost phone—wedged into a taxi seat, on the chair at a diner, in a thief's bag—you get notified of your lost thing's location on another Apple device you own.

This switch lets you opt out of joining the Find My network. But if you're going to be that way, then you won't get the benefit of the network when *you* need it, either.

Finally, **Send Last Location** means, "OK, I acknowledge that my phone's battery might be dead. But at least report to me where it was the last time it *did* have power."

- **My Location.** It's entirely possible that you own more than one Apple device. When someone's trying to locate you with Find My, which gadget's location should it report? Which one is *you*?

- **Share My Location** is the master on/off switch for all location-sharing features, including Find My.

- **Family, Friends.** These lists show who's allowed to see your location—and whose locations you're allowed to see.

Me

The final tab in this app gives you the option to find *yourself*—or the iPhone you're holding, really. If you're in an unfamiliar place and not sure how to tell a friend where to pick you up, you can check this screen to find out your current address.

Health

By the time the Health app is finished with you, there will be no greater source of medical, fitness, and health data about you on earth. This app can collect, collate, search, and present more information—and more up-to-date information—than your doctor, your insurance company, and an armful of Fitbits combined.

This app is designed to be a central data bank for health and fitness information from five sources:

- **The iPhone itself.** As you go about your day, carrying your phone on you, the sensors inside it—GPS, accelerometer, gyroscope—quietly count your steps, measure your activity, and even tally how many flights of stairs you take.

- **Apple Watch.** If you have an Apple Watch, you get even more kinds of health data, and more reliable data, because it's strapped to your wrist all day. It can measure your sleep, calories burned, heart rate, blood oxygen level, and so on.

- **Other apps and sensors.** A huge list of fitness apps and wearable fitness bands can share their data with the Health app, so you can keep all your information in one place: MyFitnessPal, Strava, MapMyRun, WebMD, 7-Minute Workout, Garmin Connect, Lark, Lose It!, Sleepio, Weight Watchers, and so on.

To hook up these apps, start on the **Summary** tab; tap your profile picture in the upper-right corner, and then tap **Apps**. Here you see a list of all the apps you already have on the phone that can share their data with Health. Tap the name of one, and then tap the data categories you wanted to share.

- **Data you log manually.** At any time, you can record a nap, a walk, your current weight and blood pressure, and so on. On the **Browse** tap, tap the category (like **Body Measurements**), tap a statistic (like **Weight**), and then tap **Add Data.**

- **Labs, hospitals, and medical networks.** You can even set things up so the records of your immunizations, lab results, prescriptions, and medical procedures are automatically fed into the Health app.

Of course, that's possible only if you belong to a "participating" health network, and not all of them are participating. A few biggies are on board, including Quest Diagnostics, AdventistHealth, Cleveland Clinic, and Kaiser Permanente, but yours may not be. You can check the list at https:/support.apple.com/en-us/HT208647.

To get started, tap **Get Started.** You're asked for your login credentials for each institution, which you may not have on hand. But that's what customer service is for.

You won't believe the scope of data that Health can track. Yes, of course, it can record steps, miles, weight, calories, body mass index, menstrual cycles, heart rate, and blood pressure. But it also has places to record data you might not think of, like electrodermal activity, waist circumference, exposure to sound, minutes of meditation, dietary details (calcium, caffeine, dietary cholesterol…), breathing rate, symptoms (abdominal cramps, acne, chills…), vital signs (blood glucose, blood oxygen, body temperature…), blood alcohol content, sexual activity, handwashing, ultraviolet sun exposure, inhaler usage, and toothbrushing.

And then there are the *minor* statistics.

Once the Health app has become a repository for all this information, it does more than just sit there like a dumb database. It actually tries to parse the data, to help you analyze it, to show you graphs and progress over time.

As a bonus, this app is also the headquarters of the iPhone's other health-related features, including Sleep Mode (page 423), your emergency Medical ID card (page 66), and your organ-donor details.

The Summary

The very first time you open Health, it asks you to fill in your name, birthday, gender, height, and weight. You can return to this screen—your profile—whenever you want (for example, as your weight goes up) by tapping your icon in the upper right and then tapping **Health Details**. Your profile is also where you can sign up to become an organ donor and create your Medical ID (page 66).

The **Summary** tab is like a daily Facebook feed where every post is about *you*. It shows your latest heart rate, how much sleep you got last night, how many steps you've taken, your weight, graphs that compare how much activity you've gotten compared with previous days, invitations to set up more of the medical features, and so on.

Scroll down far enough, and you find Highlights: graphs and observations, computed by artificial intelligence, pertaining to your status over time, and notifications of your progress (or lack thereof). One might say, for example, "Last week, you were active more minutes a day than you are this week."

Your Favorite Stats

Your interests may not match everybody else's, though, and for that reason, you can tell Health which health statistics you want it to show you. Next to

The Health app

the Favorites heading, tap **Edit**. Tap the hollow stars, turning them blue, to indicate which statistics you want to track. Tap **Done**.

The Browse Tab

Here's a massive list of the hundreds of health-related statistics that Health can track. The phone or a Bluetooth fitness gadget can input some of these numbers automatically (**Heart Rate, Exercise Minutes, Swimming Distance…**). Many others require you to enter the information manually (**Inhaler Usage, Insulin Delivery, Sexual Activity…**).

At any point, you can tap these categories, and then subcategories, to view a graph of whatever data Health has accumulated, by day, week, month, or year.

Home

For the past 20 years or so, the electronics industry has been desperately trying to convince us that we need to automate our homes. That we need "smart" or "connected" lights, thermostats, door locks, doorbells, security cameras, power outlets, ceiling fans, and sensors. That life will be perfect once we're able to control all that stuff with our phones.

Apple's home-automation technology is called HomeKit, and its front end is the Home app. It controls all the HomeKit-compatible smart-home equipment you've bought. (For want of a better noun, Apple refers to these smart-home devices as HomeKit *accessories*.)

Setting Up an Accessory

The best part of the HomeKit experience is the setup: There practically isn't any.

Introducing a new HomeKit accessory

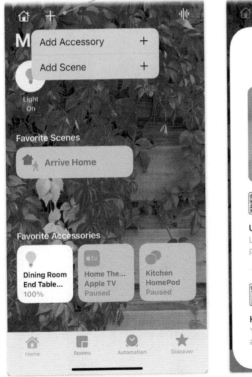
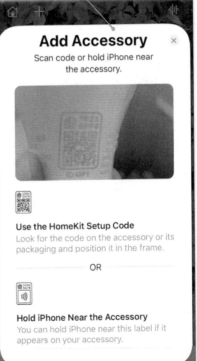

The Home app

Open the Home app on your phone, tap +→Add Accessory, aim the phone's camera at the QR bar code on your smart-home gadget or its box, give it a name, indicate what room it's in, and presto: Your "connected" smart home device really is connected. It now has an icon in the Home app that you can tap to turn on and off.

If you long-press one of these icons, you open a details screen, where you can change the accessory's name, its room assignment, and so on. Here, too, is the Include in Favorites switch, which lets you control this gadget from the Control Center (page 52) or an Apple Watch—no detour to the Home app necessary.

> **TIP:** Of course, you can also control all this stuff with Siri voice commands: "Turn on the living room lamp," "Lock the back door," and so on.

Once you've accumulated a few of these gadgets in the Home app, you can dive even deeper by creating entities like these:

- **Scenes** are groups of HomeKit gadgets that you can control simultaneously. The classic example: a scene called Goodnight, in which you say, "Hey Siri, goodnight"—and then all your doors lock and all your lights turn off.

  The less-classic example: "Hey Siri, set the mood." Your curtains close, the lights dim to half, and a Barry White playlist fills the room.

- **Rooms** are rooms. You can assign your various HomeKit accessories to these rooms, and then turn them on or off by name. "Turn off everything in the living room," for example.

- **Zones** are *groups* of rooms. If you have a zone called *Downstairs*, which contains the kitchen, living room, and dining room, you can say, "Turn off the lights downstairs."

Automation

HomeKit doodads can turn themselves on or off according to the current time or your location. Your phone, for example, always knows where it is, and so it can signal various appliances to turn off whenever you leave the house. To find these options for a certain gadget, long-press its icon to open its details screen.

You can even control your home by remote control, from across the internet— *if* you have what Apple calls a home hub. An Apple TV, a HomePod speaker, or an iPad can serve as this home hub.

At that point, the sky's the limit: You can make scenes that turn on or off or adjust themselves based on times of day, when certain people come and go, when the temperature hits certain levels, and so on.

This Home business gets complicated fast. So here's Apple's complete HomeKit user guide: support.apple.com/en-us/HT204893.

Maps

Maps is not quite as accurate as Google Maps, and there aren't quite as many features, but it certainly is pretty. Its purpose is to provide travel directions, of course, but also to give you information about any business in the civilized world: hours, phone numbers, website, reviews, photos, and so on.

In general, the Maps app likes to show the map itself at top, and an info panel below. Tap or drag that info panel to see more of it; tap again to collapse it.

Here's how to navigate (the app, that is, not the country):

- **Zoom in or out.** As always, pinch with two fingers to zoom out, and spread two fingers to zoom in. Alternately, double-tap to zoom in and then two-finger double-tap to zoom out. Or show your dexterity by double-tapping, and then with your finger still down, dragging up or down.

- **Scroll around the map** by dragging or flicking your finger.

- **Rotate the map** by twisting two fingers. (To return to north-is-up view, tap the compass at top right.)

- **Tilt into 3D view** by dragging two fingers up the screen. 3D view, which makes it look as if you're looking down at the map at an angle, doesn't become available until you've zoomed in pretty close to a city. It's really useful only once you can see the actual buildings. Tap **2D** to return to your normal flat-earth viewing.

- **Switch into satellite view**—beautiful aerial photography—by tapping ⓘ at top right and then **Satellite**. (Your other options here are **Map**—the standard, default, clean schematic of the world's roads—and **Transit**, which is great for inspecting train and subway lines.)

> **TIP:** The same little panel offers buttons for **Traffic** (which color-codes the lines of the streets, indicating how horrible the traffic is at the moment) and **Labels** (which lets you hide or show the street names in satellite view).

To pinpoint your own location on the map, tap ⊿ at top right and note the pulsing blue dot. That's you. (If you tap the ➤ until it points straight up, the map spins so the *direction you're facing* is straight up. The size of the "flashlight beam" indicates the iPhone's degree of confidence.)

> **NOTE:** Plenty of people don't appreciate having their locations tracked by apps. For that reason, the iPhone circuit that detects your geographic location has an on/off switch. If it's turned off, Maps can't find you.
>
> To flip that switch, open **Settings→Maps→Location**. (A similar option appears in Settings for every app that uses your location—weather apps, for example.) Here you can allow Maps access to your location **Never**, **Ask Next Time**, **While Using App**, or **While Using the App or Widgets**.

Searching Maps

When you tap into the search box, before you've typed a single letter, Maps offers you a vast menu of options. At the top are some of your recent searches. Next you'll find one-tap access to essential businesses you might be seeking right now, like **Fast Food**, **Gas Stations**, and **Restaurants**. When you tap one of these categories, numbered pushpin bubbles appear on the map; tap each one to pull up a list of choices in that area.

The starter list also includes Editors' Picks: nearby parks, shops, and other public attractions that Maps thinks might catch your eye. This feature draws its information from popular travel guides like Lonely Planet and All Trails.

If any of that is what you're looking for, great. The rest of the time, you'll probably want to use the search box in the more traditional way. You can type in just about anything: a street address (*200 w 79 st, new york, ny*), an intersection (*s wabash and roosevelt, chicago*), a city, a ZIP code, a point of interest (*mount rushmore*), a kind of business (*emergency rooms memphis*), and so on. You don't have to capitalize anything, and you can use standard abbreviations.

As you type, Maps displays a menu of places that match. Tap one to view its location card.

Recents, favorites

A location card, updragged to see more

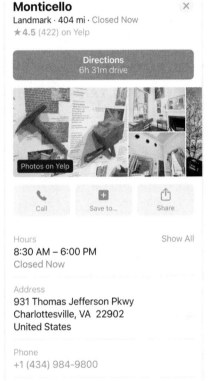

Searching Maps

The Location Card

Whenever you tap a place—or the pushpin button that represents one on the map—Maps opens the location card. It lists every conceivable action you might want to take at this point. For example:

- **Call (📞).** For a business, one tap does the trick to place a phone call.

- **Save to (⊕).** iOS 14 introduces Guides, which you can think of as saved searches: "Best Italian restaurants in Denver" or "Places to see in San Francisco." This button lets you create a new Guide—or add this place to an existing one. See the box below.

- **Share (⬆).** The Share command lets you send a little bookmark for this location either to another person (**Mail, Messages, AirDrop**) or to another app on your phone (**Notes, Reminders**).

- **Directions.** The directions panel opens, ready to go, with your current location as the starting point and the place you've selected as the destination.

- **Look Around.** If a Look Around tile appears on the location card, tap it to view a photo of the address you've been studying. See page 347.

MEET YOUR TRIP-PLANNING GUIDES

In iOS 14, Apple introduces Guides: collections of locations you assemble for your own purposes.

Imagine that you're researching a trip to San Francisco. Every time you find a promising restaurant, museum, or point of interest, you can add it to this research collection, called a Guide, for easy retrieval later when you're planning your day.

Or, if you're traveling to a conference, your Guide might include your hotel, the conference venue, the restaurants you've booked, and the local airport.

To create a new Guide, tap **Save to** on a place's location card. Tap **New Guide**; type a name for the Guide, and then tap **done**.

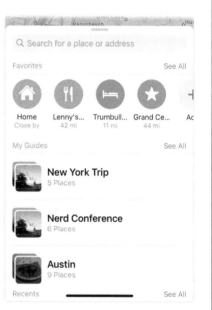

Now all you have to do is add more of the places you're researching to this Guide. That's easy enough: Find a place of interest, tap to open its location card, and tap **Save to**; tap your Guide's name to add this place to it.

You'll soon discover that your Guide acts like a folder; the places you've added appear in the menu below the search field. Yes, they're still basically bookmarks, but now they're bookmarks you can organize. ✈

- **Hours, Address, Phone, Website, Useful to Know, What People Say.** If you're looking up a business, these informational bits serve as a giant global digital Yellow Pages—with Yelp.com thrown in for customer reviews.

- **Add to Favorites.** *Favorites* are bookmarks for places you go often. The dance center where your kid takes ballet, your best friend's house, the local schools, your branch offices, whatever.

  To add a place to your list, scroll down the location card and tap **Add to Favorites**. From now on, it appears at the top of the main Maps screen before you've even tapped into the search bar. (To remove it—your favorite coffeehouse goes out of business, let's say—tap **See All** in the Favorites list, ⓘ next to the location, and **Remove Favorite**.)

> **TIP:** **Home** and **Work** are special kinds of favorites—the two addresses that presumably you use most often. Tap **Home** to see what Maps thinks your home or work address is—and if it's wrong, tap **Refine Location**. If there is no address set, tap **Add** and then **Open My Contact Card** to add it to your address card in Contacts.

- **Create New Contact** creates an address book "card" for this place in your Contacts app.

- **Report an Issue.** Behind the scenes, the Maps app and its massive database of businesses are kept updated by humans. If you discover that the information about a place is wrong—if its hours have changed, say, or if it's gone out of business—tap here to let Apple know.

The Directions Panel

Maps assumes you'll often want to use it for getting directions to a place. That's why **Directions** buttons are in your face at every turn.

When you tap **Directions**, the phone uses your current position as the starting point and fills in the ending point automatically. (To change either, tap **My Location**. Now you can edit the **From** and **To** boxes. The ↑↓ icon swaps those two locations, which is perfect for planning your return trip.)

At the top, five icons let you indicate how you'd like to travel:

- **By car.** The list offers several different routes, organized by driving time. The 🅢 marks toll roads.

  Tap a route to reveal the actual turn-by-turn directions.

> **TIP:** Farther down the screen are buttons that let you omit toll roads or highway routes.

- **On foot.** In most cases, walking takes longer than driving, although in Los Angeles, that's not a sure thing.

- **Using public transportation.** Impressively enough, Maps knows about every bus, train, and subway in every major city, complete with its schedule. That's how it's able to build complete itineraries that even include the distances you'll have to walk between, for example, bus and subway. You can tap the name of a train or subway segment to view the actual stops you'll be passing and how many minutes are between each one.

- **By bicycle.** In iOS 14, for the first time, you can choose to travel by bike. The directions include the kinds of information you'd want to know, like how steep each segment is, when you'll have to carry your bike up stairs, and what kind of road you're on (bike lane, main road, side road). Unfortunately, only a few major cities offer cycling directions.

- **By ride.** This option allows you to call a Lyft or Uber (if you've already installed and set up one of those apps). When you tap ⋔, a list of available services appears, along with the wait time and estimated cost of each ride. Tap **Next** to choose one and then **Ride** to order your car service—all without having to go into the individual app.

Once you've chosen a route and tapped **Go**, you can swipe up on the "route card" (bottom panel of stats) for more options—including a **Share ETA** button. You can now AirDrop both your ETA and your route to another person.

GUIDANCE BACK TO YOUR CAR

How's this for handy? If your phone connects to your car over Bluetooth or CarPlay, Maps can remember where you parked.

When you turn off the car, the phone checks its GPS location and lets you know, with a notification, that it's memorized the spot. (The on/off switch is in **Settings→Maps→Show Parked Location.**)

If you tap the notification, you can snap a photo of the parking spot or record notes about it. You can also see how long you've been parked, which might be useful if you have to put more quarters into a parking meter.

When you want to find your way back to the car, ask Siri, "Where's my parked car?" or even, "Dude, where's my car?" The Maps app also shows your car's location in the list of recent locations. Tap to begin your trip back to it. ✶

Navigation

Once you've tapped **Go** to begin guidance to your destination, Maps takes on a whole different look. You see a simplified map, the outlines of buildings, and next-turn banners. Siri's voice guides you, with plenty of warnings ("Go through this light; at the next one, turn right"). You don't even have to look at the screen; you can click the side button to lock the phone and rely on the voice guidance alone. (In **Settings→Maps→Navigation & Guidance**, you can adjust the volume of Siri's speaking voice as she gives you driving directions.)

> **TIP:** Once you're en route, you can say, "Hey Siri—share my ETA with Alex." Siri will text that person with your estimated time of arrival. Cool!

At bottom, you see your estimated time of arrival, the remaining time and distance, the **End** button (which, after a confirmation tap, cancels the guidance), and a **Share ETA** button. (That offers one-tap buttons for sending your arrival time to recent Messages pals.)

Maps proposes alternate routes…   *…and then guides you, turn by turn.*

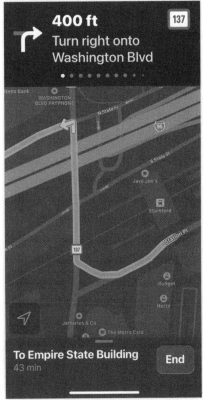

Navigating with Maps

AERIAL CITY TOURS WITH FLYOVER

Apple hasn't just been driving the planet with camera cars. It's also been flying camera-equipped helicopters.

As a result, Maps allows you to conduct your own aerial tours. The trick is to search for one of the 350 major world cities that Apple has photographed from its helicopters. Tap ⓘ to switch to satellite view, turn on **3D**, zoom in until you can see the individual buildings, and treat yourself to a self-guided virtual aerial tour of the city.

Move the map by dragging two fingers up or down. Change the camera angle by flicking two fingers up or down. Zoom in or out by spreading or pinching two fingers on the screen.

If that's too much work, you can let Maps drive the tour. This feature is called Flyover Tours, and it's pretty spectacular.

To see one, type a major city into the search field—say, San Francisco. On its location card, tap **Flyover** and then **Start City Tour**. Sit back in a slack-jawed trance and watch as Maps displays a gorgeous, preprogrammed aerial tour of that city, pointing out landmarks along the way. ✈

San Francisco
Flyover

Pause Tour

Golden Gate Bridge

You can zoom, pan, and rotate the map, which is useful when you want to peer ahead at upcoming turns or check out alternate routes. (Tap ⌁ to restore Maps' centered view.)

To see an overview map of your entire route, tap the top banner. To see a written list of directions, tap the bottom banner and tap **Details**.

Here, too, is the **Pause Spoken Audio** switch. It means "When Maps speaks an instruction, momentarily pause playback of any background recordings, like podcasts and audiobooks. Because it'd get really confusing to hear two robo-people speaking at once."

Tap the screen (or just wait) to hide these additional controls once again.

Look Around

Street View is one of the most famous features of the world's most popular maps app, Google Maps. It's a 360-degree photographic representation of any address.

Not to be outdone (usually), Apple created its own similar feature called Look Around. It's faster, smoother, and sharper than the Google feature. When you tap to move from one spot to another, it seems to morph into video mode to fly you through the streets. Unfortunately, Look Around is available in only about 15 cities.

Drag to look around in any direction. Tap to fly to a new spot.

Look Around

To try it out, search for an address in a major city like Chicago, Seattle, New York, Las Vegas, San Francisco, Boston, or Los Angeles. On its location card, tap the **Look Around** image or, once you've zoomed in close enough to see street names, tap 🔭 at top right.

Suddenly, a photo appears: what you would see if you were in that city, standing in the middle of the street. The app shows the photo and the actual map simultaneously, for your orienting pleasure; tap ↖↘ to make the photo fill the screen.

To look around you, drag your finger. Tap a more distant spot to move down the street. You can tour a whole neighborhood like this, without paying a nickel in gas (or being stuck behind the cars you'll see in traffic).

When you've finished looking around, tap Done. And send waves of encouragement to Apple to add the rest of the world's 20,000 biggest cities.

> **TIP:** In iOS 14, Maps can even show you maps of the *interiors* of 115 major airports and 375 shopping malls. (Make sure you've zoomed in close enough, and make sure you're not in satellite view.) You can see the actual store and restaurant names and whether they're before or after airport security—information that could be worth its weight in gold.

Magnifier

A few years back, Apple added a little feature that turns the iPhone into a sensational, illuminated magnifying glass. It was buried in the Accessibility settings, presumably for the benefit of the hard-of-seeing.

But Magnifier was such a hit for everyone—for reading menus in dark restaurants, for making out tiny type on pill bottles, for poring through the fine print in contracts—that in iOS 14, Magnifier is now a full-blown app. It doesn't show up, though, until you turn on Settings→Accessibility→Magnifier.

> **TIP:** And you can open it straight from the Control Center (page 52), or by triple-clicking the side button and tapping Magnifier (page 451).

Here are the basics: Open the app, point the phone at the tiny thing you want to see, use the slider to control the magnification.

> **TIP:** Swipe down on the little control panel to hide everything but the zoom slider. Or double-tap the screen to hide the controls completely.

Here's the advanced lesson. Four or five icons appear beneath the brightness slider. They're all designed to make what you're looking at even easier to see:

- **Brightness, Contrast.** Each opens a new slider that adjusts the picture.

- **Filters** apply various color modifications to the image. Each can help make your subject more legible in certain situations. (To turn off the filters, scroll all the way to the left one—called **No Filter**—and tap it.)

- **Flashlight.** Pure genius. It's always easier to see if there's good light.

- **People Detection** lets the phone detect the presence of other humans within about 15 feet. See page 426 for the full scoop.

> **TIP:** When you first fire up Magnifier, the zoom slider appears as the primary control; everything else appears in a second row. But you're allowed to customize that arrangement. Maybe you think the **Flashlight** should be a primary control instead.
>
> Tap ⚙→**Filters**. Here you can turn off the controls that don't interest you, or drag up to two other adjustments to the top row of controls—the primary ones. You may or may not include the zoom slider among them.

Magnification level   *Image adjustments*   *Shutter*

Magnifier

Unfortunately, the more you zoom in, the more your hand's jiggles get magnified, and the harder it is to keep the image steady. That's why Magnifier has a built-in freeze-framer.

At any point, tap ⊙ to snap a still image. As you take more of them, they pile up in the **View** "folder" (🗇). Tap it to scroll through the shots you've taken so far. As you go, don't miss the 🖓 button, which lets you send one of these freeze-frames to someone, save it to your Photos app, and so on.

> **TIP:** The Magnifier makes a pretty decent telescope, too. Try it on distant objects!

Measure

iOS 14 comes with only one augmented-reality app, and this is it: Measure. It's a virtual tape measure that lets you measure the sizes of things—furniture, windows, doors, paintings—just by pointing the camera at them. Its measurements aren't accurate enough for, say, nanotechnology work. But for everyday use, it's good enough.

How to Measure

When you first open Measure, a message may say, "Move iPhone to start." It wants you to move your phone through the air to help it calibrate. Once the circle with the dot appears, you're ready to measure.

WELCOME TO AUGMENTED REALITY

Everybody wants to know (and every electronics company tries to guess): What's the future of technology? What next great feature will change all our lives?

Apple believes that *augmented reality* (AR) is one answer. These are apps that treat the phone as a magical viewer for the scene around you, superimposing text or graphics over whatever the phone camera is seeing. As you move the phone around, these graphic overlays change their angle, size, and distance as though they're really in the room with you.

(In the screenshot of the IKEA Place AR app shown here, the floor, the room, and the table are all real, but the couch isn't really there.)

You may remember the Pokémon GO craze of 2016, in which millions of iPhone owners explored their neighborhoods with the intention of "capturing" Pokémon creatures that the AR app placed on sidewalks and in parks. When Snapchat adds glasses, facial hair, and antennas to your live image in video, that's augmented reality, too.

AR apps can help you with shopping (by displaying prospective furniture right in your house), games (aliens attack—in your living room!), your personal style (showing you with different hairstyles, clothing, or even tattoos), navigating cities (Google Maps superimposes huge "walk this way" arrows on the view around you), and so on. ✦

Measuring a straight distance   Measuring something rectangular

The Measure app

Here's how:

- **Rectangular object.** If you point the camera at something rectangular, like a poster, a mirror, or a window, the app recognizes it immediately. Tap inside the rectangle to see the dimensions.

- **Linear distances.** To measure a straight line, superimpose the little white dot on the starting point (the app tries to help you by detecting corners and snapping to them) and then tap ⊕ to "fasten" it there. Move the dot to the far end, and tap ⊕ again to see the measurement.

> **TIP:** After the measurement, you're welcome to adjust the positions of those starting and ending points; just drag them. The app enlarges the view to help you with precision.

- **A person's height.** The iPhone 12 Pro and Pro Max have a lidar sensor (page 22), which gives you a bonus feature: You can get an instant height measurement of a person, either standing or sitting. Just fit their whole body into the viewfinder; the height line appears automatically. You can make additional measurements on the same screen, if you like.

The measurement shows up on a tiny tab in the middle of the measurement line. If you tap it, a panel reveals the measurement in much larger type—along with the measurement in other units. If you've measured a rectangular shape, you also see the diagonal measurement here.

Feel free to tap the ◎ to take a picture of the screen, complete with measurements. You might want to send it to your builder, interior decorator, or demolition crew.

The Level

The Measure app has a secret identity as a carpenter's level.

What does that have to do with augmented reality? Nothing. But Apple figured that as long as you were measuring things, you might welcome a tool that lets you know when a picture you're hanging, or a house you're building, is perfectly horizontal or vertical.

To try it out, tap Level.

> **TIP:** If you long-press the Measure app's icon at the home screen, you get a menu that lets you jump directly into either of its modes—**Measure** or **Level**.

At this point, you can find level positioning in all three dimensions:

- **Off the vertical.** This one's useful when you're trying to hang a picture. Hold the iPhone upright—against the frame, for example. The degree markers show you, in real time, how close to perfectly upright you are. (The bottom half of the screen goes green when you've nailed it.)

- **Forward or away.** As you hold the phone upright and tip it slightly toward you or away, the readout once again lets you know when you've attained "0 degrees" level.

- **Parallel to the floor.** Set the phone flat on something, screen toward the sky. When you see the two circles perfectly superimposed (and the screen goes green), you've got level. This tool is great for trying to get a table perfectly level by stuffing something under a wobbly leg.

> **TIP:** You can designate a new "zero point" by tapping the screen. 0 degrees doesn't have to mean perfectly upright or perfectly flat.

Notes

Notes has grown, over the years, into a powerhouse of a data gobbler. It can store notes of any kind: brainstorms, recipes, phone numbers, driving directions, credit card numbers, frequent flyer numbers, and so on. But it can also store photos, documents, tables, checklists, videos, maps, web links, and just about anything else you might want to retrieve later.

Thanks to iCloud syncing, any changes you make in Notes magically appear in Notes on all your other Apple gadgets—and vice versa.

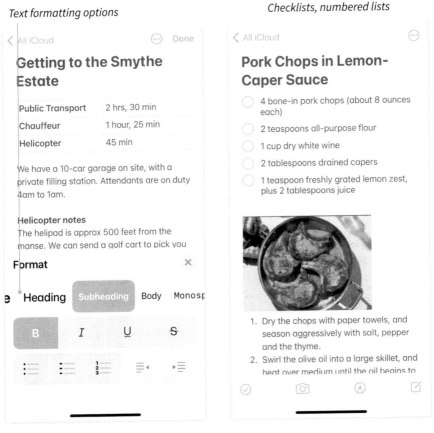

Text formatting options

Checklists, numbered lists

The Notes app

Making a Note

To add a note, tap ☑ at bottom right. You can type or paste just about any-thing into the new note.

The first line of text becomes the note's title. If you like, Notes can automatically make it stand out in big, bold type—whatever style you choose in the Settings→Notes→New Notes Start With menu.

It's fine to just type into a note. But don't miss some of your other options:

- **Paragraph styles.** Tap Aa above the keyboard to open the formatting panel. Here you can apply any of three heading styles to any selected paragraph. Or you can turn some selected paragraphs into a bulleted list (much like the one you're reading now), or a numbered list, or a list that uses dashes instead of bullets. This menu also contains buttons for **bold**, *italic*, <u>underline</u>, and ~~strikethrough~~.

- **Checklist style (⊘).** This button turns any selected paragraphs into a handy to-do list. Tap in a circle to produce a checkmark, which means "done."

- **Table.** When you tap the ⊞ button on the toolbar, you get a 2 × 2 grid: a tiny spreadsheet. (You can see an example at left in "The Notes app.")

  To add columns or rows, tap inside a cell and then tap one of the little table handles (⊶ or ⦙); you're given the option to **Paste** in text (if you have something on your Clipboard) or **Add** or **Delete** rows and columns. To jump from cell to cell as you're typing, tap next on your keyboard.

Tapping ⊶ or ⦙ also highlights a row or column—and at this point, you can drag the row or column to move it.

You can convert existing text into a table, if you like (select it and then tap the ⊞ button), or convert a table into text (tap in a cell, tap the ⊞⦿ button, and tap **Convert to Text**).

What you can't do, alas, is adjust the column widths or row heights. What do you think this is, Microsoft Excel?

- **Add a graphic.** The 📷 menu offers three options: **Choose Photo or Video** takes you into your Photos app, where you can choose an image. **Take Photo or Video** opens up your camera to do just that. And **Scan Documents** is designed to scan printed pages, automatically straightening and sharpening them. It's pretty fantastic; see page 359.

- **Add a sketch.** Tap ⊘ to use the pen, pencil, and highlighter tools to make a little drawing with your finger. Page 7 describes these tools.

> **TIP:** In iOS 14, something amazing happens when you draw a line, arc, geometric shape, heart, star, or arrow. If you leave your finger down at the end of the stroke, the iPhone turns your imperfect shape into a perfect one, with straight lines and smooth curves. Nobody will believe you drew that sketch freehand (well, freefinger).

Sharing a Note

Notes is also a *collaborative* app: You and selected friends can access the same note simultaneously, editing or consulting the same page. This is a perfect feature for brainstorming together, working on a party guest list, sharing a recipe, or whatever.

To begin, open a note and tap ⋯ in the upper-right corner. Tap **Share Note**. Before selecting your form of invitation, tap **Share Options** to specify whether these people will be able to just look at your note or make changes, too.

Now you're supposed to choose how you would like to invite all your co-conspirators to this note: by email, by text message, or by copying a link, which you can then paste into any program you like. If you chose **Mail** or **Messages**, say who you're inviting.

Once your collaborators have received and accepted your invitation, this note appears in *their* Notes apps, and they're free to edit (or just look at) the notes you've shared. At any time, if somebody misbehaves, you can rescind permission by returning to the Share Note screen and adjusting the sharing settings.

Locking a Note

Sometimes the information on one of your notes pages is best kept private. Maybe it's financial information. Maybe it's a planning list for a surprise party. Maybe it's something that's not safe for work (or family).

Fortunately, you can password-protect a note. Tap ⊙ at the top right of your note, tap **Lock**, make up a password (and/or turn on Face ID or Touch ID) and tap **Done**. From now on, this note reveals nothing until you enter the password or show your face.

> **NOTE:** A single password unlocks all your notes; there's not a different password for every note.

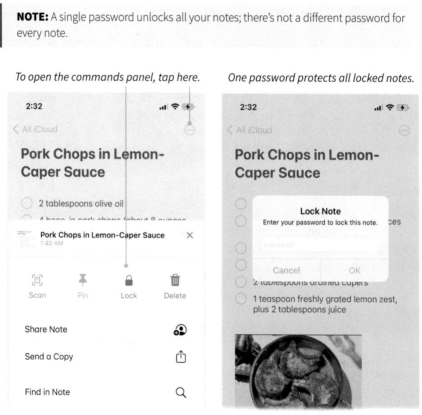

To open the commands panel, tap here.   *One password protects all locked notes.*

Locking notes

The Notes List

The ‹ at the top left of your Notes page opens a master list of your notes. It's like a table of contents, revealing the title and first line or so of every note, along with the date you last edited it. Using this list is the only way to switch from one note to another.

Notes list: Swipe to share, move, or delete   *Folders and accounts list*

Lists and folders

The notes appear chronologically by date edited, with one exception: At the top of the list is a collapsible section of *pinned* notes—the ones that you want to refer to frequently. And how did they get here? You put them here, by selecting each one, tapping ⊙, and then tapping **Pin**.

To open a note, tap it; to delete it, swipe leftward across its name in the list and then tap 🗑.

Tug down on the screen to slide the search box into view. It searches both the titles and the contents of your notes. (Notes can search only the *titles* of locked notes.)

If you tap ⊙ at this point, you get three more handy options:

- **View Attachments opens** a neat display of every photo, sketch, website, audio recording, and document that's ever been inserted into any of your notes. What's great is that you don't have to remember what you called a note; just tap a thumbnail to open it. (At that point, you can tap Show in Note to open the note that contains it.)

- **View as Gallery presents** all your notes as thumbnail images in a grid, rather than a textual list. (At that point, the command changes to say View as List.)

- **Select Notes.** In Select mode, you can select a bunch of notes all at once, just by tapping them. Why? So you can delete them en masse—or move them en masse into a different folder.

The Folders and Accounts List

If, from the Notes list, you tap ‹ Folders (or swipe right from the left edge of the screen), you'll discover a third view of your Notes world. This screen lists two groupings of your Notes:

- **Accounts.** Notes can display whatever notes you have in online accounts from Google, Yahoo, AOL, Exchange, or IMAP email accounts—and synchronize their contents in both directions. Making them appear is as easy as opening Settings→Mail→Accounts, tapping the account you want (iCloud, Gmail, AOL, or whatever), and turning on Notes.

> **TIP:** In Settings→Notes→Default Account, you can choose which Notes account should contain any new notes you create. Just don't get confused later, when notes seem to have vanished; they're just in a different account or folder.

- **Folders.** You can put your notes into folders, for your organizational pleasure. (On the Folders and Accounts screen, tap ⊟ at bottom left to create a new one.)

  You can then start moving your notes into these folders. From the Notes list, for example, swipe left on a note, tap ▬, and choose the folder name.

  Or, if the note is open, tap ⊙ and then Move Note.

And, of course, you can move a bunch of notes all at once using the **Select Notes** command described above.

> **TIP:** In **Settings→Notes**, you can turn on **"On My Phone" Account**. That's a special, hyper-private "account" that doesn't sync to anything. These notes live on your phone and *only* on your phone, and don't get synced or sent online, ever.

iPhone: The Document Scanner

Yes, your phone can make crisp, high-resolution scans of documents.

Now, you're entitled to ask: "How is that different from just taking a picture of a document?" Answer: Amazingly, Notes can recognize written words inside these photos, so you can search within them. Also, the iPhone's software helps to straighten and square up whatever you're scanning—a letter, an article, a receipt, even enormous newspaper pages that would be too big for an actual scanner. The phone can convert your scan from color to black and white, save it as a PDF file, and even let you add your signature.

In Auto mode, the phone snaps by itself.   *You can also adjust the cropping yourself.*

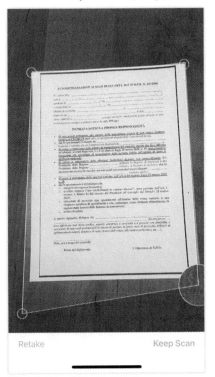

Scanning documents

The central home for the iOS scanning tools is the Notes app. (They also appear in the Files app.)

To scan a document, open a Note. Tap the 📷; from the pop-up menu, tap **Scan Documents**.

Now the phone seems to be ready to take a photo. Hold it a couple of feet above the page you want to scan—high enough, in any case, to fit the entire page image on the screen. (The document to be scanned is presumably lying on a desk, table, or even the floor. Everything works better if the surface and the page are contrasting colors.)

The camera springs into action, with the instruction that you should "Position the document in view."

The iPhone attempts to distinguish the page, which it highlights in yellow, from the background. It magically straightens the image and fixes any perspective errors that result if you're shooting the page at an angle (because you're trying to avoid shadows, for example).

If you're holding the phone steady, and the entire page image appears on the screen, the camera "scans" (snaps the photo) automatically.

(If it doesn't fire automatically—maybe it's having trouble with the lighting—you can force it to snap by tapping ◎ or pressing one of the volume keys. You can now look over the scanned image; use the round handles to crop to the page boundaries yourself. Tap **Keep Scan**.)

The scanned image collapses into the lower-left thumbnail corner. You can now scan a second page, if any. Position it, hold the phone steady, and let

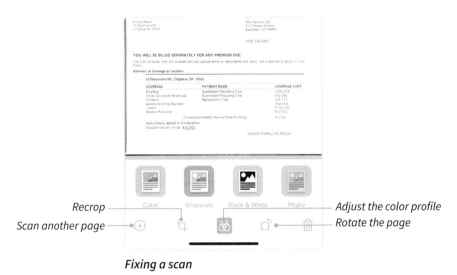

Fixing a scan

auto-snap do its thing. Repeat for as many pages as you've got. Scanning a multipage document like this is much easier than using an actual paper scanner.

When you reach the last page, tap **Save**.

You can tap the thumbnail (lower-left corner) to open one of the scanned pages. At this point, the scanned image opens so you can make adjustments: Rotate, recrop, change from color to grayscale, and so on.

Tap **Done** and then **Save**. You return to the Note that now contains the scanned images.

At this point, you can tap ⓐ to annotate the scan (to add your signature or highlighting, for example), using the tools described on page 7. Or tap ⬆→**Save to Files** or **Save to Dropbox** to export a PDF document, suitable for sending, printing, or filing on your computer.

Reminders

Reminders is a to-do app that can remind you about tasks, let you delegate tasks to somebody else, arrange your reminders in multiple lists and folders, remind you about tasks at certain times or even places, and much more.

Your to-do lists (and their checked-off status) are always up to date across all your Apple machines, thanks to iCloud syncing.

Siri, by the way, was practically *made* for Reminders. "Remind me to do my expense report when I get to work." "Remind me to watch *The West Wing*'s reunion tonight at 9." "Remind me about Alex's field-hockey game a week from Saturday."

> **NOTE:** Back in iOS 13, Apple rewrote Reminders from scratch, so it could add thrilling new features. Unfortunately, that required upgrading any existing reminders to a new data format that can't be opened on any Apple gadget *unless* it's running iOS 13 or later—or, on the Mac, macOS Catalina or later.
>
> The very first time you open Reminders in iOS 14, you'll be invited to upgrade. You can certainly decline, but you'll miss out on some nice features.

The Four Smart Lists

At the top of the screen, Reminders offers four "smart lists." Tap to open one of these intelligently collated sets of reminders that are due **Today**, that are **Scheduled** for later, that you've **Flagged** as important, or **All**.

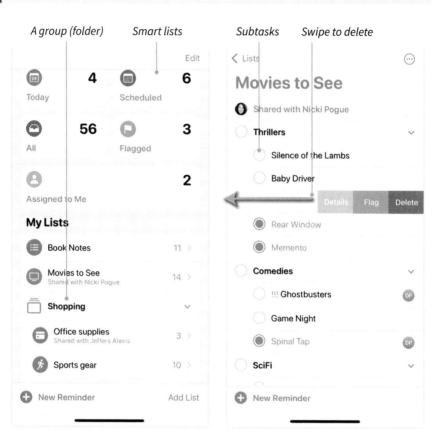

The Reminders app

You might see a bubble for **Assigned to Me** (by other people who think you could stand to shoulder some of the work).

You may even see a button called **Siri Suggestions**—a new iOS 14 feature. These are tasks that iOS itself is proposing, after analyzing your email and text messages. If someone wrote to say, "Can you send me a PDF of the manuscript?" for example, Siri might propose a Reminder called **Send me a PDF**. (If you're confused, tap it to open the original email or text message.)

My Lists

Most people somehow muddle along with only one to-do list (or none at all). The Reminders app, though, lets you create as many lists as you want, each with its own name and color code: Groceries, Kid Chores, Movies to Watch, Great Podcasts, Xmas Gift Ideas, What I Ate Today, and so on.

You can have so much fun on this list of lists:

- **Open a list** by tapping it.

- **Return to the list of lists** by tapping ‹ Lists or swiping right.

- **Make a new list** by tapping Add List. Name it, pick a color (for the list's title and also of the "checked-off" to-do circles), and, if you like, tap an icon to represent your new list.

- **Delete a list** by swiping fully left across it; then tap Delete.

- **Rearrange your lists** by long-pressing their titles and then dragging up or down.

- **Group your lists.** If you drag one list's name onto another, you make a folder in your list of lists. (Type a name for it and then tap Create.) You can drag individual lists onto this folder, thereby grouping them. For example, if you maintain to-do lists called *Things to Clean*, *Things to Fix*, *Things to Upgrade*, and *Things to Pay For*, you might want to put them all into a group folder called *House Stuff*.

> **TIP:** You can also create groups by tapping Edit→Add Group.

You can collapse a folder (to save space) by tapping it, or expand it (to see which individual lists live inside) by tapping again.

Creating Reminders

The most obvious way to enter a to-do item is to tap New Reminder at the bottom left of your screen. But there are a few other ways:

- **Tap below the last item in the list.** That's actually quicker and easier.

- **Use Siri: "Add" or "Remind me."** Quicker and easier yet: Using Siri (page 137), say, "Remind me to get my oil changed on November 7." "Add yogurt to my Groceries list." "Add 90-inch OLED TV to my Birthday Gifts list." You can use Siri for this purpose no matter what you're doing; Reminders doesn't have to be open.

- **Use Siri: "Remind me about this."** Here's a supercool option: Suppose you're using one of Apple's apps—like Calendar, Contacts, Books, Mail, Maps, Messages, Notes, Podcasts, or Safari—and you spot something that will need your attention later.

Without even leaving the app, you can say to Siri, "Remind me about this later." (Or "tomorrow." Or "at 8 p.m." Or "when I get home.")

Siri creates a new item in Reminders, named for that message or page, and attaches an icon. Later, in Reminders, you can return to the exact message, email, or web page by opening that icon.

It's like having a minion with a clipboard scurrying after you wherever you go in your work universe.

- **Select something first.** In many standard Apple apps, including Mail, Notes, Photos, and Maps, you can highlight some text, a photo, or a location and send it directly into Reminders as a new to-do. Tap your selection and then tap Share→Reminders. Specify which list you want it added to and then hit Add.

- **Subscribe to Yahoo or Microsoft Exchange tasks.** Plenty of other, non-Apple online services have their own to-do features—and Reminders can show them to you. To set that up, open Settings→Reminders→Accounts. Tap the account you want, like Exchange or Yahoo, and turn on the Reminders switch.

Anatomy of a Reminder

A reminder can be just a few words of text, like "Clean the chimney." But when you tap the ⓘ next to one, new buttons appear, which you're welcome to exploit. Among them:

- **Notes** can be anything: a phone number, random thoughts, directions.

- **URL.** A web address that might be useful.

- **Date and Time** let you specify a date and time for this reminder. Later, Reminders can sort your reminders by imminence and urgency.

- **Repeat** gives you the option to repeat your reminder Daily, Weekly, Monthly, or however often you'd like (thank you, Custom button!), as well as set an end date.

- **Location.** If you specify an address for this item, your phone can remind you when you get there (or leave there).

  For example, suppose you've dropped off your shoes for repair at the local cobbler's shop. You could set up a reminder called "Pick up shoes!" that pops up the next time you're driving by that location.

  You can type an address here, or you can use one of the choices in the menu: your current location, your home or work, or when you're getting in or out of the car (which assumes that your car connects to the phone's Bluetooth).

- **Flag.** A flag can mean whatever you want, from "super important" to "something to ask the kids to do." What it always means, though, is "Show up in the **Flagged** smart folder."

Tap for details

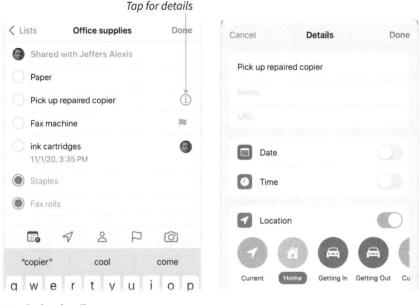

Reminder details

SHARING AND DELEGATING REMINDERS

When life starts overwhelming you, it's good to know that you can share your task list or even delegate specific items to people.

When you open one of your reminder lists, tap ⊙→**Share List**. Now you can choose how you would like to invite somebody to access this list—**Mail**, **Messages**, **Copy Link**, **AirDrop**, and so on. Address the invitation, send it, and await acceptance.

(Only people with the upgraded Reminders format—iOS 13 or later, macOS Catalina or later—are eligible.)

At this point, both of you have full access to this list. It's a fantastic way to collaborate on lists of chores, gift registries, shopping needs, and so on.

Now, though, you can take this idea a step further. You can also assign individual tasks *on* the shared list to one of your collaborators.

Tap it, tap the ⓘ next to it, and then tap **Assign Reminder→[the person's name]**. That person's initials or photo appears next to the item's name, so everybody will know who's responsible.

Delegation: It's a lifesaver. ✦

- **When Messaging** is incredibly cool. Tap Choose Person and pick somebody from your contacts list. The next time you're in Messages chatting with this person, this reminder will pop up automatically, saying, "Alex still has my bike" (or whatever you wanted to be reminded about).

Organizing Reminders

You can tweak your reminders in a thousand different ways. For example:

- **Rearrange your items.** Long-press a reminder and drag it up or down within a list.

- **Create a subtask.** If you swipe right on a task and then tap Indent, you nest the selected item *beneath* the one above it, creating a subtask. The main reminder might be "Prepare car for the drive," and the subtasks might be "Change oil," "Repair flat," and "Replace engine."

  (Later, you can swipe right on it again; this time, the button says Outdent, which returns the subtask to full task status.)

As you would hope, all this organizational effort magically reproduces itself in Reminders on all your other Apple gadgets—iPads, Watches, and Macs.

Marking To-Do Items as Done

Obviously, you check off a to-do item by tapping the empty circle next to its name. A satisfying little dot appears, and the item vanishes.

But depending on your personality type, you may prefer that checked-off items *don't* disappear, so you can look over your accomplishments with pride. In that case, tap ⊙→Show Completed.

You can also just delete items without doing them. Swipe left on the task and tap Delete.

Stocks

As you could probably guess, this app lets you track your favorite stocks (or your least favorites).

Tap the name of a stock in the list to view a graph of its price over time; it's red if the stock went down today, green if it went up. Above the graph, buttons like 3M (three months) and 2Y (two years) adjust the timescale.

To see a stock's value for a particular day, long-press the graph for a moment and then slide to the day you're interested in. If you add a second finger, the iPhone shows you the stock's performance during the time range you've

Tap to choose your stocks.

The Stocks app

highlighted in between them. Once you're done looking at a particular stock, tap ✖ to return to your watchlist.

You're not stuck with the stocks Apple starts you off with. To add a new stock to this list, type a company name or stock symbol into the search box. Tap the result you want, and then tap **Add to Watchlist**. Rearrange and delete stocks by tapping **Edit**.

Tips

This little guide to iOS doesn't quite have the charm, wit, and panache of a printed, full-color, superbly indexed book. But it does offer a set of Collections (that is, categories) of tips that cover some of iOS's marvels. Swipe leftward to see the next tip, and the next, and the next.

Translate

There's a brand-new app in iOS 14 that nobody saw coming: Translate. You can type or even *speak* in your language; the phone produces written or spoken translation. It's not quite *Star Trek*, but it's getting there.

Translate can translate to or from English (U.K. or U.S.), Arabic, Chinese, French, German, Italian, Japanese, Korean, Portuguese, Russian, or Spanish.

To get set up, hold the phone upright. Tap one of the language buttons at the top to view the list of languages. There are three features hidden in here:

- **Specify your "to" and "from" languages** by tapping their names.

- **Download a language to the phone itself,** rather than counting on having good internet service when you're traveling. Scroll down to Available Offline Languages, and tap ⊕ for the language you want.

- **Turn** Automatic Detection **on or off**—a feature for use with the spoken-translation feature of the app. When it's on, you don't have to specify which languages you're dealing with; the phone tries to figure it out.

Download a language for use offline.   *Type or dictate what you want translated.*

3:17	Languages	Done
English US		
French France		✓
German Germany		
Italian Italy		
Japanese		
Korean		
Portuguese Brazil		
Russian		
Spanish Spain		
AVAILABLE OFFLINE LANGUAGES		
Arabic		⊕
Chinese Mandarin, Simplified		⊕
English US		⊕

3:19

English US   French

English (US)

Have you seen my kids? They were over there.

French

Tu as vu mes enfants ? Ils étaient là-bas.

☆ ▯   ▶

Enter text

🎤

Translate   ★ Favorites

Translate setup

Tap **Done**.

Written Translations

Once you've specified your "to" and "from" languages, tap the **Enter text** box, type what you want to say (or tap the 🎤 button and dictate the text), and then tap **go**.

The translation appears instantly. And if you tap ▶, you get to hear the translation spoken.

> **TIP:** Apple obviously designed the Translate app for use in communication between two people. But that ▶ button turns Translate into a fantastic app for learning or practicing another language all by yourself.

You can tap ☆ to add this exchange to your **Favorites** tab, for easy reference later. You can also tap the ▭ (dictionary) and then tap any word—in either language—to see its definition, which is yet another way to improve your language skills.

Spoken Translations

This part gets nuts. Hold the phone horizontally. Tap the 🎤 and speak what you'd like translated. A split second after you stop talking, the translation appears on the right side of the screen—and the phone reads it aloud.

> **TIP:** Tap ↖↘ to make your query fill the screen. That's handy if you're trying to converse with someone across a room—or someone who's socially distant.

Enlarge to full-screen

English (US)
One of my children is carrying my passport.

French
Un de mes enfants porte mon passeport.

Real-time, spoken translation

At this point, the other-language-speaker can tap the ⦿ and respond to you—and now you hear *that* translation in your language.

Using this method, it's possible to have an entire conversation with someone who doesn't speak a word of your language.

Now, a word of expectation-setting: This is computer-based translation. It's not nearly as accurate as having a human being translate for you. There will be mistakes, sometimes hilarious ones.

But for critical exchanges when you don't speak the language—"What time do you close?" "How much is this fire extinguisher?" "Did I leave my passport here?"—it's likely to be good enough.

> **NOTE:** Translate is great to have. It is not, however, the first real-time translating app, nor the best. That would be Google Translate, whose virtues include a far longer list of available languages—and the ability to translate signs. You hold up the camera to anything written in another language; the screen somehow *erases the writing* on whatever the background is…and replaces it with ex*actly the same writing* (font, size, color) but in your language. It's some truly amazing augmented-reality mojo.

Voice Memos

This audio app is a handheld recording booth, perfect for immortalizing performances, conversations, lectures, and cute things your kids say. There's an identical app on the Mac and iPad (and even the Apple Watch), which means your recordings synchronize on all your devices.

To make a recording, tap ⦿. To end and save the recording, tap ⦿.

The new audio masterpiece appears in your list named New Recording; tap it to rename it. Now you can go to town:

- **Play it.** Tap a recording and then tap ▶ to listen to it.

- **Jump around.** Slide your finger across the scrubber "map" to skip past the boring parts. Or tap ⑮ or ⑮ to jump forward or backward by 15 seconds.

- **Delete a recording.** Tap any name in the list and then tap 🗑. Voice Memos has its own Trash (called Recently Deleted; tap the ‹ at top left to see it).

- **Send the recording** to someone else. Tap ⋯→**Share** to see your sending options.

- **Export the recording** to Files or Dropbox. That's a big deal, because these exporting methods preserve the full original quality of the recording (which, on the iPhone, is fantastic). Just tap ⋯→**Save to Dropbox** or **Save to Files**. Later, you can grab these files from your iCloud Drive or Dropbox from a computer for fancier editing.

Edit, share, export, favorite   *Map of entire recording*

The Voice Memos app

Folders

In iOS 14, for the first time, you can file your recordings into folders. From the list of recordings, tap ⟨ at top left; now you're on the folders screen. (If you're new at this, it probably lists only **All Recordings** and **Recently Deleted**.)

Tap 🗀 to create and name a new folder. Later, you can move a bunch of recordings (on the list of recordings) into it by tapping **Edit**, selecting the recordings, tapping **Move**, and then tapping the target folder.

Editing Recordings

If you tap ··· and then **Edit Recording**, you open a "map" of your audio. You can drag left or right to position the audio relative to the playhead bar (|). Now you can fix up your recording:

- **Record over a part.** When the blue playhead line is parked at the beginning or in the middle of the audio, you can tap **Replace** to record over part of your original recording. That way you can rerecord only the part where you messed up. Hit ‖ to stop.

- **Add to the end.** Position the blue playhead line at the end of the current sound waves and then tap **Resume**. Tap ‖ to stop.

- **Trim the beginning or end.** Tap ⬚. Now the map of the recording sprouts yellow crop handles. Drag them inward to isolate the part you want to keep. When you've neatly bracketed the worthy portion, tap **Trim**. If all is well, tap **Save**; to back out, tap **Cancel**.

- **Snip out the middle.** To cut a bit of audio out of the middle—that car horn that interrupted your wedding vows, say—isolate the bad spot using the trim handles as described already. This time, though, tap **Delete**. Only the outer chunks remain.

> **TIP:** If there's background noise or a room echo, tap the magic wand (✳) to try the new **Enhance Audio** command. Apple says it uses artificial intelligence to clean up static and reverberation. You may think the result is barely noticeable, or even makes things worse; in that case, you can always tap ✳ to turn it off again.

Wallet

Most people use this app, if at all, to store the electronic versions of boarding passes—and to hold credit cards for use with Apple Pay (especially its own Apple Card). Occasionally, though, you may also find an app that works with

Wallet: Major League Baseball, Starbucks, Ticketmaster, and Walgreens, for example. Some colleges offer Wallet-compatible student ID cards, too.

Usually, to make a "card" appear here, you start in the other app. For example, once you've checked into a flight using the Delta or United app, you can tap the Add to Wallet logo *there*, to place the boarding pass *here*.

And why bother? Because once you've added a ticket to Wallet, it uses the iPhone's GPS and clock to figure out when you're arriving at the airport, stadium, or theater. At that point, it displays a notification on your lock screen. That way, there's no fumbling every time somebody in security asks to see your boarding pass. It's there on your lock screen; just swipe across it to reveal the pass and its bar code.

The Wallet app

To rearrange the cards in Wallet, long-press one and then drag up or down. To purge a bunch of old tickets, scroll all the way to the bottom and then tap Edit Passes.

Tap ● in the corner of a Wallet card to read all about it. That details screen lets you delete the pass, turn Suggest On Lock Screen on or off, and open the originating app.

APP CLIPS: A GOOD SOLUTION TO A CLUMSY PROBLEM

It's great that you can download any app on demand. What's less great is dealing with the bulk and complexity of downloading an app and creating an account for it—when you're *in a hurry*.

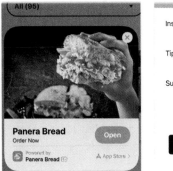

That's the situation when you're trying to rent a public bike or scooter, pay at an electronic parking meter, order fast food, rent a car, check into a hotel, and so on. In those circumstances, which often involve impatient people behind you in line, you really, really don't want to stand there waiting to download some giant app over slow internet, create a name and password, enter your credit card info, and so on.

That's why, with iOS 14, Apple invented *App Clips*. They're tiny *pieces* of the full-blown apps—under 10 megabytes, for fast downloading. They present an Apple Pay button, so there's no fussing with payment details; and they use Sign in with Apple (a one-tap, fully private, advertiser-hidden way to create a web account), so you don't waste time creating an account.

Already, software companies are making App Clips for all kinds of purposes, including letting you download demos. For example, an App Clip might give you a few minutes' worth of play for a video game whose full version is an enormous, lengthy download.

App Clips have icons—always with a dotted-line border (as though clipped,

get it?)—but they don't appear on your home screens. If you need to find one again, its icon appears in your App Library.

Once your phone is aware that an App Clip is available, a card announces its presence on the bottom half of the screen. Tap **Open** to open the clip.

But how does your phone *become* aware?

At the counter, rental station, parking meter, or front desk, you may see a QR code to scan (page 438), or one of Apple's new custom round bar codes. There may be an NFC sticker to tap with your phone. People can send you App Clips in Messages. In Safari, the App Clip card might pop up automatically for a site. In Maps, the **Order Now** button on a restaurant's location card might produce the pop-up clip.

Once you've finished your transaction, you may be prompted to **Get the full app**. Maybe you're interested, maybe you're not; but at least that's an option you can consider when you've got better internet—and fewer cranky people behind you. ✦

Watch

If you have an Apple Watch, you may come to know this app's 90 screens full of settings very well.

Just be grateful you can make all those configuration changes on the phone instead of the tiny little watch.

Weather

Hey, it's a weather app! You can see the forecast for any cities you select—and as a charming touch, Apple's artists have animated the background of the screen to depict the current sun, cloud, and precipitation conditions.

From the top of the screen, here's what you see:

- **Right now, right here.** The sky condition, huge current-temperature reading, plus today's high and low temps, of the spot where you're standing. If it's a big city, you may also see warnings here about poor air quality.

> **TIP:** You switch between Fahrenheit and Celsius on the list-of-cities screen (pinch with two fingers on any weather screen).

- **Hourly forecast.** The horizontally scrolling row of readouts shows the sky and temperature predictions for the next 24 hours.

- **Daily forecast** shows the predictions for the coming week.

- **Prose forecast.** "Clear currently. It's 55°; the high today was forecast as 61°."

- **Air quality details.** Weather shows the current air-pollution reading. You may care about that if you have asthma, children, or a distaste for breathing dirty air.

- **Meteorology fun.** At the bottom of the screen, you're treated to the sunrise/ sunset times, humidity, chance of rain, wind speed and direction, barometric pressure, visibility, UV index, "feels like" (chill or heat index), and so on.

The Multicity Weather Report

If you plan to travel, or if you're checking in on far-flung family or friends, it's handy that Weather can also show you the current conditions in other cities.

If you're already looking at the weather for some city, pinch with two fingers (or tap ☰ at lower right) to view your list of chosen cities. Tap one to see its weather; swipe left to delete it; long-press and drag to reorder them.

To enter a new city, tap Q and search for its name, ZIP code, or airport abbreviation. Tap **search** and then the city you want. Weather shows you that city's weather screen.

If you just wanted a one-time glance, tap **Cancel**. But if you tap **Add**, that city joins your list of places to track.

> **TIP:** Once you've got a couple of cities added, you can flip through their weather screens by swiping horizontally.

Swipe for the next city.

The Weather app

CHAPTER FOURTEEN

Apple's Online Stores

I n 2007, Apple Computer dropped the "Computer" from its name. The point was obvious: Its intention was to become a phone company, a tablet company, a services company, a subscription company. Today, in fact, 18% of Apple's revenue comes from selling digital goods: apps, music, movies, books, and subscription services. For the record, that's twice as much as it makes from selling *Macs*.

No wonder, then, that your phone comes with six apps dedicated to helping you find (and buy) digital multimedia goods: App Store, Books, Music, News, Podcasts, and TV.

Luckily for you, all six of these sibling apps use the same design and layout: navigation buttons across the bottom; a search button at lower right; and, in the main window, thumbnail images representing the goodies on offer.

Each store remembers what you've bought, too. You can feel free to delete stuff (books, movies, apps) from your phone, confident that you can re-download them at any time. (When you see a ☁ on a thumbnail, it means "You own this; just tap to download it again.")

The App Store

Early on, Apple decided that there would be only one source of apps for the iPhone: Apple.

That's right: The only place to find apps is the App Store.

There are some solid reasons for designing the iPhone's software ecosystem this way. For example, it's one-stop shopping. You don't waste time cruising websites trying to find just the right app, because the App Store is a single, centralized catalog of every authorized iPhone app on earth.

You also download with confidence, because Apple hand-inspects every single one of the 1.8 million apps available on its store to make sure they're clean: no malware, no viruses, nothing crashy, no porn.

There are downsides to the App Store approach, though. Software companies often resent the 30% cut that Apple takes of every app's price. And sometimes they complain of a conflict of interest, saying Apple rejects apps that threaten its own businesses.

In any case, the App Store is a vast trove of cool stuff. There are, of course, apps for all the usual suspects of the digital age: Facebook, Twitter, Yelp, Uber, Lyft, Instagram, Snapchat, TikTok, Tinder, Spotify, Netflix, *The New York Times*. There are apps that accompany all kinds of hardware products: Fitbit, Nest, Alexa, Google Home, Tesla. There are games by the thousands. And there are apps that convert the iPhone into a fax machine, radio, musical instrument, teleprompter, tip calculator, planetarium, photo retoucher, weather station, remote control, metronome, tuning fork, programming calculator, police scanner, document scanner, and on and on forever.

Really, there's so much good stuff that the hard part is knowing what to download. There are, of course, plenty of reviews and Best Of lists on the web; Apple's own editors make suggestions on the **Today** tab of the app; the **Apps** tab shows you the most popular free and paid apps at the moment. But most people, most of the time, find out about great apps by seeing other people use them.

Getting New Apps

To check out the App Store, tap its icon on your phone. As you'll discover right away, the App Store looks something like Amazon.com or any other web store—except everything listed here is an app that you can download immediately. The app-getting process goes like this:

1. **Find the app.**

   If you know what you want, tap the ⌕ icon at lower right and do a search.

   Otherwise, you can browse the categories (tap **Apps** and then scroll way down to find the category listings). You can inspect the editors' recommendations on the **Today** tab.

   In either case, whenever you tap an app's thumbnail, you open its details screen: a description, a set of screenshots, details about the software's creator, and—most important—reviews from people who've tried this app

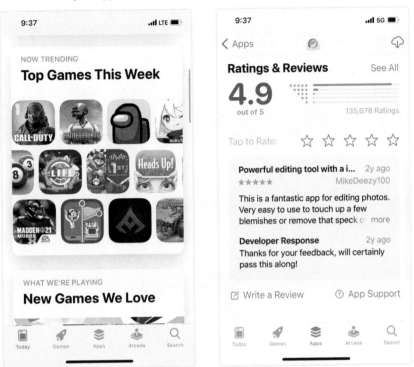

The App Store

before you. (News flash: There are 1.8 million apps on the App Store. Not all of them are amazing.)

2. Commence buying.

Every app shows one of four blue buttons. If it says **OPEN,** you already have this app on your phone! If it says ☁, you've already bought or downloaded this app. It's not on your phone at the moment, but you can tap to download.

If it shows a price or (for free apps) **GET,** tap that to begin the acquisition process.

3. Confirm your authority.

Apple has bitter corporate memories of the early days, when troublesome tweens ran up thousands of dollars' worth of charges on their parents' Apple accounts.

Therefore, the first time you try downloading something (and occasionally after that), you have to enter your Apple ID and password.

That gets old fast; fortunately, you can also set things up so your face or fingerprint is enough to prove your identity; just turn on Settings→Face ID & Passcode→iTunes & App Store. (If you have a home button, that option is in Settings→Touch ID & Passcode instead.)

From now on, you can confirm a purchase by double-pressing the phone's side button, and then supplying your Face ID (page 37) or fingerprint (page 39).

The app downloads and installs automatically, without requiring passwords, payment, or any other red tape. (Apple already has your payment information as part of your Apple account. How convenient—for both parties.)

> **TIP:** While the app is downloading, its icon on the home screen looks like a pie chart that fills in to show its progress. Tap to pause or unpause the download, or long-press the icon for a menu of choices like **Cancel Download**, **Pause Download**, and **Prioritize Download** (meaning "Download this one ahead of any other downloading apps").

When the download is complete, you'll find your new app ready to use—in the first open spot on your home screens.

Books

You've probably heard of Amazon's Kindle books, the ebooks that have taken over the world. Well, Apple wasn't about to sit on the sidelines. It has created its own version of an ebook store—and an ebook reader app. This is it.

On the Library tab (bottom of the screen), you see the thumbnails (miniature images) of all the books you've bought (if any).

Tap a book to open it on the screen and begin reading.

> **NOTE:** Thanks to the miracle of iCloud syncing, the Apple ecosystem remembers where you are in each book, no matter which Apple gadget you pick up to read it.

If no books appear here, then your first job is to shop the Book Store.

Try Before You Buy

The first few sections here (For You, New & Trending, Top Charts, and so on) are the Book Store itself. You can use the search box to seek by title or author, and you can tap one of the miniature covers to read reviews and a description. Also on the description page: the all-important Sample button, which downloads one chapter for free, so you can get a sense of whether you'd even like the book. If you love it, tap the price button to buy and download it.

The **Audiobooks** tab works exactly the same way, but for audiobooks. Listening to books read aloud by professional narrators is a great time-passer when you're working out, cleaning the kitchen, or driving.

> **TIP:** When you're listening to an audiobook, don't miss the **1x** button. It's a reading-speed control—great when the narrator is reading too slowly for your brain's superior processing power. Tap once for 2x speed (twice as fast), again for .5x (half as fast), and a third time for normal speed. The ☾ icon is a sleep timer menu. It stops playback after an interval you choose (**5 minutes**, **30 minutes**, **When current track ends**, and so on), so you can fall asleep to a book being read to you.

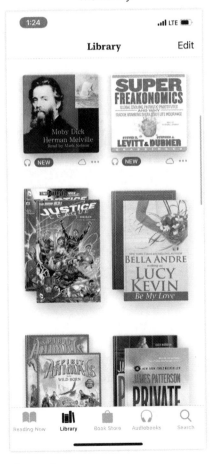

The Library   *Swipe to "turn pages."*

The Books app

Reading a Book

Once you've opened a book, you can turn pages by tapping the edge of the page—or swiping your finger across the page. (If you swipe slowly, you can actually see the "paper" curling over.) You can tap or swipe the left edge to go back a page or the right edge to go forward.

The Books app is teeming with supplementary features (to view the toolbar that contains them, tap anywhere):

- **Change the typeface or the page color** by tapping the AA button.

Tap here...   *...to change the type, size, and color.*

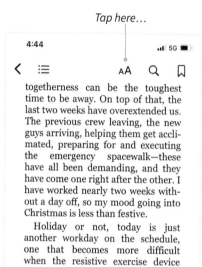

Books formatting

- **Bookmark a page** by tapping the ⬚. Later, you can view your list of bookmarks—tap ☰, then **Bookmarks**—and jump to the corresponding page by tapping its name.

- **Search the book** using the 🔍 button.

- **Highlight some text** by holding your finger down on the first word momentarily, and then dragging to the end of the text you want to highlight. If you then tap the highlighted area, you can choose a different color, add a sticky note, or remove the highlighting.

- **Listen to the book read aloud.** It's not *exactly* an audiobook, but it's pretty close, and it's free; Apple's synthesized voice is fairly realistic and even includes fake breaths.

  To set this up, open Settings→Accessibility→Spoken Content. Turn on Speak Screen.

  Now open a book in Books. Swipe down from the very top of the screen with two fingers to make the iPhone start reading the book to you.

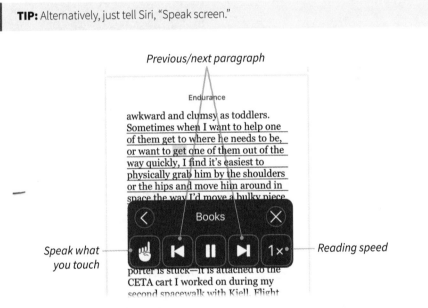

Previous/next paragraph

Speak what you touch

Reading speed

Books that read aloud

At the same time, a palette appears, offering speed controls. (After a few seconds, the palette shrinks into a transparent button at the edge of the screen to get out of your hair. Tap it to reopen the controls.)

Music

The Music app is the dashboard for Apple Music, Apple's version of Spotify. You pay $10 a month for the privilege of listening to just about any music ever recorded, streaming over the internet. Also, you can use Music to manage and play your own music files—MP3 files and whatnot that reside on your iPhone.

Apple Music

The cool thing about being a member of the Apple Music service is that you can instantly listen to any album, band, or song among the 60 million in the Apple Music catalog, or you can ask Siri to "Play the top songs of 2020" or

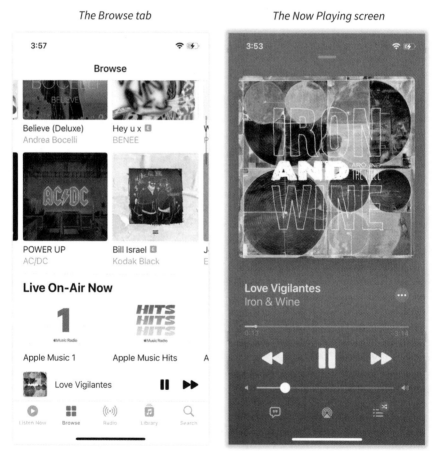

The Browse tab   The Now Playing screen

The Music app

"Play some good jogging music" or "Play Billy Joel." Or you can listen to ready-made playlists in every conceivable category.

The less cool thing is paying $10 a month forever.

In any case, the first three tabs in the sidebar are primarily intended for subscribers. Listen Now and Browse are scrolling billboards of performers, albums, and songs Apple recommends for you; Radio is a list of simulated radio stations, each of which perpetually plays a certain style or performer.

> **TIP:** Search, of course, lets you find performers, songs, albums, or lyrics, either on Apple Music or within your own library. But in iOS 14, you can *double-tap* the Q to deposit the insertion point directly into the search box. You save a couple of taps.

Your Music Collection

Even if you're not an Apple Music subscriber, the Music app can still be useful—as a jukebox to manage any music files you have on your phone.

One way to get them: Hook up an external CD drive to your Mac, open *its* Music app, and choose File→Import. Once you've got some songs on your Mac, they sync wirelessly to your phone (if Settings→Music→Sync Library is turned on). At that point, they show up in the iPhone's Music app on the Library tab.

Now you can slice and dice your collection with the categories on the Library tab: Playlists, Artists, Albums, Songs, and Recently Added.

Playlists

You can also use the Music app to organize your *playlists*—memorized lists of songs that play sequentially. You might have one for a dinner party, one for jogging, and so on. What's interesting is that you can make your playlists from either Apple Music streaming songs or music files you actually own.

Start on the Library tab; tap Playlists→New Playlist. A blank screen appears. Tap Playlist Name to rename it something descriptive; tap the 📷 icon above it to choose a photo to use as its "cover art."

Now scroll down and tap Add Music. Your job here is to find the music you want to include in the newly hatched playlist. You've got replicas of the Listen Now, Browse, and Library tabs here, as well as a search box. Drill down until you're looking at an actual album or list of songs; tap the ⊕ to add each one to this playlist. Tap Done after each deep dive into your musical world; when the playlist is complete, tap Done again.

To play the playlist, tap Library→Playlist→[the playlist's name]→Play. Or, if that's too tappy for you, just ask Siri to "Play my Romance playlist" (or whatever it's called).

Music Playback

To play a song, album, playlist, or whatever, just tap it. The music begins, and a miniature control panel appears at the bottom of the screen. It offers the name of the song, the cover art, a pause button (II), and a ▶▶ button.

> **TIP:** Tap ▶▶ to jump to the next song, or long-press to fast-forward to a later spot in *this* song.

But if you tap (or drag upward on) the mini-player, you reveal the full splendor of the Now Playing screen. It's got a lot more controls and information, including:

- **Options (●)** opens a menu of options that apply to this song, album, or playlist: **Copy, Share, Download, Delete from Library, Add to a Playlist, Rate Song,** and so on.

- **Scrubber.** The horizontal slider beneath the song's name is like a map of the song. It shows how much of the song you've heard, in minutes and seconds (at the left end) and how much time remains (at the right end). You can jump anywhere in the song by dragging the tiny round handle.

- **◀◀, ▶▶** (Previous, Next). Tap to skip to the previous or next song; hold down with your finger to rewind or fast-forward audibly.

> **TIP:** Wearing Apple earbuds? If they're the wired ones, pinch the clicker once to play or pause the music, or twice to skip to the next song.
>
> If they're AirPods, you perform these stunts by double-tapping the actual earbud in your ear (AirPods) or pinching the stem button (AirPods Pro). You can set up the right AirPod to play/pause, and the left one to skip songs, for example.
>
> To make it so, put an AirPod into your ear so it's connected to your phone wirelessly. Now, on the iPhone, open **Settings→Bluetooth→AirPods→ⓘ** to view the **Left** and **Right** control options.

- **Volume slider.** Of course, you can always press the volume buttons on the left side of the phone instead.

- **Lyrics (🗩)** displays the lyrics for the song you're listening to (usually)—and even scrolls along like a teleprompter.

- **AirPlay** (⊚) sends the music to an Apple TV, wireless receiver, or Bluetooth speaker.

- **Up Next** (▤) opens a screen that shows which songs are slated to play back next. Here's where you'll also find the buttons for Shuffle (⤭), which plays the album or playlist in random order; Repeat (⟲), which plays this playlist or album over and over; and Autoplay (∞), a new feature in iOS 14 that plays whatever you've got lined up—and then *continues* to play similar music drawn from Apple Music's catalog, either forever or until you hit **II**, whichever comes first.

Swipe down to close this screen.

> **TIP:** It's worth learning the layout of this screen. You'll encounter it again in the other Apple apps that play audio, like News and Podcasts.

Once the music is playing, you can hit the side button to turn off the screen while you work, run, cook, clean, or whatever. The music keeps right on going. (You can still adjust the volume with the buttons on the side of the phone.)

If someone calls or FaceTimes you, the music fades to a stop—and then resumes when you hang up.

> **TIP:** You've got a duplicate set of playback controls on the Control Center (page 52), so you don't have to leave the app you're using to adjust playback, and yet another set on the lock screen (page 34), so you don't even have to unlock the phone to pause or skip.

News

This app rounds up news headlines from hundreds of online newspapers, magazines, and websites. You can browse sections like **Entertainment**, **Politics**, and **Science**. You can search for a topic. Or you can tap **Following** to *tell* the app which publications and topics you care about, in effect constructing your own custom newspaper.

Over time, News is supposed to tailor the stories it's suggesting according to your own tastes, by studying which articles you actually wind up tapping to read.

There's even an **Audio** tab, which offers a selection of news stories read aloud by professional narrators. When you're driving, that's usually a better way to read the news.

NOTE: You're not getting the entire *New York Times*, the entire *Wall Street Journal*, and so on—only a subset of their stories. For $10 a month, however, Apple would be delighted to sell you a subscription to News+, which offers access to the complete current and past issues of 300 magazines and newspapers. Those, too, show up in News.

If you'd like to try it (there's a free one-month trial), tap **News+** at the bottom of the app. Once you've subscribed, that's also where you'll find the magazines and newspapers you've downloaded.

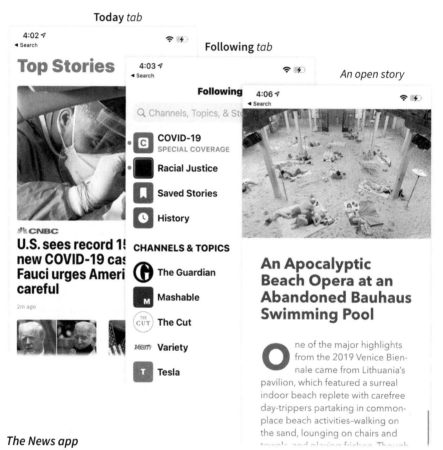

Today *tab*

Following *tab*

An open story

The News app

Podcasts

This app may look like Apple's other digital-media app stores. But this time, what you're "shopping" for is podcasts—those free "radio shows," distributed exclusively online, produced by everybody from professional radio production companies to amateurs in their living rooms.

Podcasts, as you may have heard, are spectacular audio companions when you're working out, commuting, cleaning the house, cooking, or doing anything else that requires your eyes but not the full capacity of your brain (or ears).

If you tap the "cover" of a podcast, you open up the details page for it. Here you'll find a description, reader reviews, a list of episodes—and, maybe most useful of all, a Subscribe button.

Subscribing

Most podcasts are series; there's a new episode every week or so. Subscribing tells your iPhone to download every new episode automatically.

Once you've subscribed to a podcast or two, tap Library at the bottom of the screen. Shows lists the individual podcast titles, Episodes is a list of the latest episodes, and Downloaded Episodes reveals which episodes are already on your phone. That's an important consideration if you're about to spend a long, boring time where there's no internet—like on a transatlantic flight.

> **TIP:** On the details screen for any podcast for which you have a subscription, you can tap ● and then **Settings** to reveal a vast number of controls. You can specify how many episodes you want downloaded, when you want them deleted, how often you want the phone to check for new episodes, and so on.

Handily enough, whatever podcast setup you create on your iPhone (subscribing, downloading, or listening) is magically mirrored on your Mac or iPad, so you can pick up right where you left off.

Playing Podcasts

As you might guess, tapping ▶ on any podcast or episode thumbnail image begins playback. In general, the controls—both the mini-controls at the bottom of the screen, and the full-screen version that pops up when you tap there—work exactly like the Music playback controls described on page 386.

The one big difference: Now there are ⑮ and ㉚ buttons. They let you skip backward 15 seconds or ahead 30 seconds at a time, which is ideal for repeating something you missed or zooming past ad breaks.

TIP: In **Settings→Podcasts**, you can change the skip duration for these two buttons. It can be 10, 15, 30, 45, or 60 seconds per tap.

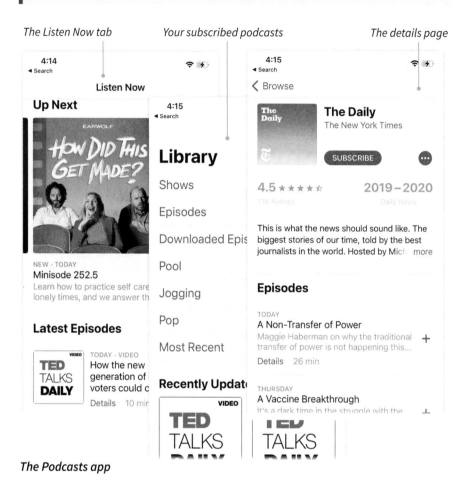

The Listen Now tab   Your subscribed podcasts   The details page

The Podcasts app

TV

This strange little app serves as housing for links to movies and TV episodes from three sources:

- **Apple's own online movie and TV-episode store.**

- **Subscription services** like Cinemax, Showtime, and Starz. Subscribing to these streaming services within the TV app has certain advantages; for example, Apple ensures that they're also available on your Mac, iPad, Apple TV, and so on. Unfortunately, the list of streaming services willing to cede 30% of their income to Apple is short. The big ones, like Netflix, Hulu, Amazon Prime Video, Disney+, and HBO Max, aren't available here.

- **Subscription services' own apps.** The TV app *can* search and display TV shows and movies from the big-name streaming services—but when you tap to play one, you get shunted off to the corresponding app on your phone (Netflix, Hulu, Prime Video, or whatever).

The Watch Now page lists a bunch of suggestions for things to watch from Apple's paid video service, any channels you subscribe to, and others.

The Library tab is where you'll find all the TV episodes and movies you've bought or rented from Apple. Not all of them are ready to play at this moment; if you see a ☁ icon, it means you have the right to watch it, but the actual video is still online. You'll have to download it, at least partially, before you can start playing it.

> **TIP:** Keep in mind that you can view whatever you're watching on your TV, which is much bigger and nicer than your iPhone's screen; see page 427.

- **Search** is handier than you might think. It can search for movies or TV shows across every channel and service at once—even ones you don't subscribe to. You can search for all George Clooney movies, or by name for a show you've heard is good on Netflix, HBO Max, Amazon Prime Video, CBS, or whatever.

Playback Fun

When you're playing videos on the iPhone, from whatever source, the controls vanish after a few seconds, so they don't ruin your view; tap to bring them back.

And feel free to switch into any other app. In iOS 14, the video continues to play, picture-in-picture style. By pinching or spreading with two fingers, you can snap to any of three window sizes for the video inset; by dragging, you can move the inset around the screen; and by swiping, you can shove it completely off the edge of the screen, where it becomes a peeky tab (❱).

> **NOTE:** If you wish that your videos would *not* keep playing when you leave the app, you can turn this behavior off in **General→Picture in Picture**.

A video can be off the screen...movable...and resizable.

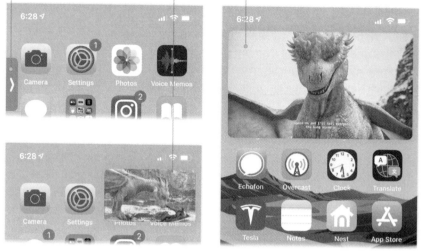

Picture-in-picture

All this convenience works with most video apps, including Apple's own TV app, Netflix, Prime Video, HBO Max, Disney Plus, and even FaceTime calls—but not, alas, YouTube or Hulu.

iCloud

Apple controls both ends of the connection between its internet services and your iPhone, iPad, or Mac. That struck somebody at the company as an amazing opportunity.

What if it could use the internet to synchronize the contents of all your gadgets? You could add somebody to your address book on the iPhone, and it would show up on the Mac and your iPad. You could take a picture with the phone, and it would appear on your iPad and your Mac. You could add a bookmark to a web page on the iPad, and it would appear in the Favorites list on your Mac and iPhone. And so on.

All that may sound like a big "duh" to you, the modern techno-citizen. But before iCloud came along, those were difficult scenarios. When you wanted to look at a photo, consult an email, or look up a phone number, you'd have to remember which device it was on.

In any case, that's only the beginning of the suite of free services that Apple calls iCloud. Yes, it does an amazing job of synchronizing your notes, reminders, appointments, email, photos, bookmarks, voice memos, passwords, and other data across all your Apple gadgets. But iCloud also includes the iCloud Drive (a backup hard drive in the sky), a free email account, Find My, the ability to share stuff you buy from Apple's online stores with family members, and much more.

The iCloud Account (Apple ID)

Like most self-respecting global tech behemoths, Apple requires that you sign up for a free account before accessing its universe of free services. Your name and password are called your Apple ID; you'll be asked to supply it many times in the coming years. Among other things, your credit card information and contact information are part of your Apple ID, which can save you all kinds of time and tedium reentering this information online.

The very first time you install iOS 14 or turn on a new iPhone, you're invited to sign up for an Apple ID. If you somehow missed that opportunity, you can make up for your tragic oversight by opening Settings→Sign in to your iPhone→Don't have an Apple ID or forgot it?→Create Apple ID. You're asked for your name and birthdate and then your email address.

If you have an existing email address, it can become your new Apple ID—but you can also tap Get a free iCloud email address, which is a useful option. Now you'll have *two* email addresses—your old one and your new "@icloud.com" one. That way you can supply one address whenever it's requested by websites; eventually, it will be harvested by the scum of the internet and become targeted with spam, ads, and hacking attempts. You can reserve the other email address exclusively for private communications, confident that it will remain pure.

> **NOTE:** The centralized hub of Apple ID info is in **Settings→[your name]**. Here, on panels called things like **Password & Security** and **Payment & Shipping**, you can examine or edit all the aspects of your Apple ID, change your password, edit your credit cards, and so on.

Synchronized Data

That business of using the internet as a coordination channel between all your Apple machines is truly a blessing—and it's part of the velvet handcuffs that keep people in the Apple ecosystem. After all: If you were to switch to some other kind of phone or computer, you'd have to deal with the hassle of recreating your calendar, address book, notes, reminders, bookmarks, voice recordings, photos, and so on.

In any case, the headquarters for data synchronization is in Settings→[your name]. If you tap iCloud, you see a list of the kinds of data the service can synchronize across your gadgets, each with its own on/off checkbox. For example:

- **Photos.** This checkbox is the on/off switch for iCloud Photos, the "all your photos are stored online" feature described on page 215.

TWO-FACTOR AUTHENTICATION AND YOU

Let's face it, people. Passwords are a bust.

Apple ID Sign In Requested
jefferdude@icloud.com
Your Apple ID is being used to sign in on the web near Norwalk, CT.

Peekskill — New Haven
New City — Bridgeport
Stamford — Norwalk
Ferry
Yonkers — Rocky Point

Don't Allow   Allow

Apple ID Verification Code
Enter this verification code on the web to sign in.

6 9 5 9 3 3

OK

Not just because they're a pain to memorize and to type—but because they're actually not very secure. How many times a year do we hear about data breaches, where a bad guy makes off with millions of customer names and passwords?

Security experts have come up with an ingenious way to keep your account protected even if somebody steals your password. This system is pretty great; it has shut down the kind of data thefts that put naked pictures of Hollywood stars online in the great iCloud hack of 2014.

Really, the only unsuccessful part of it is the name: *two-factor authentication*, a term nobody can remember and nobody understands.

In fact, it means just what it says: When you try to access your iCloud account on a new machine, you're going to need more than just your password. You also need a second factor: a six-digit code that Apple displays on all your existing Apple gadgets. If it's actually some Russian hacker pretending to be you on *their* Mac, they'll never get that code, and they'll be stopped at the gates.

Lots of Apple features don't even *work* unless you've turned on this security layer.

If you haven't already, open **Settings→ [your name]→Password & Security→ Two-Factor Authentication**. iOS asks for your phone number, which it uses to send you a verification code.

Now that "2FA" is turned on, here's what will happen the first time you try to use a new Apple device, or the first time you try to access your iCloud account in a new web browser.

On all your Apple gadgets, you see a map and the words "Your Apple ID is being used to sign in to a new device." If you don't recognize the location, somebody is trying to hack you, and you can tap **Don't Allow** to slam the door.

If you tap **Allow**, though, Apple sends a one-time, six-digit code to every one of the Apple gadgets it already knows you own. Type it into the awaiting boxes. (If you don't have more than one Apple device, Apple can send the six-digit code to your other phone number.)

You've just added a new "trusted device" to the list of machines that Apple knows are yours. They show up in a list at the left side of the **Settings→[your name]→ Password & Security** screen.

This is a one-time deal for each new web browser or Apple product you use. After that, the Apple mother ship recognizes that it's you—and thanks you for your patronage. ✦

- **Mail, Contacts, Calendars, Reminders, Notes, Messages, Voice Memos.** These apps exist in almost identical forms on Macs, iPhones, and iPads—and these checkboxes ensure that their contents will be identical, too.

- **Safari** is Apple's web browser, and this checkmark ensures that your bookmarks (Favorites) and Reading List (page 306) will appear identically across gadgets.

- **Keychain** means "memorized passwords." Not just for websites, but also for logging into other Macs or PCs on the network. Passwords are a bane of our modern existence, so it would be hard to imagine why you wouldn't want these memorized and synchronized between your Apple gadgets.

- **News, Stocks, Home, Siri.** These checkboxes sync your settings and preferences. For example, once you set up your stock portfolio in Stocks on the Mac, it will show up identically in Stocks on the iPhone. Same thing with your news-publication preferences, your home-automation setups, and what Siri has learned about the way you speak.

Continuity

If you own both an iPhone and a Mac or iPad, this one's for you. The suite of features Apple calls Continuity involve using the iPhone as an accessory to the Mac. Now you can use the Mac as a speakerphone; behind the scenes, your iPhone does the dialing. You can use your Mac to write text messages; your iPhone is actually sending them. You can write your signature on the iPhone with your finger, and drop it into a contract on the Mac. You can copy something from a web page on the Mac, and paste it into an app on the iPhone. You get the idea.

> **NOTE:** To make all this work, the Mac and iPhone must have relatively recent versions of their operating systems. They must both be signed into the same iCloud account, must have Wi-Fi and Bluetooth turned on, and must be on the same Wi-Fi network.

Calling from the Mac

Even if your iPhone is somewhere else in the house, even if it's asleep and locked, it can serve as the cellular antenna for your Mac. That's right: The Mac can be a speakerphone.

When a call comes in, your Mac rings and displays a notification. Click Accept and say hello.

And when you want to place a call, click the ☎ next to any phone number in Contacts.

Texting from the Mac

You can also send and receive text messages from your Mac, which has a glorious full-size keyboard. It's much easier to type on than glass.

To get this set up, start on the iPhone. Choose **Settings→Messages**, and turn on **Text Message Forwarding**. Turn on the name of your Mac (and any other Apple gadgets you want to get your iPhone's text messages).

From now on, your Mac shows any text messages sent to your iPhone's number. They appear as standard notification bubbles, complete with the option to reply.

You can also use Messages on the Mac to *send* standard texts to any cellphone number. (You can enter normal cellphone numbers in Messages—your correspondents don't have to be members of the Apple cult.)

You can also right-click or two-finger click any highlighted phone number in apps like Contacts or Notes on the Mac—and, from the shortcut menu, choose **Message 800-555-1212** (or whatever the number is) to commence texting.

Continuity Camera

Almost every Mac model comes with a built-in camera. Unfortunately, it's not high-quality. It doesn't do well in low light. It doesn't have a flash or a zoom. And it's not very easy to position. You can't hold it like a phone, making it higher or lower, angling it this way or that.

Fortunately, you don't care. Continuity Camera lets you use your iPhone to take a picture, which instantly appears on the Mac.

This feature works in most of Apple's built-in apps: Mail, Messages, Notes, Preview, Pages, Keynote, and TextEdit, for example. To try it out on the Mac, right-click or two-finger click a blank spot in the document. From the shortcut menu, choose **Insert from iPhone [or iPad]→Take Photo**.

Continuity Camera

> **TIP:** This shortcut menu also offers a **Scan Documents** command, which opens the iPhone's document-scanning mode (see page 359). Finally, the menu offers **Add Sketch**, which lets you draw a freehand sketch—or perhaps it should be called a *freefinger* sketch—on your phone.

Freakishly enough, the camera app automatically opens on your iPhone or iPad. Frame up the shot, snap it, and if you like the result, tap Use Photo. A second later, the photo appears on your Mac, pasted right into the document.

Continuity Markup

This one's really cool: You can sign or annotate a document on your phone, using it as though it's a wireless graphics tablet. The markings appear in real time on the Mac, thanks to Continuity.

Throughout the Mac universe—in Mail, Notes, TextEdit, Photos, Quick Look, and so on—you run across the markup tool (Ⓐ). It's a handy little drawing toolkit with markers you can use to annotate a graphic or PDF file. Which is great—if you can endure the clumsiness of drawing or writing with the mouse or trackpad.

Instead, try this. In the Finder, tap a PDF or graphic document's icon and then press the space bar to open Quick Look, the Mac's document-preview feature. Click the markup tool (Ⓐ), and then click ⬚ at the right end of the toolbar. (If you have more than one nearby iPhone or iPad, use the ⬚ as a menu to choose the one you want.)

The image or document you're editing magically shows up on your phone, with the same set of markup tools. Add your signature, cross out certain phrases, make your proofing marks, and exploit the magic of having a touch-screen. As you draw, in real time, those markings fly wirelessly through the air to the Mac version of the image. Tap **Done** to seal the deal.

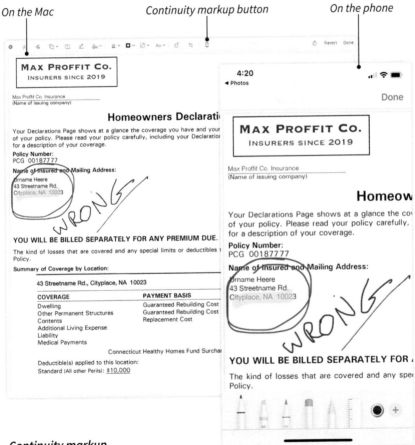

On the Mac   *Continuity markup button*   *On the phone*

Continuity markup

Continuity Clipboard

This feature lets you copy text or graphics on the Mac and paste them on the iPhone. Or go the other way. There's nothing to it, no special button or toolbar.

On the iPhone, copy something. Now, on your Mac, choose Edit→Paste. Incredibly, whatever you copied on the iOS device appears on the Mac.

> **NOTE:** You have two minutes to do the pasting. After two minutes, whatever used to be on the Mac's own invisible Clipboard reappears.

Handoff

If you need any more proof that Apple is trying valiantly to make your gadgets work smoothly together, Handoff is it. The idea is that you can begin working on something on your phone—in Notes, Mail, Calendar, Contacts, Reminders, Safari, Keynote, Numbers, or Pages—and when you sit down at your Mac, the half-finished item is right there for you to finish up.

Or vice versa.

Whatever you were editing on the Mac appears, on the iPhone, as a tiny, tap-pable banner below the app switcher (page 111). Whatever you were editing on the iPhone appears, on the Mac, near the right end of the Dock as an app icon with a tiny superimposed iPhone.

On the iPhone's app switcher screen   *On the Mac's Dock*

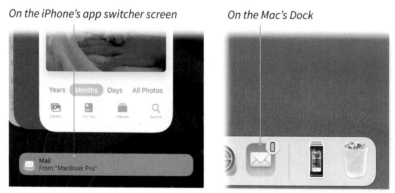

Handoff

Handoff works only if both gadgets are signed into the same iCloud account, are on the same Wi-Fi network, and have Bluetooth and Wi-Fi turned on. On the Mac, confirm that Handoff is turned on in System Preferences→General; on the iPhone, check Settings→General→AirPlay & Handoff.

Now give it a shot. Open a web page on your iPhone, or start composing an email in Mail. If you now sit down at your Mac, you'll see the Safari or Mail app icon at the right end of the Dock. Click it to open the same web page, or open Mail to view the half-finished message.

Family Sharing

In the olden days, being a member of an Apple-device family could get a little complicated. Every time your kids wanted to buy some app, movie, or book, they'd have to bug you for your credit card. Also, it made no sense that once your spouse had paid $20 for a movie, you would have to pay $20 again for the same movie, because you had a different Apple account. You're married, for heaven's sake!

Family Sharing solves all that. You identify up to six people as belonging to the same family, and just look at all you get:

- **Shared purchases.** All of you can share the same unified stash of music, books, movies, TV episodes, and apps.

- **Shared credit card.** Everyone can make charges on the same master credit card. It's not exactly Kids Gone Wild; when one of your kids tries to buy something, a notification appears on your screen. You can approve or decline the purchase on the spot.

- **Shared storage.** All of you can share an iCloud storage plan. For example, if you're paying $3 a month for 200 GB of storage, all family members can share it for use with iCloud Drive, iPhone backup, and so on.

- **Shared subscriptions.** Everybody can enjoy the family-plan versions of Apple's monthly subscription services, like Apple Music, Apple Arcade, Apple TV+, News+, and Apple One (which is a single master subscription to all the others).

- **Find one another.** You can see where your family members are on a map (with their permission, of course), using the Find My app (page 330) or by logging into iCloud.com.

- **Monitor your kids' addictions.** You can see how much time they're spending on their Apple gadgets and what they're doing.

- **Find lost gadgets.** You can also find one another's lost Apple gadgets.

- **Shared appointments, albums, and reminders.** This is a little thing, but handy: When you turn on Family Sharing, a new category called Family appears in Photos, Calendar, and Reminders. This Family category is configured so everybody in the group has access to it.

Setting Up Family Sharing

To set up Family Sharing, open Settings→[your name]→Family Sharing. Walk your way through the screens that explain how this is all going to work. Confirm that you will be the organizer, the one who has the wisdom and experience to oversee Family Sharing for everybody—and to pay for what they buy. Declaring yourself the head of the household also means specifying a credit card for all their charges. You can indicate whether or not you want *your* purchases and physical location shared with your family.

At this point, you can add up to five more people to the family. Tap **Add Member** to begin; instructions will guide you through supplying (or creating) this person's iCloud account.

> **NOTE:** Ordinarily, you must be over 13 to have an iCloud account. But the Family Sharing feature makes an exception. You, the parent, can create a special under-13 iCloud account just for this purpose by tapping **Create an Apple ID for a child who doesn't have an account**.
>
> Don't fool around with creating Child accounts just for fun, though. Once you've created such an account, you can't delete it until that "kid" is 13.

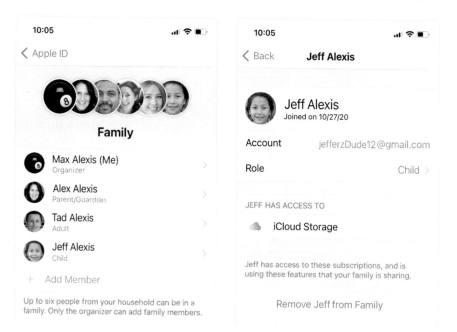

Family Sharing

Managing Family Sharing

Once you've got everybody set up, you can open Settings→[your name]→Family Sharing and look over the smiling faces (or boring first-letter initial avatars) of your immediate relatives. You can tap a person's name to make changes like these:

- **Ask To Buy.** This option, available on children's accounts, means that whenever your kid tries to buy something from an Apple store, *you* will be notified, and you'll have to give your permission. If you trust the kid, on the other hand, you can turn this option off.

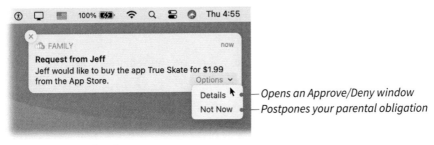

Permission notification

- **[Role].** If you tap here (adult accounts only) and then turn on Parent/Guardian, you've designated this person to be an additional approver of the kids' purchase attempts. Your spouse, for example.

- **Remove from Family.** When somebody leaves the family group—someone leaves the nest, gets disowned, gets divorced—you can remove them from the Family Sharing group. You've just opened up one of those six precious slots for somebody else.

This screen also shows the individual Family Sharing features this person can access: Location Sharing, Purchase Sharing, iCloud Storage, and so on. (For Screen Time, you're allowed to spy on family members' activities only if they're under 18.)

If you're *not* the family organizer, you can only see, not edit, most of the settings here. But these two things you can change:

- **Purchase Sharing.** You can stop sharing your purchases with the other family members. Maybe you really don't need your siblings knowing that you bought an app called *AcneKill 2000: Ten Easy Steps to Clearer Skin.*

- **Location sharing.** You can choose which family members are allowed to track you.

Finding and Hiding Purchases

In general, the whole idea of Family Sharing is to make everybody's purchases (music, movies, TV shows, apps, books) available to all family members, to save money.

The question is, where do you find these items? And the answer is "hiding within the corresponding apps":

- **TV.** In the TV app on the phone, tap Library→Family Sharing.

- **Music.** Open the iTunes Store app. Choose More→Purchased.

- **Books.** In the Books app on the phone, on the Reading Now tab, tap your round account picture (top right).

- **Apps.** In the App Store app on the phone, on the Today tab, tap your round account picture (top right). Tap Purchased.

In each case, under Family Purchases, tap the name of the family member whose stuff you want to mooch. A list of purchases appears. Tap any that you want to download to your phone.

Apple Pay

About three-quarters of all restaurants, drugstores, gas stations, big-box stores, department stores, grocery stores, and coffee shops now accept Apple Pay. That's the feature that lets you pay for things without cash or cards, just by waving your phone over a wireless reader at checkout. You don't have to open a certain app; the phone doesn't even have to be awake. It's pretty fantastic—especially when you learn that it's actually more secure than using your credit cards.

Setting Up Apple Pay

Open Settings→Wallet & Apple Pay→Add Card→Credit or Debit Card→Continue.

Now, on the Add Card screen, you can enter all the printed details of your credit card by holding it in front of the camera. Confirm that it got the digits right and tap Next. Enter the security code (CVV number) and expiration date, and hit Next.

> **NOTE:** Of course, if that's too terrifyingly high-tech for you, you can still tap **Enter Card Details Manually** and enter the digits yourself.

Proceed through whatever red-tape or legalese screens now appear. Once your bank approves the process, which may involve a phone call and/or some texts, your card appears in Settings and is ready to use.

> **TIP:** To change which card is your *default* (main) credit card, open **Settings→Wallet & Apple Pay→Default Card**.

Using Apple Pay

You're standing at checkout, ready to pay. Your phone is asleep in your hand.

Double-click the side button to make the Apple Pay screen appear. (To choose a different card at this point, tap the default card that appears.)

Authenticate by looking at the phone (Face ID). Hold the top of the phone about an inch from the reader terminal.

> **NOTE:** If you have a home button, the technique is slightly different. Rest your finger on the home button and bring the phone within an inch of the terminal.

Your iPhone buzzes, beeps, and says "Done"; you've just paid!

Apple Pay

Apple Pay on the web

When you use Apple Pay, you're making a standard credit card purchase. You get the same frequent-flyer miles, cash back, rewards points, and so on. If you have to return something to the store, it works the same way: Just when you'd swipe your card, you bring the phone near the reader until it beeps.

But Apple Pay is much more secure than a credit card. The store never touches, sees, receives, or stores your card number or your name. (Neither does Apple.) Instead, the phone provides a one-time, encoded number that means nothing to the store. It contains verification codes that only your credit card company can translate and verify.

And if someone steals your phone, no biggie. Without your face or your fingerprint, the robber can't buy anything. In the meantime, you can visit iCloud.com, click Account Settings, click your phone's name, and click Remove All to de-register your cards.

Apple Pay on the Web

Some apps and websites bear Apple Pay logos these days, too. What's nice about this payment method is that you don't enter any information. Apple already knows your name, address, phone number, and card number. It's fast and secure—you authenticate using Touch ID or Face ID. (Yes, you use your *phone* to authenticate yourself even when you're shopping on the Mac.)

Apple Cash

Ever use Venmo, Cash App, or PayPal Cash? These apps let you send cash to other people directly, for free. Pay the music teacher, tip your hairstylist, contribute your part of the restaurant bill. It's what writing checks used to be—but much faster, much easier, and much more secure.

To set this up, confirm that two-factor authentication is turned on (see page 395). Then set up Apple Pay, as described above. Whenever you pay somebody, the money will come from that credit card. (If you link to a debit card, then using Apple Cash is always free—nobody takes a cut. If you link it to a credit card, you get hit with a 3% fee each time.)

Once that's done, open Settings→Wallet & Apple Pay. Tap the picture of the Apple Cash card. The setup process will ask you to tap Add Debit Card, but if you've already got a card set up, you can skip this part.

You can tap Add Money to preload up to $3,000 onto your virtual Cash card, but you don't have to. Apple Cash will automatically pull money from your linked debit or credit card.

Now you're ready to pay someone:

- **Using Siri.** "Send 35 bucks to Alex" or "Ask Lindsay for $80." Siri shows you that she's understood your order; say "Send" to confirm.

- **In Contacts.** Tap pay on somebody's card in the Contacts app; tap their iCloud phone number or email address. You wind up in Messages, with the Apple Pay screen ready to go, as described next.

- **In Messages.** Whenever you're chatting with somebody who also has Apple Pay Cash set up, tap 🅐 and then Apple Pay (Pay).

Dial up the amount to send. *Double-click side button to confirm.*

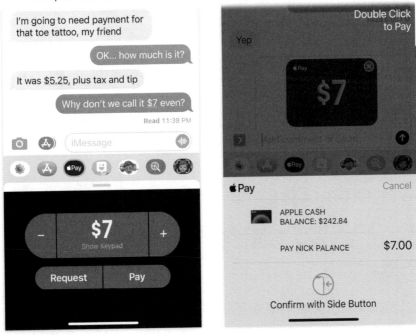

Apple Cash

> **TIP:** Cool shortcut: Tap any underlined dollar amount in a text message. The Apple Pay app opens automatically, with that amount already typed in.
>
> Similarly, if somebody texts a request for money ("Hey, do you want to split the bill? It's $30"), the QuickType bar above the keyboard offers the Pay button. Tap for insta-prep of the payment.

Enter an amount from $1 to $3,000. You can tap the + and − buttons, or tap **Show Keypad** to enter an amount.

Tap **Pay** and then ⬆. (The Send button appears in black instead of the usual blue. It's just a visual hint that you're about to send money.) Confirm with a double-click of the side button.

Your Wallet app (page 372) keeps a complete dossier of details on every transaction.

You can also use Apple Cash to request money from somebody who owes you. In a Messages chat with that person, tap an underlined dollar amount in a text message, or open the Apple Pay app. Enter the amount and then hit **Request**, and then ⬆.

If you're using Apple's virtual Cash card, then eventually incoming money piles up on it. At any point, you can dump that money into your bank account (up to $10,000 at a time, up to $20,000 a week).

In the Wallet app, tap the picture of your Cash card and then ●●●. Tap **Transfer to Bank**→**[an amount]**→**Next 1-3 Business Days**. (If you haven't yet set up your bank details, you do that now.)

If you have a Visa debit card, by the way, you can dump that money instantly instead of waiting a few days. The steps are the same, except that you tap **Instant Transfer** instead of **1-3 Business Days**—and that there might be a fee.

Everything Else

Some of iCloud's best features don't require you to learn anything or tap anywhere: They're just *there*. For example:

- **iCloud Drive.** The iCloud Drive is Apple's version of Dropbox: a simulated hard drive, apparently on your screen but actually out there on the internet, that all your Apple gadgets can access. Anything you drop into it appears instantly on all your other Apple stuff—in the iCloud Drive folders of your Macs, and in the Files app (page 327) on every iPhone or iPad. (Actually, you can even access your iCloud Drive from a Windows PC.)

All the iCloud Drive's contents are also accessible on a website—iCloud.com, of course. That's handy, because it means you can grab your files even if, in a pinch, you're forced to use somebody else's computer (::shudder::).

Thanks to iCloud Drive, you never again have to email files to yourself, copy them to a flash drive to take home, or worry that you'll lose important files forever when a lava sinkhole opens up and swallows your iPhone. Those files are still safe "in the cloud" (that is, in an internet-connected data center somewhere).

iCloud Drive also makes a fantastic backup disk—or it would, if it were bigger. Apple gives you only 5 GB of storage space for free. If you like, you can pay a monthly fee for more space: 50 gigabytes ($1 a month), 200 gigabytes ($3), or 2 terabytes ($10), which is 2,000 gigs.

- **Managing Purchases.** Apple's online stores generate millions of dollars in sales of music, movies, TV shows, books, and apps. But here's the generous part: You're allowed to download anything you've purchased to any Mac, iPhone, or iPad that's signed in with your iCloud account. Or re-download them, even if it's been years since the purchase.

  Of course, iCloud also remembers where you stopped reading each book or watching each video.

- **Automatic backup.** You can set up your iPhone to back itself up automatically to your iCloud account, wirelessly and invisibly. See page 64.

PART FIVE

Beyond the Basics

CHAPTER SIXTEEN

Seven iPhone Annoyances

T he iPhone may be the world's least annoying smartphone, but let's face it: It's still a machine, and a complicated one at that. Eventually, you'll come across things that don't work right, that frustrate you, or whose design baffles you.

May the following guidance spare you a weekend or two of Googling and swearing.

Running Out of Space

The storage capacity of your iPhone is fixed and permanently installed. You're expected, upon buying it, to somehow *anticipate* how many photos, videos, music, apps, and documents you'll ever accumulate.

Forget it. Sooner or later, almost every phone fills up, and that's a huge problem. Suddenly you can't take a photo. You can't download email. Your phone is full, and it's up to you to figure out how to free up space.

The iPhone Storage Screen

How full is your phone at this moment? Fortunately, iOS offers a handy gauge—and a bunch of ways to free up some space.

Open **Settings**→**General**→**iPhone Storage**. Here you get a handsome, color-coded bar chart that shows how full your drive is at this moment, and what's eating it: Apps, Documents, Messages (all the pictures, videos, and other stuff people have sent you), and so on.

Even sweeter, the rest of this screen offers ways to reclaim drive space. Here are the ways it might propose to tackle your capacity problem:

< General iPhone Storage 🔍

iPhone   154.2 GB of 512 GB Used

● Apps ● Messages ● Photos ● System ● Other

RECOMMENDATIONS

🅰 Offload Unused Apps   Enable

Save 9.71 GB - Automatically offload
unused apps when you're low on storage.
Your documents & data will be saved.

💬 Review Large Attachments   >

Save up to 50.04 GB - See photos,
videos, and attachments taking up
storage in Messages and consider
deleting them.

○ Messages   26.93 GB >
Last Used: Today

Overcast   17.38 GB >

< Back Attachments Edit

	Apr 23, 2020	1.36 GB
	2019	691.6 MB **Delete**
	Dec 14, 2019	691.6 MB
	Aug 18, 2015	405.1 MB
	Sep 16, 2017	285.3 MB
	Sep 16, 2017	285.3 MB
	Dec 14, 2018	239.2 MB
	Dec 14, 2018	239.2 MB
	Aug 3, 2019	223.1 MB

Freeing up space

- **Offload Unused Apps.** If you tap **Enable**, then the next time your phone
  starts getting really full, it reserves the right to delete the apps you hav-
  en't used in a while. Their icons remain on your home screen, marked by a
  ☁ badge, and you don't lose any data or settings. If you ever need that app
  again, just tap to download and open it.

  > **TIP:** You can also offload individual apps right now, on demand, on this same
  > Settings→General→iPhone Storage screen. Tap an app's name and then tap
  > **Offload App**.

  Weirdly, you can't turn this feature *off* in the same place you turned it on.
  You have to turn it off again in **Settings→App Store**.

- **Review Large Attachments** opens a list of all the email and Messages
  attachments you've received, sorted with the biggest at the top. You can
  tap one to open it and see if it's important. If not, or if you have another
  copy somewhere else, swipe left across the file to delete it. You may be
  shocked to find out how much of your phone's storage is being eaten up
  by these files.

- **Review your apps.** The bottom of the iPhone Storage screen is a list of all your apps, largest first. Tap one to open a details screen, where you can see how much room the app needs and how much its data is eating up. Some, like Messages, actually break down the storage by Conversations, Photos, Videos, GIFs and Stickers, and so on. You can tap one of *those* to review its contents—and, by swiping left, delete the ones you no longer need.

iCloud Storage Services

You should keep in mind, too, that iCloud remains at your service for hosting some of the fattest files on your phone—in exchange for a small rental fee. For example:

- **Offload your messages.** The Messages in iCloud feature stores all your text messages (and attendant attachments) on Apple's computers instead of yours. Suddenly you free up many gigs of space not just on your phone, but on your Macs and iPads, too. See page 268.

- **Offload full-resolution photos.** The iPhone's Optimize iPhone Storage feature houses all your full-resolution original photos and videos on Apple's iCloud website. Yet it leaves screen-sized versions on your phone, so you can still admire them and organize them. See page 215 for details.

- **Offload music.** In Settings→Music, you can turn on Optimize Storage. It's exactly the same deal as Optimize iPhone Storage—but this time, you're giving the phone permission to remove *music* files you haven't played in a while. You can even specify how aggressive it should be, by setting a minimum-gigabytes storage for what it *keeps* on the phone. Here again, you can still play the offloaded songs; just tap their names.

The Screen Is Always Too Dim

Apple thought this was a feature you'd *like*: The iPhone has an ambient-light sensor. Its screen brightens up in bright places and dims when it's dark. It's all supposed to improve visibility without your having to fiddle with anything.

In practice, though, the auto-dimming stuff may not produce the results you'd choose for yourself. To turn it off, open Settings→Display & Brightness and turn off Automatic. You can keep the brightness the way you like it and adjust it yourself, as needed, in the Control Center.

Lockups and Freezes

The iPhone exists to run apps, which are written by humans, who are fallible. Sometimes, they freeze, behave erratically, or lock up. (The apps, not the people.) (Actually, people, too.)

If an app is locked up, doesn't respond to taps, or otherwise acts strangely, here's your troubleshooting toolkit. Work through the following list in this sequence; the easiest and quickest fixes are listed first:

- **Quit the app.** If an app is acting glitchy, exit it and reopen it.

  To do that, bring up the app switcher (page 111). Locate the "card" for the misbehaving app, and give it a finger-flick up off the screen. You've now exited the app.

  Reopen it and see if your luck has changed.

Swipe to exit a glitchy app.

Force-quit an app

- **Restart the phone.** Problem isn't gone? Then turn the iPhone off and on again.

  On a Face ID phone, you do that by simultaneously pressing the side button and either one of the volume buttons. (On a home-button phone, hold in the side button.)

Either way, eventually, the screen says **slide to power off**. Confirm by swiping across that message; the iPhone shuts down.

Turn it back on by holding in the side button.

- **Force-restart the phone.** If you can't even shut the phone off—sometimes the operating system is so confused, it doesn't even respond to the shutoff procedure—you'll have to force a restart.

  iPhone 8 and later: Press the volume-up key, and then the volume-down key. Now hold in the side button until the Apple logo appears.

  iPhone 7: Press the side button and the volume-down key simultaneously. Keep them pressed until the Apple logo appears.

  Earlier models: Hold down the home button and the side button until the Apple logo appears.

  In every case, keep holding until you see the Apple logo. That's how you know the phone is restarting.

- **Reset the settings.** This procedure purges all your settings, which is yet another troubleshooting technique that often solves mysterious ailments. You don't lose any of your apps, data, photos, and so on, but you will have to set up your app settings again when it's over. Tap Settings→ General→Reset→Reset All Settings.

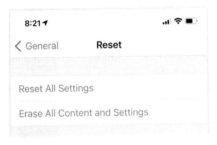

Reset buttons

- **Restore the phone.** This is the nuclear option: erasing the phone completely, back to a factory-fresh condition. When it's all over, the phone is as empty and innocent as the day you unboxed it.

> **TIP:** If the phone is healthy enough to turn on, back it up first (page 64). That way you'll be able to restore everything back to this phone (if you're troubleshooting) or your new phone (if you're upgrading).

Tap your way to Settings→General→Reset→Erase All Content and Settings. Most of the time, people wipe the phone clean because they're going to sell it or give it to a new owner, not as a troubleshooting tactic.

But once the phone is completely wiped, you can restore all your stuff to it from the backup, as described on page 66. With luck, whatever glitches were plaguing you are now gone.

Phone Won't Start Up

If you can't even get the phone to light up, its battery is probably dead. You might expect that it will spring right to life when you connect it to power (page 25)—but, in fact, a fully dead iPhone stays dark for several minutes after you begin recharging it. Once it's got enough juice to stagger to its feet, the screen lights up.

Now, if you've charged the phone for 10 or 15 minutes and it *still* won't come on, then something more serious is wrong with it. Time to make an appointment at an Apple Store.

Stuck on the Apple Logo

If the phone does come to life, but the starting-up process never makes it past the Apple logo (and there's no progress bar), the only solution is to put the phone into *recovery mode*. That's where it clings to your Mac or PC for dear life while you give it a heart transplant.

The phone may also enter recovery mode on its own. You'll know it when you see the black screen with a white Lightning cable.

Set things up like this:

- **If you have a Mac:** Open a Finder window (macOS Catalina or later) or the iTunes app (earlier Mac OS versions).

- **If you have a Windows PC:** Open the iTunes app.

Restore screen

Connect the iPhone with its white Lightning cable.

Now do a force-restart, as described on page 417. Hold the buttons in until you see the recovery-mode screen.

Click your iPhone's icon on the computer screen (in the Finder or in iTunes). The computer tells you: "There is a problem with the iPhone that requires it to be updated or restored" and offers you two buttons: **Update** or **Restore**.

Try **Update** first. It installs a fresh copy of iOS 14, but it doesn't disturb any of your apps or data on the phone. (After 15 minutes of trying to download iOS, the iPhone exits recovery mode. Just wait until the download is finished, and then start this process again.)

If updating doesn't work, you'll have to do a **Restore**, which involves completely erasing the iPhone. If you have a backup—let's hope so!—you can then restore everything on it from that.

Super-Short Battery Life

"Battery life" could mean "how many hours you get out of each charge," or it could mean "how many years your battery lasts." And they're related: The older your battery is, the less charge it holds.

See page 26 for details on making your charge last longer.

If you decide it's time for a battery swap, you can take the phone to an Apple Store (or mail it in to Apple) along with a payment of $50 or $70, depending on the model. (Of course, if the phone is under warranty or if you have AppleCare+, there's no charge.) The mail-in swap gets the phone back to you in under five business days and costs $7 for shipping (Apple mails you a box to collect the phone).

Apple says it will properly dispose of your old battery.

Cracked Screen

It's almost uncanny how many iPhones get cracked screens. It's as though gravity has a vendetta.

A phone case helps. Having one of the iPhone 12 models helps (because these phones have tougher glass). But nothing can render a screen crackproof.

If you take your phone to an Apple Store, they'll fix it on the spot for $280 to $330, depending on the model. They're not just replacing a piece of glass; it's the actual illuminating screen *and* the touch-sensitive underlayer, too.

You can also mail your phone in; Apple will send you a shipping box and ship it back in three to five business days. (If you have AppleCare+, then a replacement screen is $30, regardless of the model.)

Apple isn't the only outfit that can replace a screen, though. Lots of independent shops can do the job, usually much less expensively. (Inquire about whether you'll get an actual Apple screen as the replacement. Some companies use lower-quality, non-Apple parts.)

Finally, you can replace the screen yourself. A screen-replacement kit costs about $60 online and includes the special tools you need to open the iPhone. If you're patient and technically savvy, it's not especially hard.

Warranty and Repairs

When you buy an iPhone, you get 30 days of free tech support and a one-year warranty.

Apple cheerfully invites you, though, to add AppleCare+ insurance. That entitles you to a year of tech support, free battery repair once the capacity drops to 80%, and up to two damage repairs a year—a flat $30 for a cracked screen (the repair person comes to *you*) or $100 for any other damage.

AppleCare+ costs $80 to $200 for two years of coverage, depending on your iPhone model. For another $70, you can add theft and loss protection: They give you a new phone if your old one gets stolen or lost. There is, of course, a deductible: $150 to $270, depending on the model.

Where Else to Get Help

There is, believe it or not, an official, online, searchable Apple user's guide for iOS 14. It's at help.apple.com/iphone. (There's also a PDF version that you can open in your Books app.)

Apple also has an iPhone help website that's teeming with tips, tricks, tutorials, and troubleshooting tactics. Its crown jewel is its vast network of discussion boards, where fellow iPhoners discuss and solve one another's problems: apple.com/support/iphone.

If you need repair, here's where to find all the prices, plans, and programs Apple offers for fixing your phone and even sending you a temporary replacement while yours is in the shop: apple.com/support/iphone/service/faq.

CHAPTER SEVENTEEN

Fifteen Cool Features You Didn't Know You Had

R emember the first time a cool iPhone feature blew people away? It was the day Steve Jobs demonstrated what the very first iPhone could do, live on stage. He zoomed into a photo by spreading two fingers.

It was 2007. The audience lost its mind.

Ever since then, Apple engineers have prided themselves on adding features that are powerful, useful, and—when possible—also really cool. As in magical.

Tap the Back

You're probably under the impression that tapping the *screen* of the iPhone is what gets things done. But iOS 14 introduces a fascinating new option: You can tap the *back* of the phone.

And what, exactly, does that do for you? That's up to you. In **Settings→ Accessibility→Touch→Back Tap→Double Tap**, you can choose from a long list of functions that you trigger by double-tapping the back of the phone. You can set it up to open the Control Center. It can take you to the home screen. It can scroll down. It can lock the phone, mute it, adjust the volume up or down, or trigger Siri. It can zoom the screen, turn Voice Control on or off, or trigger any shortcut you've made (page 451).

Actually, you can set up *two* of these tap triggers—one that occurs when you double-tap the back of the phone, and another when you *triple*-tap the back of the phone.

You can tap with your fingernail or finger pad; the taps seem to work best near the middle of the phone. Two or three taps within one second or so seems to work well.

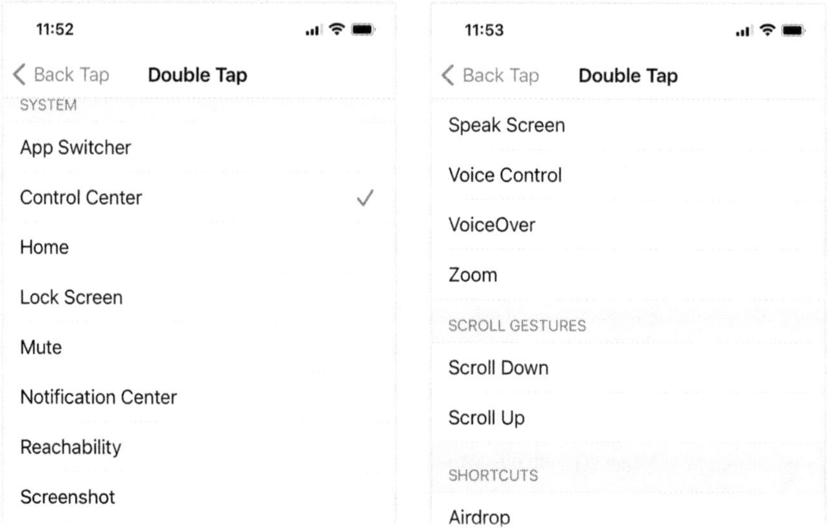

11:52 ... 🔋	11:53 ... 🔋
‹ Back Tap **Double Tap**	‹ Back Tap **Double Tap**
SYSTEM	Speak Screen
App Switcher	Voice Control
Control Center ✓	VoiceOver
Home	Zoom
Lock Screen	SCROLL GESTURES
Mute	Scroll Down
Notification Center	Scroll Up
Reachability	SHORTCUTS
Screenshot	Airdrop

Back tap shortcut options

It's a completely new way of interacting with the phone, and especially useful when your other hand is busy.

Get More Sleep

Making your *iPhone* sleep is as easy as clicking the side button. But the phone comes equipped to help *you* sleep, too.

Using the Health app, you can set up bedtimes and sleep goals for the week. The phone takes things from there: helping you wind down *before* bedtime, protecting your sleep with Do Not Disturb, and then waking you with an alarm in the morning.

In the Health app, tap **Browse→Sleep**; under Set Up Sleep, tap **Get Started→ Next**. You'll encounter these screens:

1. **Set a Sleep Goal.**

   If the phone knows how much sleep you'd like to get each night, it can calculate when it's time to wind down and then turn the lights out. Tap the + and − buttons to specify your goal—eight hours, for example. Tap **Next**.

2. **Set Your First Schedule.**

   By "first," Apple means that you, like most people, might want to set up a different schedule for use on weekends. Tap to specify which days of the week this schedule will affect, and drag the semicircular graph around the clock to indicate your desired bedtime and wake-up times. Here, too, you can set up an alarm, choose a sound for it, and permit (or disallow) the Snooze button.

Tap Add. At this point you can Add Another Schedule (for weekend nights, for example).

When you've built up your schedules, tap Next.

Click one to view the steps.

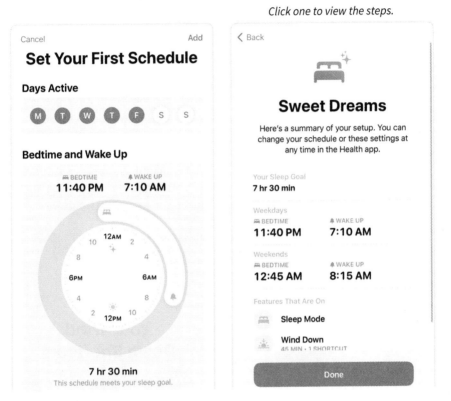

Setting up Sleep

3. Sleep Mode.

At your scheduled bedtime, the phone can dim your lock screen and hide its notifications; the idea is to help you wind down by shielding you from things that would get you riled up.

Sleep Mode also turns on Do Not Disturb While Sleeping (page 50) automatically while you're abed. It gives you a solid-black lock screen (no colorful wallpaper) and no notification bubbles—only the current time and date.

You can always get out of DND While Sleeping by long-pressing that notification and tapping **Turn Off**—if you absolutely must.

If that all sounds good, tap **Enable Sleep Mode.**

> **TIP:** You can also turn Sleep Mode on or off on demand, using its button in the Control Center (page 52).

4. Wind Down.

Using these + and − buttons, indicate how many minutes of wind-down you want before your scheduled bedtime.

To turn this feature on, tap **Enable Wind Down.** On the next screen, tap **Set Up Shortcuts**; now, on the Your Wind Down Shortcuts screen, you can choose apps or shortcuts that you want the phone to offer at the beginning of your Wind Down time. If you like to read before bed, you might choose apps like Books or Kindle; if you like to write down the day's events, you could open a journaling app; if you like to relax with some audio entertainment, you could choose a shortcut that plays a sleepy-time playlist (or sleep-inducing podcast). Mindfulness, stretching, and reviewing your next day's appointments are all good ideas, too.

Choose apps to use during Wind Down.

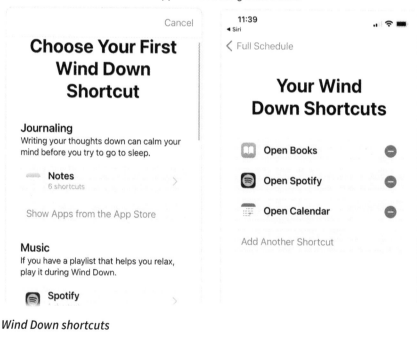

Wind Down shortcuts

For each one you'll want handy, tap the app's name and then the ⊕ below it.

Once you've set up Wind Down, tap **Next**.

5. **Sweet Dreams.**

This summary screen reveals your sleep goal, your bedtime and wakeup times, and whatever sleep-related features you've decided to use (Sleep Mode, Wind Down, Do Not Disturb). Tap **Done**.

From now on, when it's time to start winding down, you hear a sweet three-note Brahms' lullaby chime, your lock screen hides notifications, and a **Shortcuts** button appears.

> **NOTE:** You can, of course, edit any of these settings. In the Health app, tap **Browse→Sleep→Full Schedule & Options→Edit**.

Sleep History

At the top of the **Browse→Sleep** screen in the Health app, you probably noticed the graphs of your recent nights' sleep. You can hop between the **W** (week) and **M** (month) graphs, and you can swipe horizontally to see *previous* week or month graphs. You can also tap **Add Data** to record sleep manually, or hit **Show More Sleep Data** to see your cumulative scores.

And how on earth does the phone, sitting on a table beside your bed, know when you're asleep? It doesn't, really. It can only make a valiant attempt to figure out when you're *lying in bed*—by noticing when you put down the phone for the night. And if you've bought a compatible sleep-tracking gadget, these bars can also reveal when you were *actually* asleep.

> **TIP:** Alas, Fitbits are not compatible—at least not without an app like Sync Solver from the App Store.

Live Listen

What an ingenious MacGyver hack! You can use your iPhone as a remote microphone for your AirPods.

In other words, you can put your phone down next to the TV, or on the lectern next to the professor, or on the restaurant table in front of whoever's talking—and listen comfortably, loud and clear through your AirPods, from a seat up to 100 feet away.

Originally, Live Listen was a feature for use with iPhone-compatible hearing aids. But with the invention of AirPods, anybody can get in on the action.

Start by adding the Hearing button (👂) to your Control Center (page 56). Deposit the iPhone wherever it will pick up the sound you want to hear.

To use Live Listen, put on your AirPods. Open the Control Center and tap 👂. Tap **Live Listen** to start listening.

Tap to see the on/off tile.   *Tap to turn listening on...or off.*

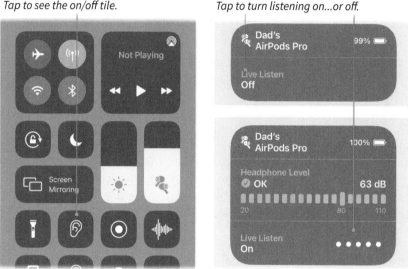

Live Listen

(Note: Do not use for evil eavesdropping purposes. That's just not fair.)

People Detection (iPhone 12 Pro)

Suppose you're blind. You're standing in a line during the time of social distancing. How are you supposed to know when it's time to shuffle forward?

People Detection, introduced in iOS 14.2, uses the lidar sensor on the iPhone 12 Pro and Pro Max (page 22) to detect objects up to 15 feet away—and let you know if they're human, and if so, how far away they are.

To try out this rather impressive feature, open the Magnifier app; tap ⚙; ensure that **People Detection** shows up in the Secondary Controls list. (If not, tap the ⊕ next to its name.)

Scroll down, tap **People Detection**, and let the app know how you'd like it to identify people's distances from you. Your choices are **Sounds** (little blips that get faster as you—or they—approach), **Speech** (announces the number of feet or meters), and **Haptics** (vibrates faster as you get closer).

> **TIP:** When you're using Sounds, the pitch of the blips get higher when you're within a certain threshold of a person—6 feet, for example, or whatever you dial in using the **Sound Pitch Distance** setting.

People Detection

When that's all set up, tap **Done**. You're ready to venture forth into the world of people.

To try the feature out, open Magnifier and tap 👤⋯👤. Hold the phone in front of you; when it sees humans, it lets you know how far away they are—sonically, verbally, visually, or vibrationally.

> **TIP:** If you're wearing AirPods Pro, the sounds from the right and left earbuds let you detect whether the person is to the left or right of you.

Your Phone on TV

Projecting your iPhone onto a big TV is the perfect way to show slideshows and movies to a group. It's super-handy for making presentations from Keynote or PowerPoint. It's great for teaching. It's fun for playing games, especially when friends or family want to watch.

To make this work, you need an Apple TV: a little black box that brings hundreds of internet "channels" (including your Netflix, Hulu, HBO, Amazon

Prime, and other online subscriptions) to the TV. But it has a stealth feature, too, called AirPlay. That's Apple's wireless video-sending technology.

> **NOTE:** You can also buy A/V receivers that have AirPlay built in. It doesn't have to be an Apple TV, although that's what most people use.

When it's time to present your iPhone's screen, open the Control Center (page 52) and tap **Screen Mirroring**. Choose the name of your Apple TV (you may also see the name of your TV or receiver if it works with AirPlay), and presto: Your TV lights up with the phone's image.

> **NOTE:** The very first time you try this, a huge four-digit number appears on the TV for you to type into the phone. That's a security thing, so your next-door neighbor can't surprise you by projecting raunchy videos onto your TV when you're trying to watch *60 Minutes*.

On the Control Center, tap...   ...to begin broadcasting to the TV.

Screen mirroring

To finish with your presenting, open the Control Center and tap **Screen Mirroring** to turn it off again. Wasn't that cool?

Use a HomePod as an Intercom

If you're the proud owner of a HomePod, one of Apple's smart speakers, you've got a nifty communication option in hand: an intercom. You can use your iPhone (or even Apple Watch) to speak through HomePods elsewhere in the house.

To set this up, open the Home app on your phone. Tap ⌂→Home Settings→Intercom. Here you have a couple of decisions to make:

- **Do you want to be notified** of intercom messages when you're out of the house?

- **Who else is allowed** to use the intercom when out of the house?

- **Which HomePods** (if you have more than one) will be part of the intercom network?

To try it out, use Siri on your iPhone or Apple Watch. Say, for example, "Dinner in five, everyone!"

Every HomePod in the house (that you've permitted to use intercom) lights up and plays your voice utterance for all to hear.

You can also specify where your voice will be heard. "Hey Siri, announce in the living room, 'Anyone seen the cat?'" Or, if you've set up rooms or zones in the Home app (page 337), you can say to Siri, "Ask upstairs, 'Who wants breakfast?'"

> **NOTE:** You can speak these queries in the Home app, too. There's a ⑈ button at top right just for this purpose.

People can reply, too. They can say "Hey Siri—reply, 'Coming down!'" Or they can direct their response to particular HomePods: "Hey Siri, reply to the kitchen, 'I already ate. Sorry!'"

AirDrop

Somebody at Apple deserves a raise for this feature. AirDrop sends files to other people's iPhones, Macs, or iPads up to 30 feet away, instantly, wirelessly, and fully encrypted. It's a breakthrough in speed and simplicity. There's nothing to set up, no accounts, no passwords, and no security risk—both sender and receiver are in total control. You don't need to be in a Wi-Fi hotspot. In fact, you don't even need an internet connection: AirDrop works when you're out at sea, in a cornfield, or in a hot-air balloon.

AirDrop is built into most of Apple's apps, so you can send a note from Notes, a photo from Photos, a map from Maps, and so on. AirDrop can even send big files, so it's ideal for sharing videos that, if you texted them, would get crunched down to a low-resolution format in the process.

Here's how to send stuff wirelessly using AirDrop:

1. **Wake both phones.**

   Everyone has to have Bluetooth and Wi-Fi turned on.

2. **Confirm that the other device can receive AirDrop shipments from you.**

   This is an annoying step, but it's required. See the box below.

3. **Call up whatever you want to send.**

   Maybe it's a photo in Photos, somebody's info card from Contacts, directions (or your current location) in Maps, a page of Notes, a web page in

HOW TO CONTROL WHO AIRDROPS YOU

Apple is fully aware that the world is full of nasty people. When creating AirDrop, Apple's engineers had to ask: How do we prevent creating a hellscape of people bombarding one another with unsolicited AirDropped raunchy photos, obscene recordings, and terrorist manifestos?

The answer: You can control who may send you things by AirDrop.

To see this setting, open your Control Center. Long-press the quartet of icons at top left; on the resulting panel, you'll see your current AirDrop setting: **Receiving Off**, **Contacts Only**, or **Everyone**. Tap to change the setting.

Receiving Off means nobody can send you things by AirDrop. **Contacts Only** means that only people in your Contacts address-book app can send you things; presumably, you trust them at least a little. Your phone is invisible to strangers. And **Everyone** means it's open season. Anyone within 30 feet of you can send you stuff by AirDrop.

Remember that you also must accept incoming AirDrop items individually; nobody can send you something without your awareness. In other words, even the **Everyone** setting doesn't mean your phone is going to fill up with nonsense. ✦

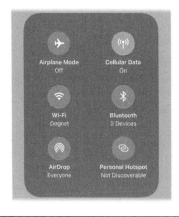

Safari, a ticket in Wallet, an app you found in the App Store, a song in Music, or whatever.

4. **Tap ⬆.**

The share sheet (page 203) appears. If only one person nearby can accept AirDrop, you see that iPhone's icon in the share sheet.

> **TIP:** When you send a *photo*, the top row of the share sheet shows nearby photos and videos from the same batch, on the premise that you might want to send several in a clump. Tap to select the ones you want to send.

If there are several AirDrop-receivable Apple gadgets nearby, the AirDrop icon indicates how many; tap **AirDrop** to see all of them.

> **TIP:** If you and the other person both have iPhone 11 or 12 models, you can *aim the top of your phone* at theirs. Their AirDrop icon appears, big and centered, at the top of the AirDrop-icons screen, making it quicker to choose.

*Tap the recipient...or tap **AirDrop** to see everyone nearby.*   *What the recipient sees*

AirDrop

5. Tap the icon of the gadget you want to share to.

Or several; you can send to multiple devices at once.

A moment later, the recipient sees a message containing a preview of whatever you're sending. It says, "[Other person's name] would like to share [the file's name]." The options are Decline or Accept.

If the receiver hits Accept, the transaction is complete. The word "Sent" appears on your screen. If you sent a file, it winds up in their Downloads folder (if it's a Mac) or in the Files app (if it's an iPad or iPhone). If you sent a note, map, web page, photo, or contact, it shows up in the corresponding app.

If they tap Decline, you see the word "Declined" on your screen, and the transfer fails.

You can also AirDrop something to yourself—from your iPhone to your Mac, for example. In that situation, there's no Accept button to click; the incoming file or data morsel goes directly into the appropriate spot or app. In theory, you're not going to send upsetting photos to *yourself.*

AirDrop. Don't send photos or videos without it.

Print

You can print from your iPhone—as long as your printer has built-in AirPrint, Apple's wireless printing feature. Most modern printers from all the major printer companies (Brother, Canon, Epson, HP, and Lexmark) have AirPrint. (A full list appears on Apple's website at support.apple.com/kb/HT4356.)

You can't print from every app. But you *can* print from the ones that display your email, notes, photos, web pages, maps, and so on. Many non-Apple apps work with AirPrint, too.

When you're ready to print something that's on the screen before you, tap ⬆ and then Print.

> **NOTE:** If you're printing from the Notes app—it could happen!—you will *never* find the **Print** command. To find it, tap the ⊙ in the corner of the note, then **Send a Copy**. Scroll up until you see **Print**.

On the Printer Options screen, tap Select Printer and then the name of your printer.

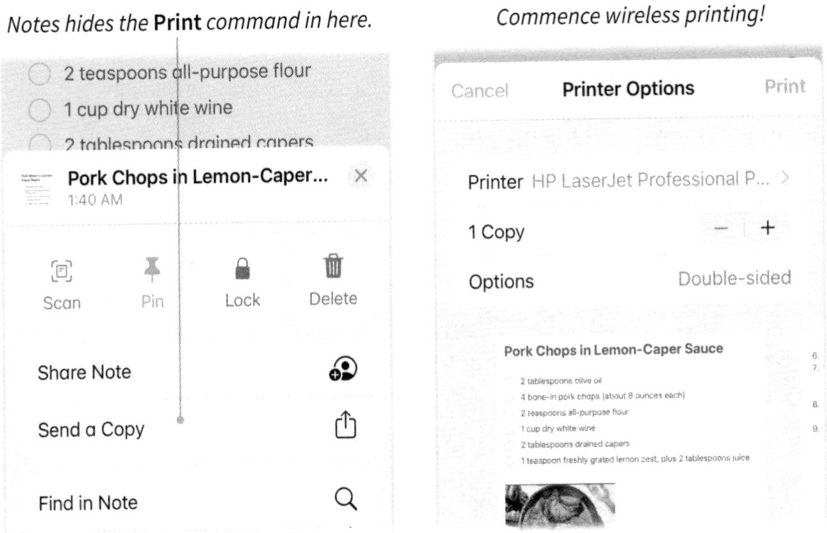

2 teaspoons all-purpose flour

1 cup dry white wine

2 tablespoons drained capers

Pork Chops in Lemon-Caper... ✕
1:40 AM

| Scan | Pin | Lock | Delete |

| Cancel | **Printer Options** | Print |

Printer HP LaserJet Professional P... >

1 Copy — +

Options Double-sided

Share Note

Send a Copy

Find in Note

Pork Chops in Lemon-Caper Sauce

2 tablespoons olive oil
4 bone-in pork chops (about 8 ounces each)
2 teaspoons all-purpose flour
1 cup dry white wine
2 tablespoons drained capers
1 teaspoon freshly grated lemon zest, plus 2 tablespoons juice

Printing from the phone

Adjust the number of copies, range of pages, and whatever printer options appear here (color or monochrome, double-sided or not, and so on). Finally, tap **Print**. The printout emerges from the printer exactly as it would if you'd printed from a computer over a cord.

> **TIP:** If your older printer doesn't offer AirPrint, you can always use Printopia ($20). It's an app for your Mac or PC that lets you send printouts from your phone *through* your computer to the printer.

Talk to Your Car (CarPlay)

Studies show that using your cellphone while driving impairs you about as much as driving drunk. And yet—come on. How are we supposed to exist without our phones for such long periods?

Apple's solution is CarPlay, a hugely simplified version of iOS that appears on your car's dashboard touchscreen. It shows only apps that are relevant to driving—Phone, Music, Maps, Messages, Podcasts, Audiobooks, Waze, Spotify, Calendar, and so on—and none of them involve reading.

Among other blessings, CarPlay lets you use Apple's Maps app (or Waze, or Google Maps) for getting turn-by-turn driving guidance instead of the (probably awful) software that came with your car.

This works, of course, only if your car *has* a touchscreen, and only if it has CarPlay built in. (About 600 car models do.) Most cars require you to connect the iPhone's USB cable; some recent ones can do CarPlay wirelessly.

To set up CarPlay, just plug the phone into the car's USB jack, turn the car on, and then tap CarPlay on the touchscreen. Or, if you have wireless CarPlay,

Swipe to see other apps.

CarPlay

WHEN YOUR PHONE IS YOUR CAR KEY

Here's a new feature that very few people can use: Car Key. It lets you use your phone (or your Apple Watch) as an electronic key to unlock and start your car.

Of course, this feature works only if your *car* also offers Car Key, and at the outset, it's a small list—a few 2021 BMW models. If all goes well, more new car models will offer Car Key. Car Key also requires a modern phone: an iPhone XR or later.

To use Car Key, just hold the phone near the door handle; it unlocks. (That's also how you lock it again.) If you have an iPhone 11 or later, you can now start the car, even if the phone is in your pocket. (Earlier iPhone models have to be in the car's wireless charging slot.)

If that sounds too simple, you can require Face ID or Touch ID before the car will unlock or start. You can share access to your car with up to five other people, provided they're all iPhone owners (you use Messages for this).

You might worry: "If my phone is dead, how can I start the car?" That's a legitimate concern, but Apple has worked things out so that even a newly "dead" phone has a five-hour reserve charge that's enough to start a car.

And if you lose your phone? Well, that's a bad day. But at least you can log into iCloud.com and turn off Car Key remotely, so the phone thieves won't also get your *car*. ✦

put the car into Bluetooth pairing mode (usually a matter of holding in the voice-command button on the steering wheel—check its manual); on the iPhone, open Settings→General→CarPlay→Available cars, and choose your car model.

Once CarPlay is set up, you can interact with your phone with Siri—use the voice-command button on the steering wheel as the talk-to-Siri button—or by tapping the big icons.

In Settings→General→CarPlay, you can choose which apps show up on the dashboard screen, and in which order. It's the standard iOS list configurator, described on page 4.

The startup screen shows you three recently used apps, the map, a couple of frequently used Maps destinations, and music playback controls. But you can swipe left to view all the rest of your CarPlay-compatible apps, exactly as you'd swipe across different home screens on the phone.

Capture the Screen

Why would you ever want to capture a picture of your screen? Maybe because you're trying to show somebody in tech support the weird glitch you're seeing. Maybe because you want to tweet about something funny you see onscreen. Maybe because you're writing a book about iOS 14.

Start by getting the screen arranged as you'll want it. Now take the shot, using any of these methods:

- **Face ID phones.** Press *both* the side button and volume-up buttons. They're directly across from each other.

- **Home-button phones.** Press both the side and home buttons, and then release both.

- **Use your voice.** If you've turned on Voice Control (page 147), you can say, "Take a screenshot."

- **Use the back tap.** If you've chosen Screenshot as the back-tap action (page 421), just double-tap the back of the phone.

The screen flashes white, and your screenshot appears as a thumbnail at the corner of the screen. If you ignore it (or swipe it away), it vanishes; you'll find your screenshot in Photos, in the Screenshots album.

But if you tap the thumbnail before it vanishes, you get a screenshot-editing window. Here are your standard markup tools (page 7), ready for

Crop handles

Standard markup tools

Editing a screenshot

annotating or drawing on the screenshot—and the ⬆ button, which lets you send the screenshot to someone, to some app, or to your Files or Dropbox virtual disks. You can also drag the white handles (corners and edges) to crop the shot, isolating the important part.

> **TIP:** If you took a bunch of screenshots within six seconds, the thumbnails stack up in the corner of the screen. When you tap, they open in the editor as a horizontally scrolling row. Now you can edit them and then send them as a batch.

If you screenshotted something taller than the screen, like a web page, email message, or a long note in Notes, you get a **Full Page** button here, too. It lets you save the screenshot as a tall, scrolling PDF document.

To exit editing mode, tap **Done**. The phone invites you to save or delete the shot.

Record Screen Video

You can capture a *video* of your screen goings-on, too. It's fantastic when you're trying to create a demo of how something works—something you love, like a cool app, or something you don't, like a bug that the software company really needs to know about.

Capturing a video is also great for snagging moments on the iPhone that aren't screenshottable, like the Setup Assistant screens (Appendix A). It's also excellent for capturing things that happen too quickly for you to grab as a screenshot.

There's no app for recording the screen. The button for creating a screen video exists in only one place: on the Control Center, and only if you install it there (page 56).

If you tap that button, you get a 3-2-1 countdown; that should be enough time to get out of the Control Center and call up whatever you want to "film." Once the capturing begins, the phone's left "ear" or status bar turns red to remind you that you're capturing. (That red spot will be part of the finished video.)

The Record Screen button on Control Center   *The long-press options*

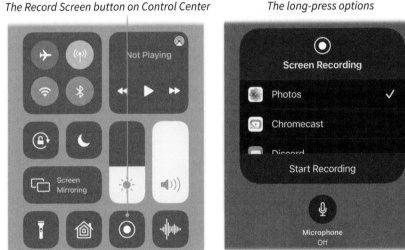

Screen video recording

To end the recording, tap the red ear or status-bar stripe. A notification appears: "Screen Recording video saved to Photos."

If you do nothing, the video winds up in Photos alongside any other videos you've captured. If you tap the notification, you jump to it and open it.

If you swipe down on the notification, two buttons appear: View, which opens the video for instant playback, and Delete, which is useful when you need a new "take."

> **TIP:** Alas, the iPhone screen recorder doesn't capture the sounds of your phone's apps. You can, however, narrate what you're doing as the phone captures the screen activity. To set that up, long-press the Screen Recording tile (◉) on the Control Center and tap the 🎙 before you start. (Also visible here: a list of apps you may prefer to send your captured video to instead of the Photos app.)

The QR Scanner

QR codes are those square bar codes. They show up a lot in marketing materials like ads and posters, and sometimes in user manuals. Usually, what they're encoding is a web address (a URL). The idea is that if you have a bar code–scanning *app,* you can point it at the QR code and go instantly to the website, without having to type in, like http://www.thisisamarketingploy.com/movieposter.html.

You, however, do not need a QR-scanning app. Just open your Camera app and point it at the code. You don't have to tap anything or choose anything. The web link appears as a notification at the top of the screen; tap it to open that page in Safari.

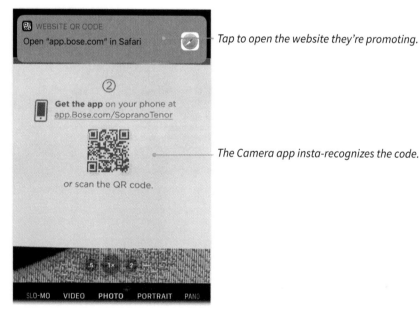

Tap to open the website they're promoting.

The Camera app insta-recognizes the code.

QR code reading

Screen Time: Protect Your Kids

On top of everything else we have to worry about, now there's the screen-time thing. We worry that too much time spent using electronics is doing long-term damage to our kids' brains—and maybe our own.

Even Apple is worried about it. That's why it came up with Screen Time, a feature that monitors what apps you're using, and for how much time, and on which Apple gadgets (Mac, iPhone, iPad). You can also set daily limits for each category—two hours a day of social media, for example—for you or for your kids.

At any time, you can open Settings→Screen Time to view a weekly report. It's a handsomely designed summary, full of graphs that show how much time you and your other family members spent staring at their screens, what apps they were using, and how many times they exceeded the limits you set.

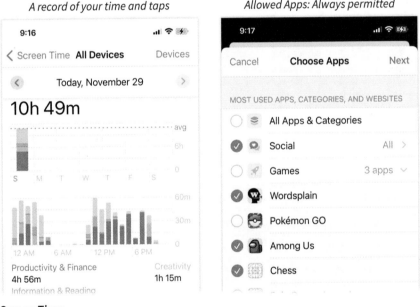

A record of your time and taps   *Allowed Apps: Always permitted*

Screen Time

Turning on Screen Time

To get started, open Settings→Screen Time. Tap Turn On Screen Time→Continue. Tap either This is My iPhone or This is My Child's iPhone.

Screen Time can, if you wish, add up the time you spend hunched in front of *all* your Apple machinery: the phone, the Mac, the tablet. Turn on Share Across

Devices if you want that kind of data horror show. (You must turn the switch on separately on each device you use.)

> **TIP:** Screen Time is a natural partner to Family Sharing (page 401). After all, you've already said you've got kids or fellow family members, right?
>
> The easiest way to fire up Screen Time for your kid is to do it on their phone. Open **Settings→Screen Time**; tap **Turn On Screen Time→Continue**. Tap **This is My Child's iPhone**. Now you'll get full reporting on your kid's activity right on your own iPhone.

The Screen Time Passcode

Of course, Screen Time won't do you or your kids much good if they can just turn it off and play 11 straight hours of Minecraft. Enter: the **Use Screen Time Passcode** option.

You're asked for a four-digit number. Without it, nobody can turn off Screen Time or change its settings.

> **TIP:** Don't lose the passcode, of course. But if you do, if you can enter your Apple ID (email address and password), you'll be allowed to make up a new one.

App Usage, Notifications, Pickups

On the **Settings→Screen Time** screen, tap **See All Activity**. Here's what you see on this brain-frying screen:

- **How much time you've spent** on Apple devices this week, broken down by app category. Tap **Day** to see a daily breakdown.

- **Which apps you've used most.** Or tap **Show Categories** to see the time you've spent in broader categories like Social, Productivity & Finance, Creativity, and so on.

- **How many times you woke and unlocked your devices ("Pickups"),** and which app you used first each time.

- **How many notifications bombarded you,** and which apps were doing it. As a handy bonus, you can tap the name of a particularly aggressive app and turn off its notifications on the spot.

> **TIP:** Tap **Devices** at top right to see these statistics for one *particular* Apple gadget—just your Mac or just your iPhone, for example.

Downtime

Downtime means "enforced no-gadgets time." You invoke it by tapping **Downtime** and then turning on the switch. Use **Every Day** to establish the no-gadgets hours—11 p.m. to 8 a.m., maybe. Or use **Customize Days** to establish separate schedules for each day, so weekends can be different from school nights.

When **Downtime** is five minutes away, a warning appears. Then, at the appointed time, the app windows all go white and their icons dim.

> **NOTE:** If this treatment seems too harsh, you can bless certain apps with exemptions to Downtime; read on.

If you've set up Downtime for *yourself*, its onset is not much more than a slap on the wrist. You can click **Ignore Limit** and then click **One More Minute, Remind Me in 15 Minutes**, or **Ignore Limit For Today**. Or you could just turn Downtime off again.

Dimmed icons: Time's up for today.   *You can, however, beg for mercy.*

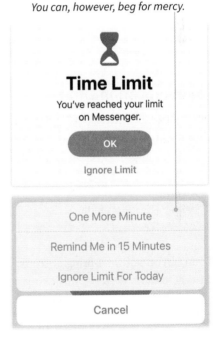

Downtime

But if you've added a passcode—for your kids, for example—there *is* no **Ignore** button. Lucky you—teachable moments every single day!

App Limits

You can also set daily time limits for individual apps or app categories.

To begin, tap **App Limits**→**Add Limit**. Turn on an app category to establish a daily time limit for that group of apps—or tap to expand a category. Now you can see the apps in that category, to limit them one by one.

Tap **Next**. Dial up a maximum amount of time per day; tap **Add**. Repeat with other apps and categories.

So what happens when your daily limit approaches for an app (or app category)? You get a five-minute warning. Here again, you can click **Ignore Limit**—*unless* you're blocked by a passcode. Which probably means you're a kid.

In that case, the kid gets a new option: **Ask For More Time**. That button sends a request to you, the all-knowing parent, wherever you are in the world. If your kid has been very, very good, you can click the notification and, using its pop-up menu, approve an extension.

> **TIP:** In **Settings**→**Screen Time**→**Always Allowed**, you can designate certain apps this person is *always* allowed to use, even during Downtime. Maybe there are apps that you don't think should ever be off-limits—for example, a guitar-practice app, a jogging-coach app, or a meditation app.

Communication Limits

This section of the Screen Time settings gives you control over who your kids are allowed to reach using the Phone, FaceTime, or Messages. You can set up one set of conditions during Screen Time and another during Downtime.

> **NOTE:** This feature requires that both of you have turned on **Settings**→**[your name]**→**iCloud**→**Contacts**. iCloud can't restrict your contacts if it can't see your address book.

If you tap one of the **During** buttons, you can limit this person's communications to **Contacts Only** (no strangers), **Contacts & Groups with at Least One Contact** (group chats where the kid knows at least one person), or **Everyone**.

And now, two items of not-so-fine print. First, your kid can always dial 911. Good.

Second, your kid can bypass all this by using WhatsApp, WeChat, Facebook Messenger, or any other non-Apple app. Oh, well.

Content and Privacy

Here's where the iPhone's parental controls live. You can use them to block stuff that might corrupt your kid's mind, like pornography and dirty words, as well as stuff that might corrupt your Visa card, like purchases of music, movies, and apps without your awareness.

In Settings→Screen Time→Content & Privacy Restrictions, tap to turn on Content & Privacy Restrictions.

You can set up restrictions in five categories:

- **iTunes & App Store Purchases** lets you keep your kid from installing apps, deleting apps, or make in-app purchases (buying new levels or weapons within a game, for example).

- **Allowed Apps** lets you block certain apps that could get a kid into trouble—Wallet, FaceTime, and the various Apple stores, for example.

- **Content Restrictions** sends you down a rabbit hole of blockades, but most of it boils down to "shielding your kid from sex and violence."

  You can block music, podcasts, news articles, music videos, movies, TV shows, books, and apps that contain raunchy material. For example, if you tap Movies→PG-13, your kid won't be able to play R-rated movies on the iPhone or rent/buy them from Apple.

  You can prevent **App Clips** from downloading, too (page 374).

  Web Content shields impressionable young eyes from online pornography. **Limit Adult Websites** consults a blacklist that Apple maintains (known naughty sites); **Allowed Websites Only** is the opposite. It's a whitelist that permits only the wholesome sites listed here.

  That doesn't mean you can't override Apple's wisdom, however. Within the Limit Adult Websites options, **Always Allow** and **Never Allow** controls let you add the addresses of websites you think should be OK (or should not be OK).

  The **Siri** options can prevent Siri from responding when your kid says, for example, "Find pictures of naked people," and **Explicit Language** lets you plug Siri's ears when your little tyke speaks crudely to her.

  The final categories create limits on playing games in Game Center, Apple's game-coordination app.

- **Privacy** offers a huge list of features, apps, and data that, if exposed to bad people on the internet, could conceivably be considered privacy violations. They're extra hatches that you can slam shut. For example, you can ensure that no app is ever allowed to detect where the phone is (turn off Location Services).

- **Allow Changes** means "various things you may not want your kids changing," like their iOS passcodes, email and social-media accounts, TV provider, and so on.

Overall, Screen Time offers a lot of controls that give you a lot of control. It can be dizzying, it should involve some frank conversations with your kids, and it may involve a few days of debugging and troubleshooting. ("Hey, Parent—you wanna know why I didn't respond to your text? Because your stupid parental controls locked me out of Messages halfway through social studies!")

But if you're at all worried about your kid's screen time and exposure to the worst of the internet, at least Screen Time gives you a fighting chance.

Large Type and Accessibility

It would be hard to imagine a phone more generous with accessibility features. If you have trouble with seeing, hearing, or mobility, this is the machine you want.

The world headquarters for these accessibility features is Settings→ Accessibility.

Help with Seeing

Some of the options in the Vision section make the iPhone more usable no matter how good your vision is. In particular:

- **VoiceOver** is the iPhone's screen reader, in which the phone speaks aloud the name of everything under your finger. It's possible, with practice, to operate the phone without being able to see anything at all, thanks to VoiceOver and its many options. A full-blown user guide for VoiceOver awaits at www.apple.com/accessibility/iphone/vision.

- **Zoom** (page 63) and **Magnifier** (page 348) are incredibly good at magnifying tiny type—on the screen and on written materials around you, respectively.

- **Display & Text Size** offers a cornucopia of useful interface tweaks. Bold Text makes all text on the screen boldface for better readability; Larger Text lets you enlarge the type in most apps.

TIP: Don't forget that you can install a Larger Text button right on your Control Center, which is far quicker to find than this control deep in Accessibility Settings. See page 56.

Fool around with the other options here to see if they make things easier to see. If you're color-blind, for example, the **Color Filters** shift everything on the screen to colors you can distinguish. (The result looks odd to anyone who *isn't* color-blind, but at least you can now distinguish colors, and maybe even pass those online color-blindness tests.)

If you have trouble reading, try **Smart Invert**; it swaps the screen's colors for better contrast like a film negative—black for white, red for green, blue for yellow. (It's smart enough not to affect photos, videos, or app icons.)

Reduce Transparency makes the backgrounds of the Dock, Notification Center, app folders, and other panels solid instead of translucent. It might be cool that you can see your wallpaper shining through those panels—but it can also make the menus harder to read.

Factory setting: Blurry translucence *Reduce Transparency on*

Reduce Transparency

- **Motion** lets you turn off motley animations, including those fancy full-screen fireworks and confetti displays in Messages (page 256). Maybe they trigger something in your neurological system, or maybe you just think they're really dumb.

- **Spoken Content** means "Your iPhone can read text aloud to you." See page 135.

Help with Mobility

If you have limited motor control, Voice Control (page 147) should have gotten your attention. It's designed to let you operate everything on the phone with voice commands.

But the options on the Touch screen are fairly epic, too. Some examples:

- **AssistiveTouch** produces a glowing white circle on your screen. You can drag it anywhere; it fades until you need it.

  When you tap it, it sprouts into a special panel that lets you trigger complex or difficult iPhone gestures with one gentle tap. Tap to open the Notification Center or Control Center without having to swipe. Rotate the screen image without having to rotate the phone. Talk to Siri without having to press the side button, and so on.

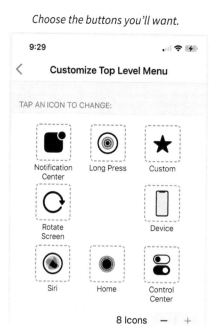

Choose the buttons you'll want.

The button waits...until you tap it.

AssistiveTouch

- **Reachability.** Some iPhone models have gotten so big that those who are small of hand have trouble reaching the top of the screen to tap things. When Reachability is turned on, you can tug down on the *bottom half-inch* of the screen to slide the *entire screen image* halfway down the glass so you can reach it. On home-button phones, touch the home button twice (touches, not full clicks).

  Once you tap anywhere, the screen jumps back to full height.

- **Haptic Touch.** This lets you customize how long a long-press has to last before the phone responds.

- **Touch Accommodations.** If you have motor-control challenges, like a tremor, these features can make it easier to perform precise taps. For example, **Hold Duration** lets you specify how long your finger has to stay in one place before the iPhone counts it as a tap, and **Ignore Repeat** screens out multiple taps within an interval (like one second).

  Similarly, **Tap Assistance** lets you indicate that the spot where you *lift* your finger is what counts (**Use Final Touch Location**), not where you initially put it down. That way, you can adjust your finger's position (during the countdown period indicated by the Gesture Delay setting here).

  Tap to Wake lets you turn off the feature that wakes the phone when you tap the screen anywhere. You can turn off **Shake to Undo**, too (page 132), which is useful if you tend to trigger it accidentally.

  And **Vibration** is the master off switch for *all* vibrations the phone makes.

  This screen is also the home of the delicious new **Back Tap** feature described on page 421.

Help with Hearing

These controls are powerful allies if you're deaf, partly deaf, or wearing fat headphones and unlikely to hear the phone.

Tap **Hearing Devices** to set up hearing aids, or **Sound Recognition** to turn on the background-sound recognition feature described in the next section.

Some of the features with broadest appeal, though, appear on the Audio/Visual screen. For example:

- **Headphone Accommodations,** new in iOS 14, lets you tune the sound of your headphones—especially if they're Apple earbuds, AirPods, or certain Beats headphones. If you tap Custom Audio Setup, you'll hear samples with different settings, and you can choose what sounds best to you.

  For example, you might want them to sound brighter, make spoken audio clearer, or boost softer sounds—maybe because you have trouble hearing, or maybe because you just like it better that way.

> **TIP:** If you turn on **Phone**, these settings affect phone calls and video calls; if you turn on **Media**, they affect music, movies, podcasts, audiobooks, Siri, voicemail, and Live Listen.

- **Mono Audio** ends the frustration of listening to stereo-mixed music if you have no hearing in one ear, you're sharing an earbud with a friend, or you've lost one AirPod. It puts the entire mix into a single channel.

- **Phone Noise Cancellation.** If your iPhone's background-noise reduction gives your ear a weird feeling—some people feel a "pressure"—you can turn it off here.

- **Balance Slider.** If one of your ears has better hearing than the other, use the slider to adjust the stereo mix.

- **LED Flash for Alerts** lets you know when you're getting a call, text, or notification by blinking the very bright LED light on the back of the phone (the flash). Useful not only if you're deaf, but also if the phone is across the room or you're jamming to loud headphone music.

Ominous-Sound Warnings

For iOS 14, Apple designed one more amazing feature for people who have trouble hearing: the Sound Recognition feature. When it's on, the phone constantly listens for sounds in your environment that you might consider important: an alarm, a baby crying, a dog barking, a cat yowling, the doorbell or a

Sound Recognition

knock, water running, and so on. When it hears something concerning, you see a notification.

Even if your hearing is fine, Sound Recognition can still be useful as a historical record—for example, if you find one of its notifications on your phone when you've been away from it for a few hours.

In any case, you set this up in Settings→Accessibility→Sound Recognition. Tap Sounds to specify which of the 13 sounds you want the phone to listen for.

Once that's done, you can switch Sound Recognition on or off using its Control Center button, once you've added it (page 56).

The only downside is that when Sound Recognition is listening for environmental sounds, you can't use "Hey Siri" (page 139). The iPhone's sonic attention span isn't that wide.

Guided Access: Help with Kids or Newbies

You've probably seen those videos of kids, even *babies*, merrily tapping away on iPhones and iPads, in full control. Well, that's great—until they tap something accidentally and wind up deleting important photos or sending gibberish emails to your employer.

You declare certain controls off-limits…

…by circling them with your finger.

Guided Access

That's why Apple created Guided Access, which you can think of as kiosk mode (or "padded-walls" mode). It locks the phone into a single app, so you can lend it to a kid or a blundering novice without worrying that they'll get into trouble.

Start by turning on Settings→Accessibility→Guided Access→Guided Access. If you'd like to protect Guided Access so the tyke can't shut it off—at least not without a six-digit passcode or your fingerprint or your face—tap Passcode Settings→Set Guided Access Passcode and make one up. Tap Time Limit if you'd like to end iPhone Fun Time after a certain period. You can even set up an alarm or a spoken warning when time is running out.

When you're ready to hand your phone over, open the app you want to lock in place. Click the side button (or home button) three times fast, and tap Guided Access.

On the Guided Access screen, here's what you can do:

- **Declare some app features off-limits.** Drag a circle around each button, slider, and control you don't want your kid to be able to activate. The phone converts your circle into a rectangle, whose size and position you can adjust.

- **Declare some phone features off-limits.** Tap Options if you'd like to prevent your little munchkin from pressing the phone's Side Button or Volume Buttons or typing on the onscreen Keyboard. If the idea is that you're handing the phone to a 2-year-old just to watch videos, you may as well turn off Touch (so taps won't register) and Motion (rotating the phone won't turn the image).

When you tap Start, the phone is locked into the app you chose, with the limited features you've requested.

Once the phone is yours again, you can exit Guided Access by triple-clicking the side (or home) button again; enter your passcode, or use your face or fingerprint, to confirm that you're the actual adult and not the 3-year-old getting lucky.

Help with Accessibility

There are a lot of people in the world, with a lot of different challenges. Clearly, accessibility is a vast area—both the topic, and the software controls. Apple maintains an entire website dedicated to its assistance features: apple.com/accessibility.

Accessibility Shortcut

Meanwhile, there are so many accessibility features that iOS offers an accessibility *shortcut*: a menu you can summon at will, listing only the features you actually use.

In **Settings→Accessibility→Accessibility Shortcut**, turn on the accessibility features you use. (**Zoom, Magnifier**, and **Voice Control** are some of the home runs for almost anyone.)

From now on, you can summon the options you've requested by *triple-clicking* the side button or home button. In any app, at any time.

> **TIP:** If you've chosen only one Accessibility feature in the list, then triple-clicking turns that one feature on or off. You don't see the little menu of buttons.

Shortcuts

Sure, sure, the iPhone can do lots of things. But most of them were dreamed up by software engineers somewhere. The beauty of Shortcuts is that *you* can to decide what the iPhone can do at your command. It's an app that builds *macros*: sequences of steps that you build in advance and then trigger, whenever it's convenient, with one command.

Most people trigger their shortcuts using Siri—you get to decide what phrase to use for each shortcut—but here are some examples of shortcuts you can build:

- **"Be quiet."** Siri turns on Do Not Disturb until you leave wherever you are.

- **"I need a nap."** The phone asks how long your nap will be, says "Happy snoozing," turns on Do Not Disturb, and rings the alarm at the end of that time.

- **"Head to work."** The phone calculates and announces your estimated arrival time, reads out your first appointment for the day, begins playing the playlist you like for commuting, and gets directions based on current traffic.

- **"Tell 'em to wait."** Siri sends a canned text that you're running late to everyone who's part of your next appointment.

- **"Take me home."** The Maps app opens and starts guiding you home from wherever you are.

- **"Log coffee."** The Health app records one cup of coffee.

- **"Top stories."** The News app opens and shows you the latest headlines.

- **"Back up the car."** Your Tesla electric car turns on, gets its bearings, and then backs out of its parking place. (Useful when someone has nearly boxed you in.)

But your shortcuts can also be icons on the home screen, looking just like other apps. You can trigger one with a double-tap on the back of the phone, as described already. They can become new entries on the share sheet (page 203).

They can even be automated, running automatically at certain times of day, when you arrive at designated places, when you get into your car (if it has CarPlay), when you change certain settings, when you open a particular app, when you snooze or shut off your alarm, when your phone connects to a Wi-Fi hotspot, and so on.

For example, a shortcut can detect that you've arrived at your home; turn on the lights; begin playing your favorite music; and text the family, "Hi, everyone! I'm home!"

> **NOTE:** And, of course, Shortcuts is also the key to creating the radical home-screen makeover designs described starting on page 108.

Ready-to-Use Shortcuts

When you open the Shortcuts app, the My Shortcuts screen offers only four demo shortcuts, just to whet your appetite.

But you're not obligated to turn yourself into a programmer. If you tap Gallery, you'll discover a universe of shortcuts that Apple has created for you. You can start using them right away; no need to understand what steps were involved in assembling them.

You can use prefab shortcuts from Apple…   *…or dive in yourself.*

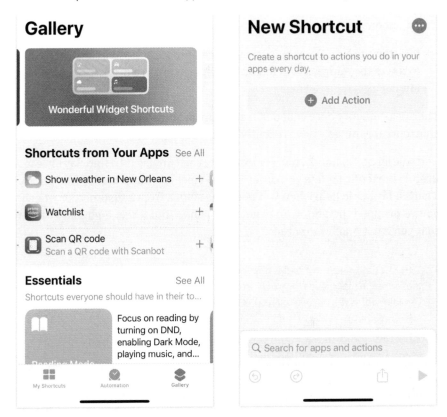

Starting up Shortcuts

The Gallery is designed to look like an app store, with shortcuts advertised in clumps called things like Essentials, Morning Routine, and Stay Healthy. Tap a tile to see the options within.

For example, if you tap See All next to Essentials, you might see options like these:

- **Home ETA** sends a text to family members with the time Maps expects you'll get home, based on current traffic.

- **Reading Mode** asks how much time you have to do some reading; turns on Do Not Disturb during that time; switches to dark mode; plays some soft instrumental music of your choice; and then asks you to choose what you want to read.

- **Make PDF** converts *anything*—an email, a note, a website—into a PDF document.

- **Directions to Next Event** inspects your calendar for the next appointment, extracts the address, and instantly displays the driving directions in Maps.

There's a category here called Shortcuts from Your Apps, too. It displays shortcuts for things you've recently performed by hand.

The search box is handy, too. For example, you can use it to find Say Cheese, a useful shortcut that lets you order Siri, by voice, to take a hands-free photo. That's a fantastic help when you're taking selfies from a distance, or when you've propped the phone up to get a cool shot—but it's so high and awkward that you can't reach the screen.

> **TIP:** The Say Cheese shortcut is designed to use the back camera. But with three taps, you can make it use the front camera for selfies instead. (Tap ⬤, then the **Back** variable, which you change to **Front**.) Or duplicate the shortcut, and have one of each!

If a shortcut looks like it could be useful to you, tap it, and then tap Add Shortcut. You've just installed its tile on the My Shortcuts tab of the app.

> **TIP:** The web is full of ready-to-roll shortcuts. Whole websites are full of them, including ShortcutsGallery.com and reddit.com/r/shortcuts.

Make Your Own Shortcuts

If you never do anything but use ready-made shortcuts by other people, you're still farther ahead than you were before. But there's nothing like the satisfaction of building a shortcut yourself.

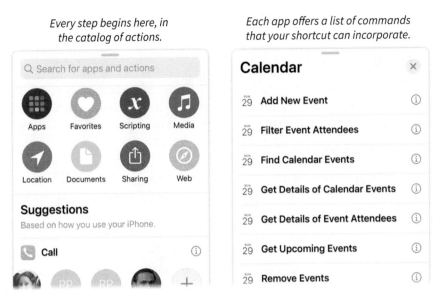

Building the shortcut

You can find a million tutorials and YouTube videos for building shortcuts online, but here's a walk-through that will show you the basics.

The one you're about to make is a classic. It texts the attendees for your next meeting that you're running late. It's only four steps long, it shows you how things work, and it's actually useful.

Here are the steps:

1. **In the Shortcuts app, on the My Shortcuts tab, tap +.**

   Welcome to the New Shortcut screen (pictured at right in "Starting up Shortcuts").

2. **Tap Add Action.**

   An *action*, in Apple's terminology, is one step of your shortcut.

   At this point, Shortcuts shows you a search screen, featuring eight icons (shown at left in "Building the shortcut"). Each represents a whole category of shortcut actions: **Apps** (a list of every Shortcuts-controllable app on your phone), **Media** (steps that make music play), **Scripting** (lets you build if/then, repeat, and **Choose from Menu** steps), and so on.

For the purposes of this "I'm running late" shortcut, you want to consult your calendar. Therefore:

3. **Tap** Apps, **and then** Calendar.

Here are all the commands your shortcut can apply to your calendar (shown at right in "Building the shortcut"). Look them over; someday, you might need one of them for a shortcut you invent yourself. But for now:

4. **Tap** Find Calendar Events.

You wind up with a little card. This is the first step your shortcut will perform. In this case, you can specify *which* calendar appointments you want the shortcut to find.

5. **Tap** Start Date **and choose** is today.

So far, you've told this shortcut: "Find my appointments today." But all you really care about is appointments for which you've entered an *address*; without that, the app has no way to calculate your travel distance.

Now, therefore, you'll limit the appointments to ones with addresses.

6. **Tap** Add Filter, **tap** Calendar, **and tap** Location.

You're back on the New Shortcut screen, where the Find All Calendar Events screen is in good shape.

UNTRUSTED SHORTCUTS

Anybody can create a shortcut and post it online for other people to download and use. Which is great—except that not everyone has your best interests at heart.

It would be technically possible, for example, for somebody to create a shortcut that requests access to information from your phone and sends it, behind the scenes, to them.

That hasn't actually happened, but Apple doesn't want responsibility if it does. So in iOS 14, Apple automatically blocks the installation of "untrusted shortcuts"—that is, any that don't come from Apple's own Gallery.

You can unblock them, though. Turn on **Settings→Shortcuts→Allow Untrusted Shortcuts**.

From now on, you're free to download other people's shortcuts (although you'll have to tap a huge red **Add Untrusted Shortcut** button first). After you add the shortcut, you'd be wise to examine it in the Shortcuts app, step by step, to see what it's doing. Hey—you might even learn a few shortcut-making tricks along the way. ⭐

Cancel	Next

New Shortcut ⚫⚫⚫

29 CALENDAR ⊗

Find All Calendar Events **where**

Start Date is today **and**

Location is anything ⊖

⊕ **Add** Filter

Sort by   None

Limit

MAPS ⊡

Get Driving **time from** Current **Location to** 🗓 Calendar Events

⊕

Details	Done

🔲 I'm Running Late ⊗

Add to Home Screen

Show on Apple Watch ⬤

Show in Share Sheet ◯

Show in Sleep Mode ◯

Import Questions ＞

ALLOW "I'M RUNNING LATE" TO ACCESS

29 Calendar ⬤

The shortcut grows

The next step will be calculating how much time it'll take you to get to this appointment.

7. Tap ⊕.

That ⊕ is always at the bottom of your chain of steps. It always means "Add another action."

At this point, the Calendar commands list from step 3 reappears, but you're done with them. Tap ⊗ to backtrack to the Apps panel.

8. Tap Maps. Choose Get Travel Time.

(If the Maps step says "Shortcuts does not have access to your location," tap **Allow Access**→**Allow While Using App**.)

This step's card already says, "Get **Driving** time from **Current Location** to **Calendar Events**," which is just what you want (see "The shortcut grows"). But you could, of course, tap **Driving** and change the travel method to **Walking**, **Transit**, or **Ask Each Time**. That's how these little cards work: You tap the variables to bend them to your will.

All right then: The shortcut now knows how long it will take you to get there. In its little head, it's referring to that number as Travel Time. Time to add a third step.

9. Tap ⊕. Tap ⊗ to return to the Apps panel, and the ⊗ again to return to the search screen.

For this step, you're going to use the Text shortcut step to compose the message.

10. Tap Documents; scroll down; tap Text.

You can construct whatever sheepish apology message you like. For example:

11. Type "I'm so sorry—running late. Will get there in about "—and then, in the Variables row just beneath, tap Travel Time. Tap Done.

So now the shortcut knows what to say; it just doesn't know whom to say it to.

12. Tap ⊕. Tap ⊗ to return to the search panel. Tap Apps and then Messages. Tap the first Send Message option.

Your final step is now in place. It says "Send 'Text' to Recipients," meaning the text from step 11.

13. Tap Next. Name your shortcut.

This is what you'll say to Siri when you want to run the shortcut. "I'm running late" will do fine.

14. Tap Done.

You're back on the My Shortcuts screen, where your shortcut is ready to use! Question is, where should it be available?

You answer that question on the shortcut's Details screen (at right in "The shortcut grows"). (On the My Shortcuts tab, tap ● to open the editing screen; tap ●.) Your options include these:

- **On the home screen.** You can add a shortcut to your home screen as an app icon, if you like. Tap **Add to Home Screen**. You'll have the chance to choose a name and icon for it.

- **In the share sheet.** Some shortcuts are designed to process text, graphics, or other data—the **Convert Anything to PDF** shortcut comes to mind—so you want them to show up as an option on the share sheet of any app.

The bottom line: To run your new shortcut, tell Siri, "I'm running late." Or tap the shortcut's home-screen icon. Or open Shortcuts and tap the tile. Or set up "I'm running late" to run when you double-tap the back of the phone (page 421).

With time, interest, and enough curiosity to study how other people have made their shortcuts, you can get really good at shortcutting. You can create really elaborate ones, dozens of steps long, with branching outcomes and conditional responses.

Automated Shortcuts

You can also create shortcuts that run automatically based on a time, a place, or what you're doing on your phone.

Start on the **Automation** tab in Shortcuts. Tap ⊕.

A shortcut can auto-run when any of these events occur—like connecting your charger.

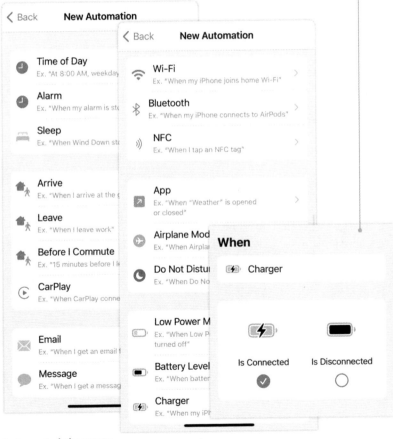

Automated shortcuts

You have a choice here: **Create Home Automation** shortcuts control HomeKit accessories, as described on page 338. (These shortcuts also appear in the Home app.)

But **Create Personal Automation** shortcuts are ones that run automatically based on *events*: You arrive or leave an address, your phone connects to CarPlay or a certain Wi-Fi hotspot, an email arrives from someone in particular, a certain app opens or closes, and so on. Take a look at the list of possibilities; it'll blow your mind.

You could create a shortcut that makes a certain album play automatically when you come near your Bluetooth speaker. Another one could text your assistant to bring you a power bank when your phone enters Low Power Mode. A third could turn on all the lights whenever your alarm goes off, so you can't *possibly* oversleep.

PART SIX

SIX

Appendixes

Installing iOS 14

There's always a little thrill when you unbox a new iPhone. Everything's shiny and new and filled with promise. And nothing has fingerprints on it yet.

Even upgrading to a new iOS version has its pleasures. You get a few minutes of the thrill of exploration before you start encountering any bugs.

Getting going is straightforward; if you can tap a **Next** button, you can manage it.

New iPhones

In almost every way, each year's new iPhone is better than the last. The camera's better, the chip is faster, the feature list is longer. So if you can afford it, and you're the kind of person who likes to have the latest, dive right in.

But the arrival of a new year's model is also good news if you're *not* a cutting-edge sort, because it means instant price cuts on *previous* years' models. For example, the very day the iPhone 12 debuted for $700, the previous year's model, the iPhone 11, dropped to $600. Even older models are still around for *really* low prices.

You can buy a new phone either from Apple or from a carrier, like Verizon, AT&T, or T-Mobile/Sprint.

> **TIP:** Check coverage areas before you choose a carrier! It's deeply frustrating to not be able to make cellphone calls from your home because you have no service there.

If you're changing cellular plans, or getting your first one ever, the carrier's website makes the pricing *somewhat* clear. The carriers all encourage you to sign up for an unlimited plan, so you don't have to worry about using up a monthly allotment of data—but the "data-bucket" plans are still available. For example, you might sign up for a plan that gives you 4 gigabytes of data per month. (You can change plans whenever you want.)

> **TIP:** If your plan has a data limit each month, don't miss the statistics at **Settings→Cellular** or **Settings→Mobile Data**. Here you can see how much data you've used so far in this carrier's billing period, broken down by the apps that have used it. That's important, because if you go over your monthly allowance, additional gigabytes' worth of data are very expensive, on the order of $15 per gig.

If you have an older iPhone or Android phone, switching your phone number to your new phone is as easy as swapping the SIM card (page 31) from the old one to the new one (assuming you're keeping the same cell carrier).

If this is your very first smartphone, if your previous phone used a different style of SIM card, or if you're changing carriers, you may have to ask the carrier to transfer your number for you.

Upgrading to iOS 14

If you've just bought a new iPhone, then you've already got iOS 14 (or iOS 14.2, 14.3, or whatever). Well done!

If you have an earlier iPhone that's running an earlier software version—for example, iOS 13—then one fine day, you'll see a notification that a newer version is available. You'll also see a red number badge (❶) on the **Settings** app, and on the **General** line in the list of panels. It's trying to tell you that a free upgrade is available. You'll get new features, but you won't lose any of your apps, files, or settings.

Software update badge

NOTE: The chance of something going wrong during one of these updates is very tiny. Still, you should back up your phone—or any kind of computer, actually—before you begin. See page 64.

Here's another very slim chance: that you're still running that classic 2016 hit, iOS 10, and one of your favorite apps is an ancient *32-bit* app. The iPhone hasn't been able to run those since iOS 11. It's best to come to terms with the app's loss (or see if there's an updated version) now, before the update.

To see if any of your apps are 32-bit, open the App Store, tap your icon at top right, tap **Purchased** (and then **My Purchases**, if you see that). Any apps whose **Open** or ☁ buttons are dimmed are the ones you won't be able to run in iOS 14.

To download iOS 14, open Settings→General→Software Update to see the note about iOS 14. Tap Download and Install. Provided you're on a Wi-Fi network and have at least 50% battery charge, the installation begins immediately.

NOTE: For faster service, you may prefer to upgrade your phone to iOS 14 by connecting it to your Mac or PC. Open a Finder window (macOS Catalina or Big Sur) or the iTunes app (Windows or earlier Mac versions) and click the iPhone's icon. Click the **Summary** or **General** tab, click **Check for Update**, and then click **Download and Update**.

The Setup Assistant

If you'd already been using your iPhone with iOS 13, then the upgrade process to iOS 14 is as simple as waiting about 10 minutes. A welcome screen appears, and you're in. Start exploring the new features, as described in this book.

But if you've bought a new iPhone, or your first iPhone, or you've installed iOS 14 by *restoring* your phone (that is, erasing it and starting from scratch—see page 417), then you'll encounter the Setup Assistant. It's a sequence of screens, each inviting you to turn on another setting.

The screens you see depend on your phone model and other factors, but here's the gist of it:

- **Hello** in various languages. Swipe up to continue.

- **[Choice of Language].** What language would you like the iPhone to speak?

- **Select Your Language and Country.** Where in the world do you live? (The iPhone takes a guess: the same country where you bought the phone.)

- **Quick Start** is an extremely slick way to import important settings from an old phone onto a new one. It requires that the older phone be running iOS 11, 12, 13, or 14.

  If you don't have such a phone, tap **Set Up Manually** and skip to the next step.

  If you do, though, make sure Bluetooth is turned on for both devices. Now bring your old phone *physically near* the new phone. On the old phone, on the Set Up New iPhone screen, tap **Continue**.

  At this point, on the new phone, you see the *craziest* bar code you've ever seen: a swirling cloud of animated pixels. Hold the old phone a few inches over the new phone's screen; make the animation fit inside the brackets on the old phone's screen. Suddenly, the two phones are paired, and the old one starts sending encrypted information to the new one.

Quick Start animated bar code

Now you're supposed to enter the old phone's passcode on the new phone. That's now the passcode for the new phone, too.

Over the next few minutes, the old phone transfers some basic settings to the new phone: your Wi-Fi password, your iCloud password, and your iPhone passcode.

Now you're treated to an interlude, during which you're supposed to set up Face ID (page 37) or Touch ID (page 39).

But at this point, the phone offers to bring over everything else from the old phone: all your apps, data, and settings. If you tap **Transfer from iPhone**, that stuff comes over wirelessly. It's a slow process—a couple of hours, maybe—but when it's over, your new phone has *everything* on it that your old phone did. Your apps are there, in the same positions as before, and you don't have to reenter any passwords or data.

If you tap **Download from iCloud** instead, the new phone downloads all the old phone's contents from your iCloud backup over Wi-Fi. You can start using the new phone in about 15 minutes, although you'll see only place-holders for most of the apps on your home screens. They'll continue to download from the App Store in the coming hours, and some may require you to sign into them again.

Once your new phone is fully loaded, you can skip most of the other steps described here.

> **NOTE:** Backing up your iPhone doesn't actually back up any of the stuff you've bought from Apple, like apps, songs, books, movies, and TV shows. Instead, Apple keeps track of which ones you own. Once you've restored your backup, your phone just re-downloads everything from the corresponding online stores—no charge. The dimmed app icons on your home screen gradually fill up as they arrive.
>
> This scheme means your backups don't eat up much time or disk space. Unfortunately, it also means waiting quite a while to restore all your stuff to a new (or erased) phone.

- **Apps & Data** offers to reload all your apps (and data!) from a recent backup, from a Mac or PC backup, from another phone, or from an Android phone. All of this can save you a ton of time and hassle.

If you tap **Restore from iCloud Backup**, the phone downloads all your apps and data from your most recent iCloud backup (page 64). You have to sign in with your Apple ID. Now, on the Choose Backup screen, tap the most recent backup—or, if you have some weird reason to believe that it's bad news, you can choose one of your older backups. You whip through a few of the settings screens described below, and then the Restore from iCloud screen appears.

Hint: The download takes a while. Start a movie.

Choose **Restore from Mac or PC** if your backup was on your computer—in the iTunes app or the Finder in macOS Catalina or Big Sur.

You can use **Move Data from Android** if your old phone was an Android phone; see Appendix B. Finally, tap **Don't Transfer Apps & Data** if this is your very first iPhone, or if you want to start fresh.

- **Choose a Wi-Fi Network.** You won't get far with your iPhone without getting online. Here's where you choose the Wi-Fi hotspot you want to join. If there isn't one available, or you don't want to deal with it now, tap **Use Mobile Connection** (meaning your cellular antenna).

Now iOS takes a moment to *activate* your phone—to get connected with your cell carrier.

- **Data & Privacy.** Here's where Apple lets you know about its philosophy on data collection: to collect as little data as possible.

- **Touch ID or Face ID.** Teach the phone to recognize your face (page 37) or fingerprint (page 39) for the purpose of unlocking it.

If you're in a hurry now, tap **Set Up Later in Settings.** When you've got more time, you can always set up face recognition or fingerprint recognition in **Settings→Face ID & Passcode** [or **Touch ID & Passcode**].

- **Create a Passcode.** Even if you've taught the phone to recognize your finger or face, you must also make up a passcode for unlocking your phone, as a backup. See page 36 for some notes on choosing a passcode.

The iPhone proposes a six-digit passcode, but that's not a requirement. If you tap **Passcode Options**, you can choose options like **Custom Alphanumeric Code** (any password, any length, any characters), **Custom Numeric Code** (any number of digits), or **4-Digit Numeric Code** (like an ATM).

If you don't intend to use Face ID or Touch ID, you can even tap **Don't Use Passcode.** The phone's security conscience will scold you—but if you're confident that nobody would ever have the opportunity to steal your phone or peek at what's on it, you're free to leave your phone unprotected. Since you'll probably be waking it (and unlocking it) dozens of times a day, you'll save yourself a lot of time and hassle.

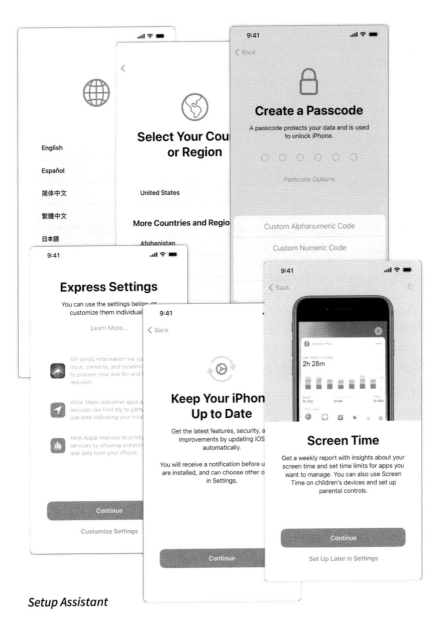

Setup Assistant

- **Apple ID.** An Apple ID is an email address and password. You need one if you ever intend to buy anything from Apple, like an app, a song, a book, or a computer. You also need one to access the features of iCloud and Apple Pay (Chapter 15), to play games against other people online, to use iMessages, to make an Apple Store appointment, and so on.

If you already have an Apple ID, enter it here. If not, tap **Forgot password or don't have an Apple ID?** and then **Create a Free Apple ID**. You're asked to provide your name, birthday, email address (or you can create a new iCloud email address), and a password of your choice. You also get to opt out of junk email from Apple.

Apple now sends you a two-factor authentication code, as described on page 395.

- **Terms and Conditions.** Tap **Agree**. Otherwise, you've got yourself an $800 paperweight.

- **Express Settings.** Here, by tapping **Continue**, you're saying, "OK, fine, skip over all the individual questions about what settings I'd like and give me the usual setup."

You're agreeing to get automatic updates when Apple releases updates to iOS; to let other people reach you using FaceTime (page 239) and iMessages (page 245); to let navigation apps like Maps monitor your location using Location Services; to use Siri; to let Apple collect anonymous audio recordings of your Siri requests (for quality-assurance purposes, of course); and to give Apple data about how you use the phone.

If you'd rather approve or deny each of these settings individually, tap **Customize Settings**.

> **NOTE:** If you decided to use Quick Start (page 466), and your new phone is downloading the backup from your old phone, you don't get Express Settings. Instead, you're offered **Settings from Your Backup**. This option, too, skips the Siri, location, and analytics questions—but it grabs the settings from your backup.

- **Keep Your iPhone Up to Date.** At this stage, Apple used to offer you the *option* of using Automatic Updates, which installs any new iOS versions automatically when Apple releases them. Usually, that's what you want, since new versions tend to have more features and fewer bugs.

    But in iOS 14, all you get here is a **Continue** button. You can't opt out—at least not here. (You'll get a notification before a new version is installed.)

    You *can* turn Automatic Updates off later, in **Settings→General→Software Update→Automatic Updates**.

- **Location Services** is the circuitry and software that determine where you are. Some of Apple's apps, like Maps and Find My—and other apps, like weather apps—don't work without it. On the other hand, some people are creeped out by the phone knowing where they are, even though it's an inanimate hunk of glass and silicon. This is your chance to turn Location Services on or off.

- **Siri.** Here's where you turn on the Siri voice-assistant feature, and set up the hands-free "Hey Siri" magic. (That involves speaking a few sample commands; the screen walks you through it.) You can always set this stuff up later, in Settings→Siri & Search.

THE SMARTPHONE THAT DOES CONTACT TRACING

Now that you've got your phone set up, one of the next settings to consider is one that could save lives.

Contact tracing means tracking the spread of a virus from each person to everybody they've been in contact with—and informing them so they can quarantine. It's an incredibly important tool in containing a pandemic—but it's incredibly hard to pull off. How could you, the ordinary citizen, possibly remember every single person who's been near you? Including total strangers in the store or on the bus?

In early 2020, Apple and Google, usually archrivals, collaborated on an ingenious solution called exposure notification. It works like this:

If you decide to participate, your phone begins continuously broadcasting a Bluetooth beacon—a random number that changes every few minutes—to any iPhones or Android phones within about 15 feet. Meanwhile, *your* phone is picking up the beacons from all *other* phones nearby. It remembers these interactions for 14 days.

Suppose that a few days later, some guy who stood close to you tests positive for COVID-19. If he reports his diagnosis—anonymously, of course—then everybody he's exposed to the virus in the past two weeks gets notified on their phones. The notification doesn't say anything about the interaction with this person; you'll never know who, where, or when, because none of that information is collected. You do know *that* it happened, though, and you should get tested and avoid other people.

The genius of this system is that it's completely anonymous and completely voluntary. Above all, there's no centralized database! Nobody can get their hands on the list of infected people—not the government, not Apple, not Google, not Russia—because there is no such list. No *person* is tracked, no locations are ever collected, and nobody knows *who* exposed them.

Apple and Google don't even make the app; that's up to individual states and countries. If your state offers such an app—and Settings→Exposure Notifications→Availability Alerts is turned on—your phone will invite you to join the program.

For your health and the country's, it's an excellent idea to accept. ✦

- **Screen Time** is basically parental controls. It lets you set time limits for certain apps (or categories of apps); see page 439. You can tap Continue to set it up now, or Set Up Later in Settings.

- **iPhone Analytics.** If you tap Share with Apple, your iPhone will send activity logs to the mother ship, including your location and what you're doing on your iPhone. Apple uses this data to fix bugs, locate dead spots in the cell network, and so on. None of this information is associated with *you* or your account; it's stripped of all identity information and pooled into a vast database of aggregated customer data.

- **App Analytics.** Same thing, but for your non-Apple apps.

- **True Tone Display.** The iPhone 8 and later models can adjust the screen colors to keep them looking consistent under different lighting conditions. Here's your chance to try it out. If you like the effect, tap Continue. If you don't, you still have to tap Continue—but you can then turn it off in Settings→Display & Brightness.

- **Appearance** is your chance to choose either light mode or dark mode for your phone (page 87).

- **Display Zoom.** Some iPhone models have pretty big screens, especially the Plus and Max models. On this screen, if you choose Standard, your phone will use that extra screen space by displaying more stuff per screenful than smaller iPhones can. If you choose Zoomed, then you see the same thing you'd see on the smaller phones—but *bigger*.

  You can always change your mind in Settings→Display & Brightness.

- **Meet the New Home Button.** This screen appears only if you're setting up an iPhone 7, 8, or SE model. It lets you adjust the feeling of its clickless home button.

- **Welcome to iPhone.** Enough with this entry interview! Swipe up (or tap Get Started) to reveal the home screen and start using iOS 14.

> **NOTE:** When you next open Settings, a strange message greets you: **Finish Setting Up Your iPhone**. Really? After all that, you're *still* not finished?
>
> If you tap that link and then tap **Finish Setting Up**, Settings walks you through any steps you may have skipped during the setup process. This is also where Apple tries to persuade you to set up Apple Pay (page 404).

Switching from Android

There are, as you know, two major phone religions: iPhone and Android.

Since you're here, reading about switching, you don't need to be reminded of the iPhone's advantages. You can skip hearing about the fragmented state of the Android world, where every phone is running a different version, some phones can never be updated, and only 10% are ever running the latest software. (By contrast, about 75% of iPhones are generally running the very latest iOS version.)

There's no need to point out how chaotic, unsupervised, and virusy the Android app store can feel. You don't need reminding that every iPhone has a quick-flick silencer switch.

And you don't need anyone to note the smoothness and depth of integration across Apple's phones, tablets, computers, and watches. AirDrop. iMessages. FaceTime. AirPlay. Free tech support at Apple Stores. Mmmmm.

Making the Switch

Fortunately, it's very easy to transfer your life from an Android phone to an iPhone. Here's how:

1. **Get everything ready.**

   Plug both your iPhone and your old Android phone into power. Make sure both are on Wi-Fi. Consider whether your iPhone has *room* for everything on your Android phone (including on its SD memory card, if any).

On the Android phone, download the app called Move to iOS from the Google Play store.

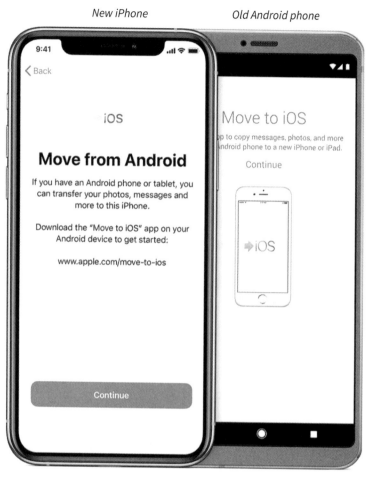

New iPhone   Old Android phone

Move to iOS

2. **Start setting up your iPhone. On the Apps & Data screen, tap** Move Data from Android.

You have one opportunity to import everything from your old phone, and that's the Apps & Data screen that appears during the Setup Assistant process (page 465). In other words, there's no automated way to bring over your Android stuff *after* you've already set up your phone.

If it's too late, you can use the Settings→General→Reset→Erase All Content and Settings option to start over from scratch—or just use one of the methods described in the next sections.

3. **On the Android phone, open the Move to iOS app.**

Tap **Continue**; on the legalese screen tap **Agree**; on the Find Your Code screen, tap **Next**.

4. **On the iPhone (still on the Move to Android screen), tap Continue.**

Your iPhone now displays a numeric code.

> **NOTE:** If the Android phone gives you some notification about a weak internet connection or "Internet not available," you can ignore it; tap **Keep Wi-Fi Connection**. The iPhone is creating a tiny Wi-Fi hotspot just for this purpose, which may confuse the Android phone.

5. **Enter the code on the Android phone.**

After a moment, the Transfer Data screen appears.

6. **Choose what you want to bring over.**

You can choose your email account (which generally includes your cal-endar), text messages, web bookmarks, and so on. Hard to imagine why you'd want to exclude any of your own data, but here's your chance.

What to transfer

7. **Once you've made your choices on the Android phone, tap** Next.

The transfer begins!

It can take a very long time.

It's also a little temperamental. If a call comes in on the Android phone, the process stops. If you switch to another Android app, the process stops. If you have apps that mess with the Wi-Fi connection, like Sprint Connections Optimizer or Smart Network Switch, the process stops.

Don't touch the phones until the *iPhone* indicates that everything is done. (The Android phone may say that it's done, but it's not.)

If you can't make it work, try deleting ("forgetting") any stored Wi-Fi networks on your Android phone (in Settings→Wi-Fi on the iPhone). Try restarting both phones. Try turning off cellular on the Android phone. Try holding a small incense ceremony.

8. **Tap** Done **on the Android phone.**

On the iPhone, tap Continue Setting Up iPhone. Now you can continue with the standard iOS setup process, as described in Appendix A.

Once it's all over, you may see, on the iPhone, a message: "Add your Android Device Apps from the App Store?" It's offering to auto-download all the free apps that are available on both the iPhone App Store and the Google Play store.

You'll also be asked to enter the passwords for the email accounts you transferred. That's OK; it's a one-time thing.

Cloud Transfers

The Move to iOS app is a good start. But it has two huge limitations.

First, it doesn't transfer *everything* from your Android phone. For example, it doesn't attempt to move music, ebooks, or PDF files.

Second, you have only one chance to use the Move to iOS app: during the setup process for a brand-new (or freshly erased) iPhone. If you want to bring stuff over later, it's of no use to you.

There are, however, ways to bring things over manually—without either of those restrictions.

The good news is that these days, a lot of the stuff you might want to transfer lives in the *cloud*—that is, on online services, accessible equally easily by Android and iPhone. For example:

- **Apps.** Every brand-name app you've ever used on Android is also available on the iPhone. Facebook, Twitter, LinkedIn, Spotify, Pandora, Flickr, Yelp, Netflix, YouTube, Wikipedia, Google Maps, Waze, Uber, Lyft, Skype, Hipmunk, *The New York Times*, Kindle, TED, Mint, Snapchat, TikTok, OpenTable, Fandango, Ticketmaster, Tripadvisor, and every national chain of restaurants, stores, movie theaters, airlines, and banks. Open your App Store app and download away.

- **Mail, contacts, notes, and calendars.** If you use one of the biggie email/calendar services, like Google, Microsoft Exchange, AOL, or Yahoo, you're all set. You can set them up on the iPhone at any time, as described on page 271.

- **Photos and videos.** If you keep your photographic output on a site like Amazon Photos, Google Photos, Flickr, or SmugMug, no big deal. Download the corresponding app on the iPhone, log in, and boom: You've got exactly the same photo and video collection on your new phone.

> **TIP:** Using iCloud Photos for this is especially easy, because the pictures and movies appear *automatically* on your iPhone, in Photos. To get your photos from the Android phone to iCloud Photos, sign up for an Apple ID (page 393), log into iCloud.com, and click **Photos.** Upload your photos and videos right from the phone. Make sure you have enough iCloud storage to hold your collection (otherwise you'll have to transfer the stuff in chunks).

- **Music.** If you listen to music through a streaming service like Spotify, Amazon Music, YouTube Music, Google Play Music, Pandora, or (of course) Apple Music, you're all set. Download the app onto your new iPhone, sign in, and start listening to all your existing favorites and playlists.

- **Ebooks.** If we're talking about ebooks from the Kindle, Nook, or Google Play bookstores, this will be easy. Download the corresponding app from the iPhone App Store, log in with your account, and boom: There are your books, ready to download onto your new iPhone.

Manual Transfers

Those cloud-based services sure make life easy! But not everyone keeps their photos, music, and ebooks online. Some of it is *physically* located on your Android phone.

If you missed your chance to transfer this stuff from your Android phone to your iPhone during the setup process, as described earlier, you can still do the job.

Use a Shared Online Drive

Dropbox, Google Drive, and Microsoft OneDrive are fantastic ways to transfer almost anything from Android phones to iPhones. Sign up for a free account; download the corresponding app on each phone.

Then copy any files on the Android phone that you want to transfer onto this "virtual hard drive." Once you've signed into the same app on the iPhone, those files show up automatically after a few minutes.

Of course, you get only limited space with the free accounts: 2 gigs (Dropbox), 5 gigs (Microsoft OneDrive or Apple iCloud Drive), or 15 gigs (Google Drive). You can pay a small amount to increase that storage limit—and if your only goal is to transfer stuff from Android to iPhone, you can sign up for the generous account, do the transfer, and then cancel, having paid for only one month's worth of service.

Use the Mac or PC as an Intermediary

If you've got too many files for the Dropbox method, you can also use your Mac or PC as a way station between the old and new phones. The process goes like this:

1. **Connect your Android phone to the computer.**

   Use the USB cable it uses for charging.

   If you have a Windows PC, the Android phone shows up in an Explorer window as though it were an external drive. Now you can browse the Android phone to find the stuff you want to copy over.

   If you have a Mac, you'll need a little help with this. Go to www.android.com/filetransfer and download the app called Android File Transfer. It lets you browse the contents of an Android phone you've connected to a Mac.

2. **Find the folder that contains the files you're looking for.**

   For photos and videos, that's usually the DCIM→Camera folder on the phone. For music, it's the Music folder. For documents, ebooks, and PDF files, it's usually Documents.

3. **Drag the files you want to transfer into a folder on the Mac or PC.**

They might include photos, music, PDF files, whatever.

Android File Transfer

4. **Disconnect the Android phone from the computer. In its place, hook up the iPhone with its white Lightning cable.**

Now your job is to tell the computer to copy the Android files over to the iPhone.

5. **Open the iTunes app for Mac or Windows.**

You can download the free iTunes app from www.apple.com/itunes.

If you have a Mac running macOS Catalina or Big Sur, you don't need iTunes. Apple has built its features right into the desktop. Open any Finder window.

6. **Click the iPhone's icon. Click the appropriate tab, and choose the stuff to transfer.**

For photos, click the Photos tab of iTunes or the Finder window. Turn on Sync photos to your device from, and choose the folder that contains your Android photographic work. At this point, you can also specify a subset of photos or albums to copy to the iPhone, and you can omit videos, if you like.

For music files, click the Music tab (of iTunes or the Finder window). Find that folder you created to hold your Android tunes, and drag it into the Songs section of the screen. Finally, click your iPhone's icon and then click Music. Specify how much of the music you want to copy to the iPhone: your whole music library, or only the music you just added.

For free, not-copy-protected PDF files and ePub books, find the files on your Mac or PC; drag them into iTunes (on Windows) or into the Books app (on a Mac).

Click Sync to begin the transfer process.

About the Author/Illustrator

David Pogue was the weekly tech columnist for *The New York Times* from 2000 to 2013. Today, he's a correspondent for *CBS Sunday Morning*, a *New York Times* bestselling author, host of 20 science specials on PBS's *NOVA*, and host of the podcast *Unsung Science with David Pogue*.

He's won five Emmy awards, given five TED talks, and written or co-written more than 120 books, including dozens in his *Missing Manual* series; six in the *For Dummies* line (including *Macs*, *Magic*, *Opera*, and *Classical Music*); two novels (one for middle-schoolers); and three books of essential tips and shortcuts: *Pogue's Basics: Tech*, *Pogue's Basics: Life*, and *Pogue's Basics: Money*.

In his other life, David is a magician, a funny public speaker, and a former Broadway show conductor. He lives in Connecticut and San Francisco with his wife, Nicki, and a blended family of five fantastic kids.

You can find a complete list of his columns and videos, and sign up to get them by email, at authory.com/davidpogue. His website is davidpogue.com; on Twitter, he's @pogue. He welcomes feedback about his books by email at david@pogueman.com.

About the Creative Team

Julie Van Keuren (editing, indexing, interior book design) spent 14 years in print journalism before deciding to upend her life, move to Montana, and work from home before it was cool. She now provides skilled editing, writing, book layout, and indexing to a variety of terrific clients. She and her husband have two adult sons. Email: JulieVanK@gmail.com.

Kellee Katagi (proofreading) has devoted most of her 20-plus-year writing and editing career to covering technology, fitness and nutrition, travel, and sports. A former managing editor of *SKI* magazine, she now smiths words from her Colorado home, where she lives with her husband and three kids. Email: kelkatagi@gmail.com.

Diana D'Abruzzo (proofreading) is a Virginia-based freelance editor with more than 20 years of experience in the journalism and book publishing industries. More information on her life and work can be found at dianadabruzzo.com.

Judy Le (proofreading and symbol font) is a magazine editor and school board member in Virginia, where she lives with her husband and their son in a near-constant state of amazement.

Sandy Writtenhouse (technical editing) is a freelance writer and editor for technology-related topics. She has a decade of experience in the online publishing world and holds a B.S. in information technology. You can see where she currently works at www.sandywrittenhouse.com.

Julie Elman (design consulting) teaches courses in publication design, editorial illustration, and the creative process at a university in Ohio. She has more than 15 years' experience working as a visual journalist. Examples of her work can be found at julie-elman.com.

Jason Heuer (cover design) is an art director, designer, and artist living in New York City. He holds a BFA in advertising and design from the School of Visual Arts and was associate art director and recruiter at Simon & Schuster from 2004 to 2015. Currently he is principal at Jason Heuer LLC, a design and art studio in Rockaway. Find his work at jasonheuer.com.

Acknowledgments

Creating this book from scratch—concept, philosophy, style, graphics, layout, design, covers—was a thrill. It began with my agent, Jim Levine, and the team at Simon & Schuster—Richard Rhorer, Priscilla Painton, and Jonathan Karp—who offered their full support for launching a new tech-book series in 2020. They said from the start they believed in the idea, and, wow, did they come through.

I was delighted that Jason Heuer (who designed the cover of my book *How to Prepare for Climate Change*) was willing to try his hand at a tech book; that Sandy Writtenhouse was available to take on the technical editing; and that Diana D'Abruzzo, Julie Elman, Kellee Katagi, and Judy Le contributed their design and editing talents.

My most active collaborator, though, was Julie Van Keuren, who was responsible for basically every aspect of the book except for the writing and the figures. She came up with the title; she did the editing; she oversaw the design, proofing, and tech editing; she did the layout; she created the index; she chased down all the cross-references; she designed the custom symbol font; and she was my partner in creation on a thousand decisions, large and small.

Writing a book on a tight deadline always means withdrawing, to some degree, from the rest of the world. I'm grateful to my kids, Kell, Tia, and Jeffrey, for their tolerance during this period, and above all, to my bride, Nicki. Without her support, ideas, and belief in me, you'd be using Google searches right now to learn about your iPhone. —*David Pogue*

Index

E

earbuds and headphones. *See also* **AirPods**
Bluetooth pairing 69–70
Headphone Accommodations 448
headphone adapter 24
mono audio 448
not included in box 29
playback controls 386
status bar icon 83
ebooks. *See* **Books app**
editing
button lists 4–5
by voice 148–150
Calendar appointments 321
cut, copy, paste 129–132
home screens 89–93, 94–95
Look Up command 131
moving the insertion point 127–128
phone options 4–5
photos and videos 210–220
selecting text 128–129
standard markup tools 7–9
typed text 127–132
undo and redo 132
voice recordings 372
widgets 102–103
effects
in Messages app 256–260
on FaceTime calls 241–242
email. *See* **Mail app**
emergencies 66–69
auto-calling 911 69
Emergency SOS 66–69
Medical ID 68
emoji 119
Animoji 266–267
Memoji 264–266
new in iOS 14 1
search box 119
stickers 263
texting giant emoji 257–258
erasing the phone 417–418
eSIM 32
exposure
dedicated control 163–166
defined 161

exposure lock 162–163
in video mode 186
making permanent 166
exposure notifications 471

F

Face ID 37–39
Attention Aware features 37
Face ID vs. home-button phones 23
how it works 38
passcode requirement 35
setting up 39
FaceTime 239–244
audio-only calls 244
Eye Contact setting 240
group chats 242–244
using Animoji 267
video calls 239–244
facial recognition
Face ID 37–39
in Photos app 198–199
security cameras 12
Family Sharing 401–404
defined 401–402
managing 403
privacy 403
setting up 402
sharing purchases 404
Fantastical app 102
favorites
in Files app 330
in Health app 335–336
in Home app 56, 338
in Maps app 343
in Phone and Contacts apps 223–225
in Photos app 192
in Safari 293
Mail app VIPs 289
pinning notes 357
pinning text conversations 248
file attachments. *See* **attachments**
Files app 327–330
favorites 330
managing files 328–329
tagging files 329–330
widget 98